ADVANCED MECHANICS OF MATERIALS

ADVANCED MECHANICS OF MATERIALS

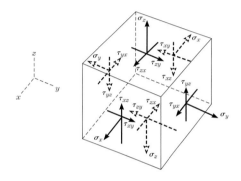

WILLIAM B. BICKFORD

ARIZONA STATE UNIVERSITY

An imprint of Addison Wesley Longman, Inc.

Menlo Park, California • Reading, Massachusetts • Harlow, England
Berkeley, California • Don Mills, Ontario • Sydney • Bonn • Amsterdam • Tokyo • Mexico City

Senior Acquisitions Editor: Michael Slaughter
Assistant Editor: Susan Slater
Production Manager: Pattie Myers
Senior Production Editor: Teri Hyde
Art and Design Supervisor: Kevin Berry
Composition: G&S Graphics
Artist: Precision Graphics
Cover Design: Yvo Riezebos
Cover Image: William B. Bickford
Text Design: Lesiak/Crampton Design Inc.
Text Printer and Binder: MapleVail
Cover Printer: Lehigh Press

Copyright © 1998 Addison Wesley Longman, Inc.

All rights reserved. No part of this publication may be reproduced, or stored in a database or retrieval system, or transmitted, in any form or by any means, electronic, mechanical, photocopying, recording, or otherwise, without the prior written permission of the publisher. Printed in the United States of America. Printed simultaneously in Canada.

Many of the designations used by manufacturers and sellers to distinguish their products are claimed as trademarks. Where those designations appear in this book, and Addison-Wesley was aware of a trademark claim, the designations have been printed in initial caps.

Library of Congress Cataloging-in-Publication Data
Bickford, William (William B.)
    Advanced mechanics of materials / William B. Bickford.
      p.   cm.
    Includes bibliographical references and index.
    ISBN 0-673-98195-9 (std. ed.)
    1. Strength of materials.   2. Mechanics, Applied.   3. Finite element method.   I. Title.
TA405.B497   1998
624.1'7—dc21   97-44789
                                                                      CIP

The full complement of supplemental teaching materials is available to qualified instructors.

Instructional Material Disclaimer
The programs presented in this book have been included for their instructional value. They have been tested with care but are not guaranteed for any particular purpose. Neither the publisher or the authors offer any warranties or representations, nor do they accept any liabilities with respect to the program.

ISBN 0-673-98195-9

1 2 3 4 5 6 7 8 9 10—MA—02 01 00 99 98

Addison Wesley Longman, Inc.
2725 Sand Hill Road
Menlo Park, California 94025

CONTENTS

Preface   vii

CHAPTER 1   Basic Ideas of Solid Mechanics   1

1–1   Problem Statement   1
1–2   Basic Idea I: Equilibrium and Stress   2
1–3   Basic Idea II: Deformation and Strain   15
1–4   Basic Idea III: Material Behavior: Stress-Strain Relations   25
1–5   Boundary Value Problems: Combination of Basic Ideas   28
1–6   Modeling and the Design Process   31
1–7   Failure and Failure Criteria   35
  1–7–1   Failure Criteria for Ductile Materials   40
  1–7–2   Failure Criteria for Brittle Materials   45
1–8   Factor of Safety and Working Stresses   47
1–9   Approximate Formulations: Mechanics of Materials   50
1–10 Summary   56
References   56
Exercises   56

CHAPTER 2   Variational and Approximate Methods   64

2–1   Introduction   64
2–2   Work and Potential Energy   65
2–3   Strain Energy and Stiffness Formulations   69
  2–3–1   Strain Energies for Structural Members   72
  2–3–2   Stationary Potential Energy   74
  2–3–3   Structural Analysis: Castigliano's First Theorem   77
2–4   Approximate Methods   81
  2–4–1   Method of Ritz   81
  2–4–2   Finite-Element Methods   87
2–5   Complementary Energy Ideas   98
  2–5–1   Complementary Energies for Structural Members   100
  2–5–2   Dummy Loads   105
  2–5–3   Statically Indeterminate Systems: Principle of Least Work   107
2–6   Summary   112
References   114
Exercises   114

Chapter 3 Torsion 124

3–1 Introduction 124
3–2 Torsion of Circular Cross-Section Bars 124
3–3 Torsion of Noncircular Cross-Section Bars 128
    3–3–1 The Membrane Analogy 136
    3–3–2 Thin Rectangular Cross Section 137
    3–3–3 Thin-Walled Open Sections 139
    3–3–4 Finite-Element Models 142
3–4 Torsion of a Single Thin-Walled Tube 144
3–5 Torsion of Multicelled Tubes 151
3–6 Warping Restraint in Thin-Walled Open Sections 158
3–7 Alternative Formulation for Warping Restraint in Thin-Walled Open Sections 166
3–8 Summary 171
    References 172
    Exercises 172

Chapter 4 Transverse Loading of Unsymmetrical Beams 182

4–1 Introduction 182
4–2 Unsymmetrical Bending 185
    4–2–1 Stresses in Unsymmetrical Bending 191
    4–2–2 Displacements in Unsymmetrical Bending 198
4–3 Shear Stresses in Thin-Walled Open Sections 204
4–4 Shear Center for Thin-Walled Open Sections 209
4–5 Bending of a Single Thin-Walled Tube 218
4–6 Thin-Walled Multitube Bending 222
4–7 Combined Bending and Torsion 227
4–8 Summary 233
    References 233
    Exercises 233

Chapter 5 Plane-Curved Beams 244

5–1 Introduction 244
5–2 Development of the Governing Equations 245
5–3 Bending Stresses 249
5–4 Transverse Shear Stresses 254
5–5 Displacements 260
5–6 Summary 266
    References 267
    Exercises 267

Chapter 6 Thick-Walled Cylinders 276

6–1 Introduction 276
6–2 Development of the Theory 277
6–3 Internal and External Pressures 282
    6–3–1 Shrink-Fit Problems 285
6–4 Long Cylinders 288
6–5 Thermal Problems 292
6–6 Rotating Disks 299
6–7 Variable-Thickness Rotating Disks 305
6–8 Finite-Element Solutions 309
6–9 Summary 313
    References 314
    Exercises 314

Chapter 7 Plates 322

PART I—RECTANGULAR PLATES 322
7–1 Introduction 322
7–2 Force Resultants and Equilibrium 323
7–3 Deformations—Strain-Displacement Relations 326
7–4 Stress-Strain and Combination 328
7–5 Solutions to Rectangular Plate Problems 334

PART II—CIRCULAR PLATES 339
7–6 Introduction 339
7–7 Force Resultants and Equilibrium 340
7–8 Deformations: Strain Displacement Relations 342
7–9 Stress-Strain and Combination 344
7–10 Solutions to Circular Plate Problems 347
7–11 Summary 359
  References 360
  Exercises 360

Chapter 8 Membrane Shells of Revolution 366

8–1 Introduction 366
8–2 Geometry of Shells of Revolution 369
8–3 Governing Equations for Membrane Theory 371
  8–3–1 Equilibrium 371
  8–3–2 Deformations 373
  8–3–3 Material Behavior and Combination 375
8–4 Axisymmetrically Loaded Membrane Shells 377
  8–4–1 Cylindrical Shells 379
  8–4–2 Spherical Shells 381
  8–4–3 Conical Shells 384
8–5 Asymmetrically Loaded Membrane Shells 386
  8–5–1 Cylindrical Shells 386
  8–5–2 Conical Shells 390
  8–5–3 Spherical Shells 393
8–6 Summary 397
  References 398
  Exercises 398

Chapter Appendices 403

A Calculus of Variations 403
B Finite-Element Models and MATLAB Codes 409
C Sectorial Area 447
   Index 452

PREFACE

The first course in solid mechanics generally involves an elementary treatment of each of the topics of stress, strain and stress-strain relations in one form or another. These basic ideas are then combined to form individual elementary theories for investigating the axial, torsional and bending deformations of straight prismatic bars. Such a course can serve as a prerequisite for the topics covered in this text.

In the first course there is generally insufficient time to carefully consider the assumptions that are made in developing the theories and to delve into the corresponding approximate nature of the theories. The second course in solid mechanics is intended to carefully extend the elementary theories to make them applicable in more general settings as regards the geometry of the region and the type of loadings allowed. Additionally, several important technical topics not covered in a beginning course, as well as several exact solutions of the equations of solid mechanics, are presented.

The basic underlying philosophy of the second course involves making simplifying assumptions about the kinetic and kinematic variables, which essentially reduces the partial differential equations of the general theory of solid mechanics to one or more ordinary differential equations. Solutions to these ordinary differential equations generally constitute approximate solutions in solid mechanics. Force resultants are generally used in place of stresses to represent the kinetics. Simplifying assumptions are made regarding the kinematics to reduce the number of independent variables on which the displacements depend to either one or two, that is, to reduce it to either a one- or a two-dimensional problem. Whenever possible, the results from the approximate theories are compared to the corresponding exact solution of the problem in question. When appropriate, the finite-element method is used to discuss approximate solutions to boundary value problems, which would otherwise be difficult or impossible to solve analytically.

Chapter 1 is a brief introduction providing an overview of the mechanics of deformable bodies, that is, a discussion of the quantities that are considered as given and those that are to be determined from solutions of the equations of solid mechanics. Chapter 1 introduces the three basic ideas of solid mechanics, namely, (1) equilibrium or kinetics, (2) deformation or kinematics and (3) material behavior or constitution. Definition and properties of stress, stress transformations and the equations of equilibrium in terms of the stresses are also presented. Strain is defined, after which the components of strain are expressed in terms of the displacements. Stress-strain relations are defined for a general linearly elastic solid and specialized for the technically important cases of plane stress and plane strain. The results from the three basic ideas are then combined to yield the governing equations in terms of the displacements. Stress and displacement boundary value problems are discussed. Also presented are the commonly used theories of failure along with the ideas of factor of safety and working stresses.

Chapter 2 is devoted to variational formulations in solid mechanics. The primary emphasis is on using variational principles as a basis for analyzing idealized structural systems. The ideas of work and potential energy are presented. Strain energy ideas and their application to stiffness formulations are discussed in connection with Castigliano's first theorem. The approximate method of Ritz is discussed and applied to a specific one-dimensional problem. Within the context of the Ritz method, the basics of the finite-element method are presented using the same one-dimensional example. Complementary energy ideas and their application to flexibility formulations are discussed in connection with Engesser's first theorem. The use of the principle of least work is discussed in connection with flexibility formulations for statically indeterminate structural systems.

Chapter 3 begins with a treatment of the exact solution for the torsion of a circular bar. The general problem for the torsion of a prismatic bar of arbitrary cross section is then formulated in terms of the warping function and in terms of the Prandtl stress function. Exact solutions for the ellipse and rectangle are discussed. The membrane analogy is introduced and used in developing several solutions for bars having thin-walled open and thin-walled closed cross sections. The torsion of a single thin-walled tube and of a multitube section are covered in detail. The problem of the additional stresses associated with restraint against warping is discussed.

Chapter 4 begins with a brief review of the results from the theory of bending for symmetric sections. This is followed by the development of the general equations for bending stresses and transverse deflections of beams having unsymmetrical cross sections. Shear stresses in thin-walled open sections and the idea of the shear center are treated in detail. Bending of beams for which the cross section is a single thin-walled tube as well as the location of the corresponding shear center are presented. Also covered in the same sense is the bending of multitube cross-section beams. The chapter ends with an application of the basic idea of superposition to the case where the loading does not pass through the shear center, that is, combined bending and torsion.

Chapter 5 is devoted to the theory of plane-curved untwisted bars. It is assumed that the cross section of the bar possesses a plane of symmetry so that deformations are restricted to the plane of curvature. Bending and shear stresses are covered. Displacements are determined using energy methods and by integrating the differential equations.

Chapter 6 covers thick-walled pressure vessels from the standpoints of both plane stress and plane strain. Interference fits, thermal inputs and rotation effects are treated. Finite-element models are used to analyze the variable-thickness rotating disk problem with the results compared to the analytical solution.

Chapter 7 covers the problem of the small deflection of thin plates. The basic ideas are discussed and combined to develop the governing equations of classical thin-plate theory for both rectangular plates and symmetrically loaded circular plates. Solutions are presented for rectangular plates that are simply supported on two or more edges and for several solid and annular circular plates.

Chapter 8 treats membrane theory for thin shells of revolution. The governing equations are developed for loadings and supports consistent with membrane theory. Stresses and displacements for cylindrical, conical and spherical shells subjected to axisymmetric and simple asymmetric loadings are presented.

There are approximately 400 end-of-exercise problems. These problems are of several types. When possible and appropriate, the student is asked to verify some aspect of the theory developed in the text of the chapter. In the same vein, the student is asked to verify that the results from a specific problem actually satisfy the requirements of the corresponding theory. The verification problems are intended to help the student realize that checks of this sort can and should be made in order to minimize the occurrence of errors, either those resulting from misapplication of the theory or those associated with the calculations. There are also numerous problems that ask the student to demonstrate his or her understanding of the theory in the text by application to specific practical situations.

Finally, there are three appendices. Appendix A considers an elementary treatment of the ideas from the calculus of variations necessary to investigate the stationary value of an integral. Appendix B presents a development of the finite-element model for the torsion problem using linearly interpolated triangular elements. Appendix B also contains several MATLAB codes that can be used to investigate the following:

1. Torsion problems from Chapter 3.
2. Thick-walled pressure vessel problems from Chapter 6.
3. Rectangular plate problems from Chapter 7.
4. Circular plate problems from Chapter 7.
5. Shell problems from Chapter 8.

The solutions manual available to instructors will include a diskette containing the MATLAB codes. Finally, Appendix C gives a brief account of the topic of sectorial areas needed in Chapters 3 and 4 for thin-walled sections.

The author and the publisher wish to thank the following reviewers for their comments and suggestions during the development of the manuscript: Prof. James R. Barber, University of Michigan; Prof. Francesco Costanzo, Pennsylvania State University; Prof. Marijan Dravinski, University of Southern California; Prof. Mica Grujicic, Clemson University; Prof. Phil McLaughlin, Villanova University; Prof. Colin P. Ratcliffe, United States Naval Academy; Prof. Hassan Rejali, California State Polytechnic University at Pomona; Prof. Alan S. Wineman, University of Michigan.

Basic Ideas of Solid Mechanics

1-1 Problem Statement
1-2 Basic Idea I: Equilibrium and Stress
1-3 Basic Idea II: Deformation and Strain
1-4 Basic Idea III: Material Behavior: Stress-Strain Relations
1-5 Boundary Value Problems: Combination of Basic Ideas
1-6 Modeling and the Design Process
1-7 Failure and Failure Criteria
      1-7-1   Failure Criteria for Ductile Materials
      1-7-2   Failure Criteria for Brittle Materials
1-8 Factor of Safety and Working Stresses
1-9 Approximate Formulations: Mechanics of Materials
1-10 Summary
      References
      Exercises

1-1 Problem Statement

Given a solid body acted upon by contact forces applied to the surface of the body and/or body forces acting throughout the body, such as that shown in Figure 1-1, the two basic questions that are posed in the study of solid mechanics can be stated as follows:

1. How are the forces transmitted throughout the interior of the body?
2. What are the resulting deformations?

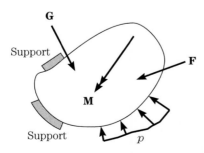

FIGURE 1-1 Solid body acted upon by contact and body forces.

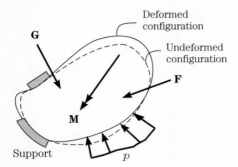

FIGURE 1-2 Undeformed and deformed configurations.

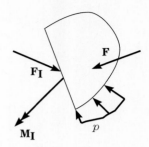

FIGURE 1-3 Section showing internal force resultants.

Deformations refer to the differences in the shape and size of the body before and after the loads are applied, that is, the natural shape and size of the body are in general changed by the application of the loads, as indicated in Figure 1-2.

Transmission of force refers to the *internal force distribution* resulting from the application of loads together with the types and locations of supports. The resultants associated with this internal force distribution can be determined by taking a section through the body, as indicated in Figure 1-3.

Requiring equilibrium would allow us to determine the resultant internal force $\mathbf{F_I}$ and a resultant internal moment $\mathbf{M_I}$ that are statically equivalent to the applied external forces on the section. There are, however, an infinite number of force *distributions* on the face of the section that would be statically equivalent to $\mathbf{F_I}$ and $\mathbf{M_I}$. The basic goal is to determine the internal force distribution corresponding to a specific set of loads and supports.

> The initial shape of the body, the location and type of supports and the loading are known quantities. The deformations and internal force distribution are unknown quantities. Roughly, the formulation of a solid mechanics problem involves establishing the framework within which the relationship(s) between the loads, the internal forces and the deformations can be determined.

In this chapter the three basic ideas common to all solid mechanics problems will be introduced. It will be demonstrated in general how these ideas enter into the formulation and solution of such problems. Failure and theories of failure as well as the idea of a factor of safety will be discussed as they apply to problems in stress analysis. In addition, the formulation and solution of approximate theories of solid mechanics will be investigated.

1-2 Basic Idea I: Equilibrium and Stress

The formulation of a problem in solid mechanics invariably involves three basic ideas or ingredients. One of these is the requirement that the body or any part of the body be in a state of equilibrium or that the motion be in accordance with Newton's second law.

1-2 BASIC IDEA I: EQUILIBRIUM AND STRESS

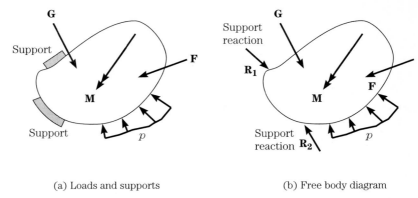

(a) Loads and supports (b) Free body diagram

FIGURE 1-4 Loads, supports and equilibrium of the entire body.

This is one of the basic physical laws that governs any and all solutions to problems in solid mechanics. This basic idea is frequently referred to as *kinetics*. As indicated in Figure 1-4b, this basic physical law can be used to establish relationships between the loads and the reactions by drawing a free body diagram (FBD) and requiring the entire body to be in equilibrium

As mentioned above (see Figure 1-3 and the accompanying discussion), equilibrium equations can be written for portions of the body obtained by sectioning to establish the force resultants acting on a section through the body. In the remainder of this section, the tools necessary to determine the internal force distributions will be investigated.

Stress at a Point. Consider a body subjected to contact forces and body forces, as shown in Figure 1-5. Section the body by a plane with normal **N** pointing away from the plane as shown in Figure 1-5a. On a small differential area ΔA containing the point P

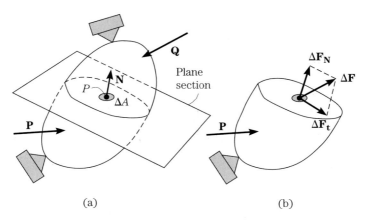

(a) (b)

FIGURE 1-5 Plane section for defining stress at a point.

there will be in general a resultant force $\mathbf{\Delta F}$, as shown in Figure 1-5b. Define the *stress vector at P for the plane whose normal is in the N direction* as

$$\mathbf{t_N}(P) = \lim_{\Delta A \to 0} \frac{\mathbf{\Delta F}}{\Delta A}$$

Note that the symbol $\mathbf{t_N}(P)$ denotes the choice of a point P in the body and then the choice of one of the infinite number of planes—namely, the particular plane whose normal is $\mathbf{N}$—that passes through P. Relative to the plane the stress vector $\mathbf{t_N}(P)$ can have any direction. If the portion of the body on the other side of the section is considered, the normal would be $-\mathbf{N}$ and the resultant force $-\mathbf{\Delta F}$ such that

$$\mathbf{t_{-N}}(P) = \lim_{\Delta A \to 0} \frac{-\mathbf{\Delta F}}{\Delta A} = -\lim_{\Delta A \to 0} \frac{\mathbf{\Delta F}}{\Delta A} = -\mathbf{t_N}(P)$$

That is, the stress vector on the opposite face would have the same magnitude but the opposite direction. This is of course Newton's third law as it applies to the stress vector.

With $\mathbf{N}$ and $\mathbf{t}$ as unit vectors normal to and in the plane, respectively, resolve the resultant force $\mathbf{\Delta F}$ into components $\mathbf{\Delta F_N} = \mathbf{N} \Delta F_N$ in the direction normal to the plane and $\mathbf{\Delta F_t} = \mathbf{t} \Delta F_t$ in a direction tangent to the plane. The force $\mathbf{\Delta F_N}$ can act either toward or away from the plane, and the force $\mathbf{\Delta F_t}$ can have any direction in the plane. Define the normal stress at P for the plane whose normal is in the N direction as

$$\sigma_N(P) = \lim_{\Delta A \to 0} \frac{\Delta F_N}{\Delta A}$$

As indicated, a normal stress is positive when it produces tension. Define the shear stress at P as

$$\tau_N(P) = \lim_{\Delta A \to 0} \frac{\Delta F_t}{\Delta A}$$

Clearly, stress has dimensions F/L^2. In the US customary system the units of stress are usually expressed in lbf/in^2, often abbreviated as psi. The units ksi = 1000 psi are also frequently used. In the SI system the units of stress are expressed in newton/m^2 or in Pascals, where 1 Pa = 1 N/m^2. It is very often the case that 1 MPa = 10^6 newton/m^2 is a useful measure of stress in the SI system.

Stress Notation. Consider the stresses acting at a point on an infinitesimal block referred to a rectangular coordinate system, as shown in Figure 1-6. There are two faces perpendicular to the x axis. The face whose normal points in the positive x direction is termed the positive face; the other, the negative face. For the positive x face the stresses σ_x, τ_{xy} and τ_{xz} act in the positive x, y and z directions, respectively. Hence for the shear stresses the first subscript denotes the direction of the normal to the face, and the second subscript denotes the direction of the stress component. Then, by virtue of Newton's third law the stresses σ_x, τ_{xy} and τ_{xz} act in the negative x, y and z directions, respectively, on the negative face. Similar statements apply for the components acting on the other faces.

1-2 BASIC IDEA I: EQUILIBRIUM AND STRESS

The collection of stresses

$$\mathbf{S} = \begin{bmatrix} \sigma_x & \tau_{xy} & \tau_{xz} \\ \tau_{yx} & \sigma_y & \tau_{yz} \\ \tau_{zx} & \tau_{zy} & \sigma_z \end{bmatrix}$$

shown as an array is referred to as the stress tensor. The elements of the first, second and third rows of $\mathbf{S}$ correspond to the stresses acting on planes perpendicular to the x, y and z axes, respectively. The student is asked in the exercises to show that moment equilibrium of the block shown in Figure 1-6 leads to the equalities

$$\tau_{yx} = \tau_{xy}$$
$$\tau_{zx} = \tau_{xz}$$
$$\tau_{zy} = \tau_{yz}$$

which is referred to as the symmetry of the stress tensor.

Stress Transformations. Consider an infinitesimal tetrahedron extracted from the interior of a stressed solid, as shown in Figure 1-7. The stresses on the negative x, y and z faces, respectively, are denoted by $-\mathbf{t_x}$, $-\mathbf{t_y}$ and $-\mathbf{t_z}$, where

$$\mathbf{t_x} = \mathbf{i}\sigma_x + \mathbf{j}\tau_{xy} + \mathbf{k}\tau_{xz}$$
$$\mathbf{t_y} = \mathbf{i}\tau_{yx} + \mathbf{j}\sigma_y + \mathbf{k}\tau_{yz}$$
$$\mathbf{t_z} = \mathbf{i}\tau_{zx} + \mathbf{j}\tau_{zy} + \mathbf{k}\sigma_z$$

The stress on the inclined face whose positive normal is given by $\mathbf{N} = \ell\mathbf{i} + m\mathbf{j} + n\mathbf{k}$ is denoted by $\mathbf{t_N}$. Requiring equilibrium of the tetrahedron leads to

$$-\mathbf{t_x}A_x - \mathbf{t_y}A_y - \mathbf{t_z}A_z + \mathbf{t_N}A_N = 0$$

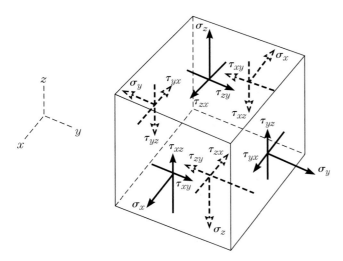

FIGURE 1-6 Notation for stresses in rectangular coordinates.

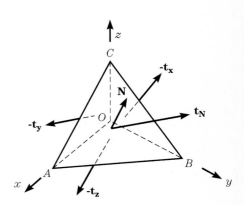

FIGURE 1-7 Equilibrium of a tetrahedron.

or, since $A_x/A_N = \ell$, $A_y/A_N = m$ and $A_z/A_N = n$, to

$$\begin{aligned}
\mathbf{t_N} &= \ell \mathbf{t_x} + m \mathbf{t_y} + n \mathbf{t_z} \\
&= \mathbf{i}(\ell \sigma_x + m \tau_{yx} + n \tau_{zx}) \\
&\quad + \mathbf{j}(\ell \tau_{xy} + m \sigma_y + n \tau_{zy}) \\
&\quad + \mathbf{k}(\ell \tau_{xz} + m \tau_{yz} + n \sigma_z) \\
&= \mathbf{i} t_{Nx} + \mathbf{j} t_{Ny} + \mathbf{k} t_{Nz}
\end{aligned}$$

where

$$\begin{aligned}
t_{Nx} &= \ell \sigma_x + m \tau_{yx} + n \tau_{zx} = \ell \sigma_x + m \tau_{xy} + n \tau_{xz} \\
t_{Ny} &= \ell \tau_{xy} + m \sigma_y + n \tau_{zy} = \ell \tau_{xy} + m \sigma_y + n \tau_{yz} \\
t_{Nz} &= \ell \tau_{xz} + m \tau_{yz} + n \sigma_z = \ell \tau_{xz} + m \tau_{yz} + n \sigma_z
\end{aligned}$$

for the components of the stress vector $\mathbf{t_N}$ on the inclined plane in terms of the stress components on the coordinate faces. The normal component of the vector $\mathbf{t_N}$, which amounts to the normal stress on the plane whose normal is in the N direction, is given by

$$\begin{aligned}
\sigma_N &= \mathbf{N} \cdot \mathbf{t_N} \\
&= \ell^2 \sigma_x + \ell m(\tau_{yx} + \tau_{xy}) + \ell n(\tau_{zx} + \tau_{xz}) \\
&\quad + mn(\tau_{zy} + \tau_{yz}) + m^2 \sigma_y + n^2 \sigma_z
\end{aligned}$$

or, upon using the symmetry of the stresses, by

$$\sigma_N = \ell^2 \sigma_x + m^2 \sigma_y + n^2 \sigma_z + 2\ell m \tau_{xy} + 2 mn \tau_{yz} + 2\ell n \tau_{xz} \qquad (1\text{-}1)$$

Viewed along a line perpendicular to the plane of $\mathbf{N}$ and $\mathbf{t_N}$, as shown in Figure 1-8, it is easily seen that the total shear stress $\tau_N(P)$ is given by

$$\tau_N = \sqrt{\mathbf{t_N} \cdot \mathbf{t_N} - \sigma_N^2} = \sqrt{t_{Nx}^2 + t_{Ny}^2 + t_{Nz}^2 - \sigma_N^2}$$

Thus, on a plane whose normal is in the N direction the stress vector can be expressed in terms of the quantities σ_N and τ_N as

$$\mathbf{t_N} = \mathbf{N} \sigma_N + \mathbf{t} \tau_N$$

where $\mathbf{t}$ is a unit vector perpendicular to $\mathbf{N}$ and lying in the plane of $\mathbf{N}$ and $\mathbf{t_N}$.

Given the stresses $\mathbf{S}$ at a point P expressed in the xyz coordinate system, it is frequently useful to be able to determine the corresponding stresses in another orthogonal coordinate system at P. Figure 1-9 shows the geometry of the situation along with

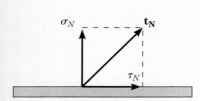

FIGURE 1-8 Normal and shear stresses.

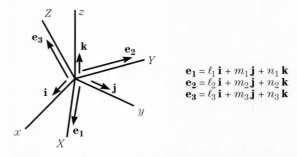

FIGURE 1-9 Direction cosines for the transformation of stress.

the mutually orthogonal unit vectors $\mathbf{e_1}$, $\mathbf{e_2}$ and $\mathbf{e_3}$ expressed in terms of the unit vectors $\mathbf{i}$, $\mathbf{j}$ and $\mathbf{k}$ and the three sets of direction cosines (ℓ_1, m_1, n_1), (ℓ_2, m_2, n_2) and (ℓ_3, m_3, n_3) given by

$$\ell_1 = \cos(x, X), \qquad m_1 = \cos(y, X), \qquad n_1 = \cos(z, X)$$
$$\ell_2 = \cos(x, Y), \qquad m_2 = \cos(y, Y), \qquad n_2 = \cos(z, Y)$$
$$\ell_3 = \cos(x, Z), \qquad m_3 = \cos(y, Z), \qquad n_3 = \cos(z, Z)$$

The normal stress σ_X can be computed as follows: First compute the stress vector $\mathbf{t_1}$ (or $\mathbf{t_X}$) acting on the face whose normal is $\mathbf{e_1}$ (or X) according to

$$\mathbf{t_1} = \ell_1 \mathbf{t_x} + m_1 \mathbf{t_y} + n_1 \mathbf{t_z}$$

The normal stress σ_X is then

$$\sigma_X = \mathbf{e_1} \cdot \mathbf{t_1} \qquad (1\text{-}2a)$$
$$= \ell_1^2 \sigma_x + m_1^2 \sigma_y + n_1^2 \sigma_z + 2\ell_1 m_1 \tau_{xy} + 2m_1 n_1 \tau_{yz} + 2\ell_1 n_1 \tau_{xz}$$

with the two shear stresses given similarly as

$$\tau_{XY} = \mathbf{e_2} \cdot \mathbf{t_1} = \ell_1 \ell_2 \sigma_x + m_1 m_2 \sigma_y + n_1 n_2 \sigma_z$$
$$+ (\ell_1 m_2 + \ell_2 m_1)\tau_{xy} + (m_1 n_2 + m_2 n_1)\tau_{yz} \qquad (1\text{-}2b)$$
$$+ (\ell_1 n_2 + \ell_2 n_1)\tau_{xz}$$

and

$$\tau_{XZ} = \mathbf{e_3} \cdot \mathbf{t_1} = \ell_1 \ell_3 \sigma_x + m_1 m_3 \sigma_y + n_1 n_3 \sigma_z$$
$$+ (\ell_1 m_3 + \ell_3 m_1)\tau_{xy} + (m_1 n_3 + m_3 n_1)\tau_{yz} \qquad (1\text{-}2c)$$
$$+ (\ell_1 n_3 + \ell_3 n_1)\tau_{xz}$$

Then with $\mathbf{t_2} = \ell_2 \mathbf{t_x} + m_2 \mathbf{t_y} + n_2 \mathbf{t_z}$ and $\mathbf{t_3} = \ell_3 \mathbf{t_x} + m_3 \mathbf{t_y} + n_3 \mathbf{t_z}$, it follows in a completely similar fashion that

$$\tau_{YX} = \mathbf{e_1} \cdot \mathbf{t_2} = \ell_1 \ell_2 \sigma_x + m_1 m_2 \sigma_y + n_1 n_2 \sigma_z$$
$$+ (\ell_1 m_2 + \ell_2 m_1)\tau_{xy} + (m_1 n_2 + m_2 n_1)\tau_{yz} \qquad (1\text{-}2d)$$
$$+ (\ell_1 n_2 + \ell_2 n_1)\tau_{xz}$$
$$= \tau_{XY} = \mathbf{e_2} \cdot \mathbf{t_1}$$

$$\sigma_Y = \mathbf{e_2} \cdot \mathbf{t_2} = \ell_2^2 \sigma_x + m_2^2 \sigma_y + n_2^2 \sigma_z \qquad (1\text{-}2e)$$
$$+ 2\ell_2 m_2 \tau_{xy} + 2m_2 n_2 \tau_{yz} + 2\ell_2 n_2 \tau_{xz}$$

$$\tau_{YZ} = \mathbf{e_3} \cdot \mathbf{t_2} = \ell_3 \ell_2 \sigma_x + m_3 m_2 \sigma_y + n_3 n_2 \sigma_z$$
$$+ (\ell_3 m_2 + \ell_2 m_3)\tau_{xy} + (m_3 n_2 + m_2 n_3)\tau_{yz} \qquad (1\text{-}2f)$$
$$+ (\ell_3 n_2 + \ell_2 n_3)\tau_{xz}$$

and

$$\tau_{ZX} = \mathbf{e_1} \cdot \mathbf{t_3} = \ell_1 \ell_3 \sigma_x + m_1 m_3 \sigma_y + n_1 n_3 \sigma_z$$
$$+ (\ell_1 m_3 + \ell_3 m_1)\tau_{xy} + (m_1 n_3 + m_3 n_1)\tau_{yz} \qquad (1\text{-}2g)$$
$$+ (\ell_1 n_3 + \ell_3 n_1)\tau_{xz}$$
$$= \tau_{XZ} = \mathbf{e_3} \cdot \mathbf{t_1}$$

$$\begin{aligned}
\tau_{ZY} = \mathbf{e_2} \cdot \mathbf{t_3} &= \ell_3\ell_2\sigma_x + m_3m_2\sigma_y + n_3n_2\sigma_z \\
&+ (\ell_3m_2 + \ell_2m_3)\tau_{xy} + (m_3n_2 + m_2n_3)\tau_{yz} \\
&+ (\ell_3n_2 + \ell_2n_3)\tau_{xz} \\
&= \tau_{YZ} = \mathbf{e_3} \cdot \mathbf{t_2}
\end{aligned} \quad (1\text{-}2\text{h})$$

$$\begin{aligned}
\sigma_Z = \mathbf{e_3} \cdot \mathbf{t_3} &= \ell_3^2\sigma_x + m_3^2\sigma_y + n_3^2\sigma_z \\
&+ 2\ell_3m_3\tau_{xy} + 2m_3n_3\tau_{yz} + 2\ell_3n_3\tau_{xz}
\end{aligned} \quad (1\text{-}2\text{i})$$

In the *XYZ* coordinate system the stresses are thus represented as

$$\mathbf{S}' = \begin{bmatrix} \sigma_X & \tau_{XY} & \tau_{XZ} \\ \tau_{YX} & \sigma_Y & \tau_{YZ} \\ \tau_{ZX} & \tau_{ZY} & \sigma_Z \end{bmatrix}$$

Equations (1-2a–i) are referred to as the *stress transformation equations* in that given the components **S** of stress in an *xyz* coordinate system, and a rotational transformation from the *xyz* system to the *XYZ* system, the components **S'** in the *XYZ* coordinate system can be determined. The set of transformation Equations (1-2a–i) are precisely the equations that govern the behavior of a second-order cartesian tensor upon rotation of the coordinate system, hence the designation *stress tensor*.

Principal Stresses. With the stress components at a point being in general different for each plane or direction, the question as to the direction(s) for which the stresses assume maximum values naturally occurs. It can be shown that the maximum values of the normal stresses occur on planes where there is no shear stress, that is, where the stress vector is normal to the plane. Alternatively, principal stresses are frequently *defined* as the normal stresses that act on planes on which there are no shear stresses. In any event, with the normal given by $\mathbf{N} = \ell\mathbf{i} + m\mathbf{j} + n\mathbf{k}$ this condition can be characterized as $\mathbf{t_N} = \mathbf{N}\sigma$ or

$$\begin{aligned}
&\mathbf{i}(\ell\sigma_x + m\tau_{xy} + n\tau_{xz}) \\
&+ \mathbf{j}(\ell\tau_{xy} + m\sigma_y + n\tau_{yz}) \\
&+ \mathbf{k}(\ell\tau_{xz} + m\tau_{yz} + n\sigma_z) = \sigma(\ell\mathbf{i} + m\mathbf{j} + n\mathbf{k})
\end{aligned}$$

where σ is the unknown value of the normal stress on such a face. Equating coefficients yields the three equations

$$\begin{aligned}
\ell(\sigma_x - \sigma) + m\tau_{xy} + n\tau_{xz} &= 0 \\
\ell\tau_{xy} + m(\sigma_y - \sigma) + n\tau_{yz} &= 0 \\
\ell\tau_{xz} + m\tau_{yz} + n(\sigma_z - \sigma) &= 0
\end{aligned}$$

that is, the classic linear algebraic eigenvalue problem, which can be written as

$$(\mathbf{S} - \sigma\mathbf{I})\mathbf{N} = \mathbf{0}$$

showing that the principal stresses at a point are the eigenvalues of the stress tensor at that point. The corresponding eigenvectors represent the normals to the planes on which the principal stresses act. The symmetry of the tensor $\mathbf{S}$ guarantees that the principal stresses (eigenvalues) are real and that the principal directions (eigenvectors) are mutually orthogonal.

The values of the principal stresses are determined by requiring that the determinant $\det(\mathbf{S} - \sigma \mathbf{I}) = 0$, leading to the characteristic equation

$$\sigma^3 - I_1 \sigma^2 + I_2 \sigma - I_3 = 0$$

where

$$I_1 = \sigma_x + \sigma_y + \sigma_z$$
$$I_2 = \sigma_x \sigma_y + \sigma_y \sigma_z + \sigma_z \sigma_x - \tau_{xy}^2 - \tau_{yz}^2 - \tau_{xz}^2$$
$$I_3 = \det(\mathbf{S})$$

are invariants of the stress tensor, meaning in terms of I_1, for example, that the sum of the normal stresses on any three mutually orthogonal planes is a constant, that is,

$$\sigma_x + \sigma_y + \sigma_z = \sigma_X + \sigma_Y + \sigma_Z$$

Factoring the characteristic equation in terms of the principal stresses σ_1, σ_2 and σ_3 as

$$\sigma^3 - I_1 \sigma^2 + I_2 \sigma - I_3 = (\sigma - \sigma_1)(\sigma - \sigma_2)(\sigma - \sigma_3)$$

it is easily shown that the invariants can be expressed in terms of the principal stresses as

$$I_1 = \sigma_1 + \sigma_2 + \sigma_3$$
$$I_2 = \sigma_1 \sigma_2 + \sigma_2 \sigma_3 + \sigma_3 \sigma_1$$
$$I_3 = \sigma_1 \sigma_2 \sigma_3$$

◆**EXAMPLE 1-1** Consider the stresses at a point given by

$$\mathbf{S} = \begin{bmatrix} 2 & -\sqrt{3} & 0 \\ -\sqrt{3} & 4 & 1 \\ 0 & 1 & 2 \end{bmatrix}$$

The principal stresses are determined by solving $\det(\mathbf{S} - \sigma \mathbf{I}) = 0$, leading to the characteristic equation $(2 - \sigma)[(4 - \sigma)(2 - \sigma) - 4] = 0$ with roots $\sigma_1 = 0.764, \sigma_2 = 2.000$ and $\sigma_3 = 5.236$. Note that $\sigma_1 + \sigma_2 + \sigma_3 = 8$ as required. The corresponding directions are determined by solving, one at a time, the dependent set of equations

$$(\mathbf{S} - \sigma_i \mathbf{I})\mathbf{N_i} = \mathbf{0} \quad i = 1, 2, 3$$

For $i = 1$ there results

$$1.236 \, N_1 - 1.732 \, N_2 + 0 \, N_3 = 0$$
$$-1.732 \, N_1 + 3.236 \, N_2 + 1 \, N_3 = 0$$
$$0 \, N_1 + 1 \, N_2 + 1.236 \, N_3 = 0$$

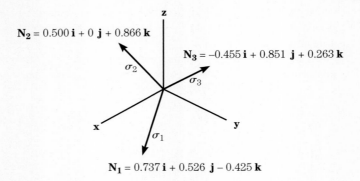

FIGURE 1-10 Principal directions.

with solution $\mathbf{N_1} = [0.737\ 0.526\ -0.425]^T$. Repeating these calculations for the other two principal stresses results collectively in

$$\sigma_1 = 0.764 \quad \mathbf{N_1} = [0.737\ 0.526\ -0.425]^T.$$
$$\sigma_2 = 2.000 \quad \mathbf{N_2} = [0.500\ 0.000\ 0.866]^T.$$
$$\sigma_3 = 5.236 \quad \mathbf{N_3} = [-0.455\ 0.851\ 0.263]^T.$$

with the geometry of the principal directions shown in Figure 1-10. The student should check that the unit vectors indicating the principal directions are mutually orthogonal. ◆

Denoting $\sigma_1 \geq \sigma_2 \geq \sigma_3$ as the principal stresses, it can be shown that the absolute maximum shear stress occurs on the plane that bisects the two principal planes on which σ_1 and σ_3 act, and has the value $\tau_{max} = (\sigma_1 - \sigma_3)/2$. The values of the shear stresses on the other two planes that bisect principal planes are $|\sigma_3 - \sigma_2|/2$ and $|\sigma_2 - \sigma_1|/2$. This situation is pictured in Figure 1-11 where the radii of the three circles pictured represent the values of the three corresponding shear stresses.

Plane Stress. There are frequently important technical applications where the body in question is a thin plate subjected to loadings parallel to the plane of the plate. Thin means that the thickness h is very small compared to the in-plane dimensions of the plate, as shown in Figure 1-12.

Assuming that the top and bottom surfaces of the plate are stress-free, that is, that the stresses σ_z, τ_{xz} and τ_{yz} are zero there, it is argued that these components are small everywhere in the plate and can hence, as a first approximation, be ignored. The remaining stress components, namely, σ_x, σ_y and τ_{xy}, are assumed to depend only upon the in-plane coordinates x and y. The plane $z = 0$ is taken to be located midway between the top and bottom surfaces of the plate. Contact stresses applied around the boundary of the plate as well as body forces throughout the volume of the plate are assumed to be constant over the thickness. The class of problems resulting from these assumptions is referred to as *plane stress*.

At a point P in the plate shown in Figure 1-13a, the stresses on a typical infinitesimal block are as shown in Figure 1-13b. On an inclined face, representing a differ-

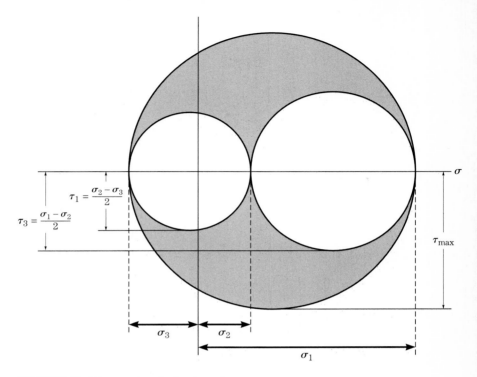

FIGURE 1-11 Extreme values for the shear stresses.

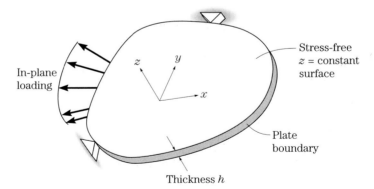

FIGURE 1-12 Plane stress geometry and loading.

ent coordinate system at the point P, the normal and shear stresses are as shown in Figure 1-13c. The reader should show by considering the equilibrium of the wedge pictured in Figure 1-13c that

$$\sigma_X = \sigma_x \cos^2 \theta + \sigma_y \sin^2 \theta + 2\tau_{xy} \sin \theta \cos \theta \tag{1-3a}$$

$$\tau_{XY} = -(\sigma_x - \sigma_y) \sin \theta \cos \theta + \tau_{xy}(\cos^2 \theta - \sin^2 \theta) \tag{1-3b}$$

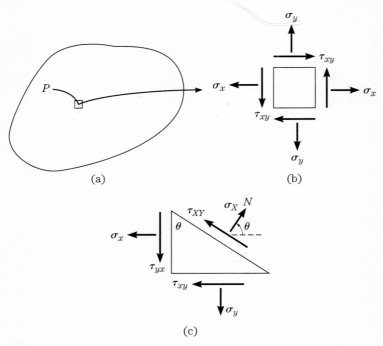

FIGURE 1-13 Stress transformation for plane stress.

By taking $\theta = \theta + \pi/2$ there also results

$$\sigma_Y = \sigma_x \sin^2 \theta + \sigma_y \cos^2 \theta - 2\tau_{xy} \sin \theta \cos \theta \qquad (1\text{-}3c)$$

The reader should also show that these same results obtain by specializing the transformation Equations (1-2a–i).

If Equations (1-3a–c) are put in double-angle form, the results are

$$\sigma_X = \frac{1}{2}(\sigma_x + \sigma_y) + \frac{1}{2}(\sigma_x - \sigma_y) \cos 2\theta + \tau_{xy} \sin 2\theta \qquad (1\text{-}4a)$$

$$\sigma_Y = \frac{1}{2}(\sigma_x + \sigma_y) - \frac{1}{2}(\sigma_x - \sigma_y) \cos 2\theta - \tau_{xy} \sin 2\theta \qquad (1\text{-}4b)$$

$$\tau_{XY} = -\frac{1}{2}(\sigma_x - \sigma_y) \sin 2\theta + \tau_{xy} \cos 2\theta \qquad (1\text{-}4c)$$

These equations can be used for a graphical construction called Mohr's circle as follows: Transfer the first term on the right side of Equation (1-4a) to the left side, square it and add it to the square of Equation (1-4c) to obtain

$$\left[\sigma_X - \frac{1}{2}(\sigma_x + \sigma_y)\right]^2 + \tau_{XY}^2 = \frac{1}{4}(\sigma_x - \sigma_y)^2 + \tau_{xy}^2$$

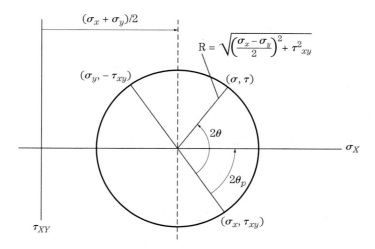

FIGURE 1-14 Mohr's circle for plane stress.

With σ_X and τ_{XY} as the abscissa and ordinate, respectively, this is the equation of a circle whose center has coordinates $[(\sigma_x + \sigma_y)/2, 0]$ and whose radius is

$$R = \sqrt{\frac{1}{4}(\sigma_x - \sigma_y)^2 + \tau_{xy}^2}$$

as shown in Figure 1-14.

With τ_{xy} positive down as shown, the stresses on the positive x face are plotted as (σ_x, τ_{xy}) and the stresses on the positive y face as $(\sigma_y, -\tau_{xy})$. The stresses on a face of the block rotated through an angle θ from the x face are then determined from the coordinates of the point (σ, τ) obtained by rotating the diameter of Mohr's circle through an angle 2θ in the same direction. The principal stresses, occurring on faces where there is no shear stress, are clearly given by

$$\sigma_{max,min} = \frac{\sigma_x + \sigma_y}{2} \pm \sqrt{\left(\frac{\sigma_x - \sigma_y}{2}\right)^2 + \tau_{xy}^2}$$

The corresponding angles made by the principal planes are given by

$$\tan(2\theta_p) = \frac{2\tau_{xy}}{\sigma_x - \sigma_y}$$

The maximum in-plane shear stresses, equal in magnitude to the radius of the circle, occur on planes that bisect the principal planes. The absolute maximum shear stress, discussed in detail in section 1-7, turns out to be $\max(|\sigma_1/2|, |\sigma_2/2|, |(\sigma_1 - \sigma_2)/2|)$ where $\sigma_1 = \sigma_{max}$ and $\sigma_2 = \sigma_{min}$.

Stress Equilibrium. Decisions as to the strength of a structure are made primarily on the basis of comparing the maximum stresses in the structure to allowable stresses in materials from which the structure is composed. For this reason, the basic kinetic or force variable in solid mechanics is generally taken to be stress.

By the very nature of the problem, equilibrium equations are written with respect to the deformed configuration of the body, that is, the configuration that results after the loads are applied. Hence implicit in the development of the equations of equilibrium is the assumption of small deformations, that is, we do not distinguish between undeformed and deformed coordinates. In situations where large deformations occur it is necessary to alter the equilibrium equations to account for them. This results in a set of nonlinear equations that are beyond the scope of this text.

In any event, it seems logical that the kinetic equations that govern the behavior of a structure should be expressed in terms of the stresses. To this end consider the equilibrium of a typical plane stress differential element in the interior of a body, as shown in Figure 1-15. Shown are the stresses, the changes in the stresses for the case of plane stress, and body forces B_x and B_y per unit volume. Multiplying the stresses by the appropriate areas, multiplying the body forces by the volume of the element and summing forces in the x and y directions, respectively, yields

$$\frac{\partial \sigma_x}{\partial x} + \frac{\partial \tau_{xy}}{\partial y} + B_x = 0$$

$$\frac{\partial \tau_{yx}}{\partial x} + \frac{\partial \sigma_y}{\partial y} + B_y = 0$$

(1-5)

It is left for the student to show by summing moments about an axis perpendicular to the plane that, as mentioned above without proof, $\tau_{xy} = \tau_{yx}$, that is, the stress ten-

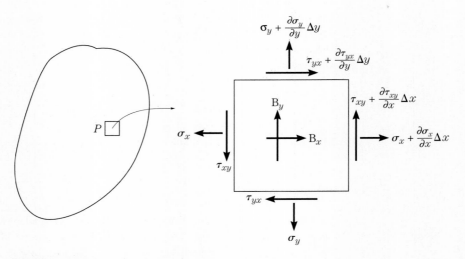

FIGURE 1-15 Differential equations of equilibrium for plane stress.

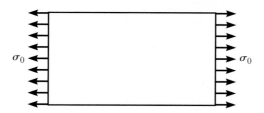

FIGURE 1-16 Stress boundary conditions.

sor is symmetric. Developed in an analogous manner, the corresponding fully three-dimensional equations appear as

$$\frac{\partial \sigma_x}{\partial x} + \frac{\partial \tau_{xy}}{\partial y} + \frac{\partial \tau_{xz}}{\partial z} + B_x = 0$$

$$\frac{\partial \tau_{xy}}{\partial x} + \frac{\partial \sigma_y}{\partial y} + \frac{\partial \tau_{yz}}{\partial z} + B_y = 0 \qquad (1\text{-}6)$$

$$\frac{\partial \tau_{xz}}{\partial x} + \frac{\partial \tau_{yz}}{\partial y} + \frac{\partial \sigma_z}{\partial z} + B_z = 0$$

where the symmetry of the stresses has already been assumed.

As functions of position, any stresses must satisfy these equations, or their counterparts in other coordinate systems, in order to be considered valid for a solid mechanics problem. Equilibrium equations in other coordinate systems are left for the homework exercises.

The stresses that satisfy equilibrium in the interior of the body must also satisfy any force-type conditions at the boundary. As an example consider a rectangular region acted upon by constant surface tractions on the x faces, as shown in Figure 1-16. The stresses in the interior of the rectangle would have to be such that as the two vertical faces are approached $\sigma_x \to \sigma_0$ and $\tau_{xy} \to 0$ and that as the horizontal faces are approached $\sigma_y \to 0$ and $\tau_{xy} \to 0$. The more general case will be discussed in the section on boundary value problems toward the end of the chapter.

1-3 Basic Idea II: Deformation and Strain

The second basic idea of solid mechanics arises in connection with the need to be able to describe the deformations that are associated with the differences in the undeformed and deformed configurations. This basic ingredient is sometimes referred to as *kinematics*. In the study of statics and dynamics bodies are generally assumed to be rigid and hence undergo what are termed rigid body motions. The kinematics that is appropriate for discussing the motion of a rigid body deals with translations and rotations.

A translation, such as that indicated in Figure 1-17a, is a motion such that the displacement **u** of each point in the body is the same. On the other hand, a rotation about a point O results in a different displacement **u** for each point in the body, that is, $\mathbf{u}(P) \neq \mathbf{u}(Q)$, as indicated in Figure 1-17b. Any rigid body motion can be considered a

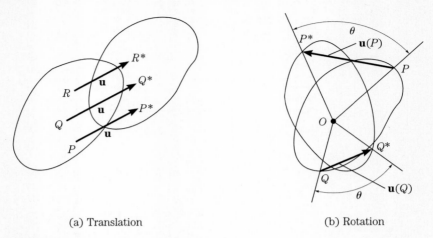

(a) Translation (b) Rotation

FIGURE 1-17 Rigid body translation and rotation.

suitable combination of a translation and a rotation. Rigid body translations and rotations change only the position and orientation of the body.

The distance between any two points in the body is unchanged as a result of any rigid body motion.

When formulating the kinematics for a deformable body it is necessary to be able to describe changes in the size of the body and changes in the shape of the body in addition to any rigid body motions that may result. A typical deformation indicating both of these changes is shown in Figure 1-18. A rectangle $ABCD$ located within the body deforms into the quadrilateral $A^*B^*C^*D^*$. In connection with the deformation there are generally changes in lengths ($\ell_{A^*B^*} \neq \ell_{AB}$) and changes in angles (angle $B^*A^*C^* \neq$ angle BAC) between line segments. These changes in size and shape can be related to quantities termed *displacement gradients*. Strains in turn are defined in terms

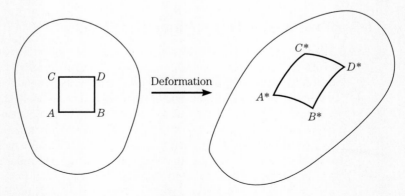

FIGURE 1-18 Deformation resulting in changes in size and shape.

1-3 BASIC IDEA II: DEFORMATION AND STRAIN

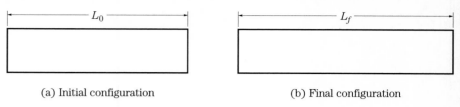

(a) Initial configuration (b) Final configuration

FIGURE 1-19 Extensional deformation.

of displacement gradients by what are generally referred to as *strain displacement relations*.

A very simple example of a change in the size of a body is shown in Figure 1-19, where a rectangle is deformed into another rectangle with one of the dimensions of the rectangle increased uniformly from L_0 to L_f. Such a deformation is referred to as a *uniform elongation* or a *uniform extension*.

A quantitative measure of the elongation is the change in length divided by the original length, termed the engineering strain, and is defined as

$$e = \frac{L_f - L_0}{L_0}$$

Note that the engineering strain e is dimensionless.

A very simple example of a change in shape is shown in Figure 1-20, where a rectangle has been deformed into a rhombic shape. This is referred to as a shear deformation. The shear strain γ is defined as the decrease in the right angle DAB, that is,

$$\gamma = \frac{\pi}{2} - \text{angle } D^*A^*B^*$$

measured in radians, so that shear strain is also dimensionless. In the following section deformation and strain will be treated for more general situations.

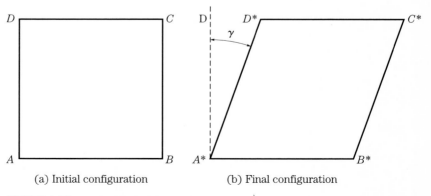

(a) Initial configuration (b) Final configuration

FIGURE 1-20 Shear deformation.

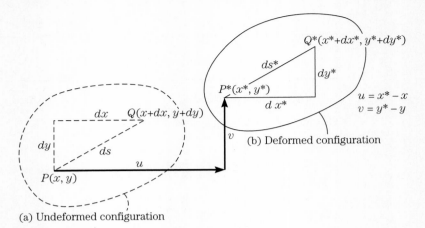

(a) Undeformed configuration

FIGURE 1-21 General two-dimensional deformation.

Strain-Displacement Relations. Consider the thin, arbitrarily shaped plate indicated in Figure 1-21. Typical undeformed and deformed configurations are pictured. Before deformation P is located at the point (x, y) and Q at the point $(x + dx, y + dy)$ with the square of the length of the infinitesimal line segment PQ as

$$ds^2 = dx^2 + dy^2$$

After deformation the coordinates of P^* are (x^*, y^*) and those of Q are $(x^* + dx^*, y^* + dy^*)$. The line P^*Q^* generally has a different length and orientation with the deformed length as

$$ds^{*2} = dx^{*2} + dy^{*2}$$

The displacements $u = x^* - x$ and $v = y^* - y$ in the x and y directions, respectively, are indicated in Figure 1-21. It then follows that $dx^* = dx + du$ and $dy^* = dy + dv$ so that

$$ds^{*2} = dx^2 + dy^2 + 2dx\,du + 2dy\,dv + du^2 + dv^2$$

and

$$ds^{*2} - ds^2 = 2dx\,du + 2dy\,dv + du^2 + dv^2$$

With u a function of x and y it follows that

$$du = \frac{\partial u}{\partial x}\,dx + \frac{\partial u}{\partial y}\,dy$$

and

$$dv = \frac{\partial v}{\partial x}\,dx + \frac{\partial v}{\partial y}\,dy$$

where $\partial u/\partial x$, $\partial u/\partial y$, $\partial v/\partial x$ and $\partial v/\partial y$ are the displacement gradients for the plane problem. It then follows that

$$ds^{*2} - ds^2 = 2\left[\frac{\partial u}{\partial x} + \frac{1}{2}\left\{\left(\frac{\partial u}{\partial x}\right)^2 + \left(\frac{\partial v}{\partial x}\right)^2\right\}\right]dx^2$$
$$+ 2\left[\frac{\partial v}{\partial y} + \frac{1}{2}\left\{\left(\frac{\partial u}{\partial y}\right)^2 + \left(\frac{\partial v}{\partial y}\right)^2\right\}\right]dy^2$$
$$+ 2\left[\frac{\partial u}{\partial y} + \frac{\partial v}{\partial x} + \frac{\partial u}{\partial x}\frac{\partial u}{\partial y} + \frac{\partial v}{\partial x}\frac{\partial v}{\partial y}\right]dx\,dy$$

The extensional strain for the line segment PQ is then *defined* as

$$\epsilon_n(P) = \frac{ds^{*2} - ds^2}{2ds^2} = \ell^2\epsilon_x + m^2\epsilon_y + \ell m \gamma_{xy} \tag{1-7}$$

where

$$\epsilon_x = \frac{\partial u}{\partial x} + \frac{1}{2}\left[\left(\frac{\partial u}{\partial x}\right)^2 + \left(\frac{\partial v}{\partial x}\right)^2\right] \tag{1-8a}$$

and

$$\epsilon_y = \frac{\partial v}{\partial y} + \frac{1}{2}\left[\left(\frac{\partial u}{\partial y}\right)^2 + \left(\frac{\partial v}{\partial y}\right)^2\right] \tag{1-8b}$$

are the extensional strains of line segments along the x and y axes, respectively, and

$$\gamma_{xy} = \frac{\partial u}{\partial y} + \frac{\partial v}{\partial x} + \frac{\partial u}{\partial x}\frac{\partial u}{\partial y} + \frac{\partial v}{\partial x}\frac{\partial v}{\partial y} \tag{1-8c}$$

is the shear strain between the x and y axes. In Equation (1-7), $\ell = dx/ds = \cos(n,x)$ and $m = dy/ds = \cos(n,y)$ are the direction cosines of the line segment PQ with the n subscript of $\epsilon_n(P)$ indicating the direction of the line segment being considered at point P. The quantities ϵ_x, ϵ_y and γ_{xy} are called the *components of strain* with Equations (1-8a–c) referred to as the *strain-displacement relations*.

Unless specifically stated to the contrary it will be assumed in what follows that the displacements are continuous differentiable functions of position and that they are small compared to characteristic dimensions of the body. For most engineering solids the values of ϵ and γ are very small with maximum values on the order of 10^{-3}. Thus it will also be assumed that the derivatives are uniformly small so that squares and products can be neglected. The strain-displacement relations for the two-dimensional case can then be written as

$$\epsilon_x = \frac{\partial u}{\partial x}, \quad \epsilon_y = \frac{\partial v}{\partial y} \quad \text{and} \quad \gamma_{xy} = \frac{\partial u}{\partial y} + \frac{\partial v}{\partial x}$$

and are referred to as the *linear* strain-displacement relations.

A simple interpretation of the extensional strain ϵ_x can be developed in connec-

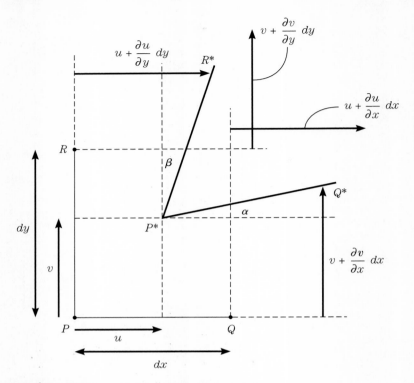

FIGURE 1-22 Geometric interpretation of shear strain.

tion with Figure 1-22 as follows. The length of the line segment P^*Q^* can be represented as

$$ds^* = \sqrt{\left(dx + u + \frac{\partial u}{\partial x}dx - u\right)^2 + \left(\frac{\partial v}{\partial x}dx\right)^2}$$

which, after using the binomial expansion and neglecting the products of the derivatives, can be expressed as

$$ds^* \approx (dx)\left(1 + \frac{\partial u}{\partial x}\right)$$

so that with $ds = dx$

$$\epsilon_x = \lim_{ds \to 0} \frac{ds^* - ds}{ds} = \frac{\partial u}{\partial x}$$

showing that the linear part of the extensional strain can be interpreted as a limit of a change in length divided by the original length. Similarly, and also in connection with Figure 1-22, a simple geometric interpretation of the linear portion of the shear strain γ_{xy} can be developed as follows.

The small angles α and β are given, respectively, by

$$\sin \alpha \approx \alpha = \frac{\partial v}{\partial x} \quad \text{and} \quad \sin \beta \approx \beta = \frac{\partial u}{\partial y}$$

so that the decrease in the right angle QPR is

$$\gamma = \alpha + \beta = \frac{\partial v}{\partial x} + \frac{\partial u}{\partial y}$$

which coincides with the small deformation shear strain γ_{xy}. γ_{xy} is clearly a measure of the decrease in right angle between the originally perpendicular x and y axes.

The student is asked in the exercises to show that, collectively, the strain transformation equations for the two-dimensional case can be expressed in terms of the direction cosines $\ell = \cos\theta$ and $m = \sin\theta$, shown in Figure 1-23 as

$$\begin{aligned}
\epsilon_X &= \epsilon_x \cos^2\theta + \epsilon_y \sin^2\theta + \gamma_{xy}\sin\theta\cos\theta \\
\epsilon_Y &= \epsilon_x \sin^2\theta + \epsilon_y \cos^2\theta - \gamma_{xy}\sin\theta\cos\theta \\
\gamma_{XY} &= \gamma_{xy}(\cos^2\theta - \sin^2\theta) - 2(\epsilon_x - \epsilon_y)\sin\theta\cos\theta
\end{aligned} \quad (1\text{-}9)$$

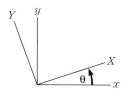

FIGURE 1-23 Direction cosines $\ell = \cos\theta$ and $m = \sin\theta$.

Equation (1-9) is an example of a *strain transformation equation*. Given the strains ϵ_x, ϵ_y and γ_{xy} in the xy system, the strains in any other orthogonal system at the point in question can be determined.

Note that if in the corresponding stress transformation equations developed in section 1-2, σ_x and σ_X are replaced by ϵ_x and ϵ_X, σ_y and σ_Y are replaced by ϵ_y and ϵ_Y, and τ_{xy} and τ_{XY} are replaced by $\gamma_{xy}/2$ and $\gamma_{XY}/2$, the above strain transformation equations result. Thus it follows that Mohr's circle can be used in exactly the same manner as for plane stress if the points $(\epsilon_x, \gamma_{xy}/2)$ and $(\epsilon_y, -\gamma_{xy}/2)$ are used to plot the circle.

It also then follows that the collection of strains

$$\boldsymbol{\epsilon} = \begin{bmatrix} \epsilon_x & \epsilon_{xy} \\ \epsilon_{xy} & \epsilon_y \end{bmatrix}$$

behaves as a two-dimensional cartesian tensor. The quantity $\epsilon_{xy} = \gamma_{xy}/2$ is referred to as the *tensorial component of shear strain* with the quantity γ_{xy} commonly called the *engineering shear strain*.

The student is asked to show in the exercises that the strain-displacement relations for the three-dimensional case are

$$\begin{aligned}
\epsilon_x &= \frac{\partial u}{\partial x} + \frac{1}{2}\left[\left(\frac{\partial u}{\partial x}\right)^2 + \left(\frac{\partial v}{\partial x}\right)^2 + \left(\frac{\partial w}{\partial x}\right)^2\right] \\
\epsilon_y &= \frac{\partial v}{\partial y} + \frac{1}{2}\left[\left(\frac{\partial u}{\partial y}\right)^2 + \left(\frac{\partial v}{\partial y}\right)^2 + \left(\frac{\partial w}{\partial y}\right)^2\right] \\
\epsilon_z &= \frac{\partial w}{\partial z} + \frac{1}{2}\left[\left(\frac{\partial u}{\partial z}\right)^2 + \left(\frac{\partial v}{\partial z}\right)^2 + \left(\frac{\partial w}{\partial z}\right)^2\right] \\
\gamma_{xy} &= \frac{\partial u}{\partial y} + \frac{\partial v}{\partial x} + \frac{\partial u}{\partial x}\frac{\partial u}{\partial y} + \frac{\partial v}{\partial x}\frac{\partial v}{\partial y} + \frac{\partial w}{\partial x}\frac{\partial w}{\partial y} = 2\epsilon_{xy} \\
\gamma_{xz} &= \frac{\partial u}{\partial z} + \frac{\partial w}{\partial x} + \frac{\partial u}{\partial x}\frac{\partial u}{\partial z} + \frac{\partial v}{\partial x}\frac{\partial v}{\partial z} + \frac{\partial w}{\partial x}\frac{\partial w}{\partial z} = 2\epsilon_{xz} \\
\gamma_{yz} &= \frac{\partial w}{\partial y} + \frac{\partial v}{\partial z} + \frac{\partial u}{\partial y}\frac{\partial u}{\partial z} + \frac{\partial v}{\partial y}\frac{\partial v}{\partial z} + \frac{\partial w}{\partial y}\frac{\partial w}{\partial z} = 2\epsilon_{yz}
\end{aligned} \quad (1\text{-}10)$$

where w is the displacement in the z direction. The corresponding linear strain-displacement relations for the three-dimensional case are clearly

$$\epsilon_x = \frac{\partial u}{\partial x}, \quad \epsilon_y = \frac{\partial v}{\partial y}, \quad \epsilon_z = \frac{\partial w}{\partial z} \tag{1-11}$$

$$\gamma_{xy} = 2\epsilon_{xy} = \frac{\partial u}{\partial y} + \frac{\partial v}{\partial x}, \quad \gamma_{xz} = 2\epsilon_{xz} = \frac{\partial u}{\partial z} + \frac{\partial w}{\partial x}, \quad \gamma_{yz} = 2\epsilon_{yz} = \frac{\partial w}{\partial y} + \frac{\partial v}{\partial z}$$

In either case the strain tensor is given by

$$\boldsymbol{\epsilon} = \begin{bmatrix} \epsilon_x & \epsilon_{xy} & \epsilon_{xz} \\ \epsilon_{xy} & \epsilon_y & \epsilon_{yz} \\ \epsilon_{xz} & \epsilon_{yz} & \epsilon_z \end{bmatrix}$$

◆EXAMPLE 1-2 Consider the uniform deformation of a rectangle, as shown in Figure 1-24. The original and deformed configurations are indicated. The constant values of the displacement gradients are given by

$$\frac{\partial u}{\partial x} = \frac{3.0025 - 3}{3} = 0.000833$$

$$\frac{\partial u}{\partial y} = \frac{0.002 - 0}{2} = 0.001000$$

$$\frac{\partial v}{\partial x} = 0$$

$$\frac{\partial v}{\partial y} = \frac{2.0013 - 2}{2} = 0.000650$$

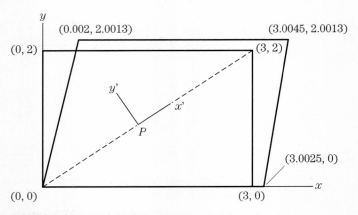

FIGURE 1-24 Uniform deformation of a rectangle.

1-3 BASIC IDEA II: DEFORMATION AND STRAIN

The linear strains are given by

$$\epsilon_x = \frac{\partial u}{\partial x} = 0.000833$$

$$\epsilon_y = \frac{\partial v}{\partial y} = 0.000650$$

$$\gamma_{xy} = \frac{\partial u}{\partial y} + \frac{\partial v}{\partial x} = 0.001000$$

Using Equations (1-9), the strains at a point P in the middle of the rectangle for directions oriented along and perpendicular to the diagonal are

$$\epsilon_{x'} = 0.000833\left(\frac{3}{\sqrt{13}}\right)^2 + 0.000650\left(\frac{2}{\sqrt{13}}\right)^2 + 0.001000\left(\frac{3}{\sqrt{13}}\right)\left(\frac{2}{\sqrt{13}}\right) = 0.001238$$

$$\epsilon_{y'} = 0.000833\left(\frac{2}{\sqrt{13}}\right)^2 + 0.000650\left(\frac{3}{\sqrt{13}}\right)^2 - 0.001000\left(\frac{3}{\sqrt{13}}\right)\left(\frac{2}{\sqrt{13}}\right) = 0.000245$$

$$\gamma_{x'y'} = 0.001000\left(\left(\frac{3}{\sqrt{13}}\right)^2 - \left(\frac{2}{\sqrt{13}}\right)^2\right) - 2(0.000183)\left(\frac{3}{\sqrt{13}}\right)\left(\frac{2}{\sqrt{13}}\right) = -0.000217$$

The negative value for $\gamma_{x'y'}$ indicates that as a result of the deformation, the angle between the original x' and y' axes decreases.

The principal strains can easily be determined by constructing Mohr's circle, as shown in Figure 1-25. The radius R is given by

$$R = \sqrt{0.0005^2 + 0.000092^2} = 0.000508$$

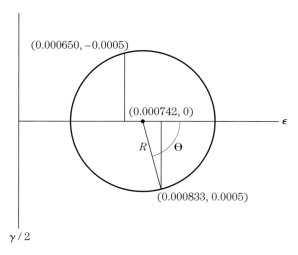

FIGURE 1-25 Mohr's circle for strain.

so that

$$\epsilon_{max} = 0.000742 + 0.000508 = 0.001250$$
$$\epsilon_{min} = 0.000742 - 0.000508 = 0.000234$$

The angle Θ is given by

$$\tan \Theta = \frac{0.000500}{0.000092} = 5.435$$

from which $\Theta = 79.6$ degrees. Shown in Figure 1-26 are the deformations associated with the principal strains and maximum shear strain. Figure 1-26a depicts a block whose sides are oriented along the principal directions. This block experiences no shear strain. Figure 1-26b depicts a block whose sides are oriented along the directions between which the maximum shear strain occurs. The corresponding extensional strains for those directions are also pictured.

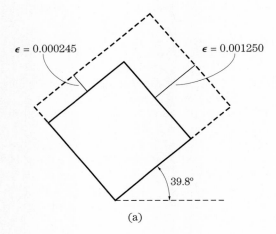

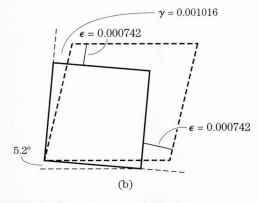

FIGURE 1-26 Deformations of blocks oriented along principal strain directions.

1-4 Basic Idea III: Material Behavior: Stress-Strain Relations

In the above discussions of stress and equilibrium, six unknowns—namely, the stresses $\sigma_x, \sigma_y, \sigma_z, \tau_{xy}, \tau_{xz}$ and τ_{yz}—were introduced. In considering deformation and strain the additional unknown variables that were involved were the three displacements u, v and w and the six strains $\epsilon_x, \epsilon_y, \epsilon_z, \gamma_{xy}, \gamma_{xz}$ and γ_{yz} for a total of 15 unknowns. The equations relating these unknowns are the six strain-displacement relations and the three equilibrium equations for a total of nine equations, that is, an insufficient number compared to the number of unknowns. Even more to the point, there is no relationship whatsoever between the kinetic variables (stresses) and the kinematic variables (displacements and strains). Even if it were possible, for instance, to solve the equations of equilibrium for the stresses, there would be no way to then determine displacements. Thus additional information that bridges the gap between the kinetics and the kinematics and, at the same time, brings into balance the number of equations and unknowns, is needed.

Logically for any specific application, this additional information will be related to the properties of the material(s) in question. It is expected, for instance, that two bodies that are identical geometrically and loaded in precisely the same fashion, but that are made of different materials, will respond differently. The relationships between the force and displacement variables for a given material are generally referred to as *material behavior*. This discussion of material behavior will characterize, both qualitatively and quantitatively, the basic responses of different materials to mechanical and thermal inputs by specifying the form of, as well as the constants appearing in, the relations between the force and displacement variables.

EXAMPLE 1-3   As an example of some of the ideas of material behavior, consider two geometrically identical bars, one of a steel and one of a rubber material, both loaded identically, as shown in Figure 1-27. In each case, consider establishing a relationship between the load P and the corresponding end displacement Δ, that is,

$$\Delta_S = f(P)$$

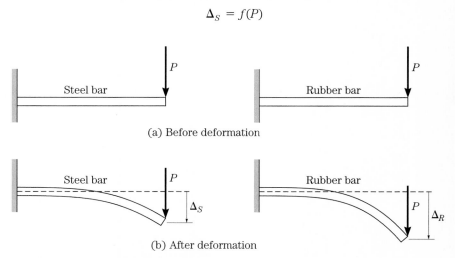

FIGURE 1-27 Deformations of bars of steel and rubber.

for the steel bar and

$$\Delta_R = g(P)$$

for the rubber bar. The two functions f and g could be different, say, $f(P) = kP$ (a linear relationship) and $g(P) = k_1 P + k_2 P^3$ (a nonlinear relationship). Or it could also turn out that both relationships are linear, that is, $f(P) = k_1 P$ and $g(P) = k_2 P$, but that $k_1 \neq k_2$. There are obviously many other possibilities. ◆

In general, the *form* ($f \neq g$ in the previous example, for instance) of the relationship between the forces and deformations can be different, for different materials. It is also possible that the form of the relationship varies for a given material depending on the environment. For example, the form of the relationship for a typical steel might depend on the temperature and/or the rate of loading. The values of the constants appearing in the relationships between the force variables and the displacement variables are referred to as *material constants*. Material constants must be determined by experiment. It will be seen throughout our study of solid mechanics that this basic ingredient, referred to loosely as *material behavior*, is ultimately necessary to bridge the gap between kinetics (force variables) and kinematics (displacement variables).

For a problem in solid mechanics the relationships that describe the material are usually taken to be relationships between stress and strain. Such relationships can be fairly general, such as $f(u, \epsilon, \dot{\epsilon}, \sigma, \dot{\sigma}, T) = 0$, indicating that there is some relationship between displacement, strain, strain rate, stress, stress rate and the temperature T. For many engineering applications it is sufficient to assume the solid to be characterized by Hooke's law, which asserts that the relationship between stress and strain is *linear* with the zero stress state coinciding with the state of zero strain, that is, there are no initial stresses in the undeformed body. For a completely anisotropic material, the most general form of Hooke's law is characterized by the relations

$$\sigma_x = C_{11}\epsilon_x + C_{12}\epsilon_y + C_{13}\epsilon_z + C_{14}\gamma_{xy} + C_{15}\gamma_{xz} + C_{16}\gamma_{yz} - \beta_1 T$$
$$\sigma_y = C_{21}\epsilon_x + C_{22}\epsilon_y + C_{23}\epsilon_z + C_{24}\gamma_{xy} + C_{25}\gamma_{xz} + C_{26}\gamma_{yz} - \beta_2 T$$
$$\sigma_z = C_{31}\epsilon_x + C_{32}\epsilon_y + C_{33}\epsilon_z + C_{34}\gamma_{xy} + C_{35}\gamma_{xz} + C_{36}\gamma_{yz} - \beta_3 T$$
$$\tau_{xy} = C_{41}\epsilon_x + C_{42}\epsilon_y + C_{43}\epsilon_z + C_{44}\gamma_{xy} + C_{45}\gamma_{xz} + C_{46}\gamma_{yz} - \beta_4 T$$
$$\tau_{xz} = C_{51}\epsilon_x + C_{52}\epsilon_y + C_{53}\epsilon_z + C_{54}\gamma_{xy} + C_{55}\gamma_{xz} + C_{56}\gamma_{yz} - \beta_5 T$$
$$\tau_{yz} = C_{61}\epsilon_x + C_{62}\epsilon_y + C_{63}\epsilon_z + C_{64}\gamma_{xy} + C_{65}\gamma_{xz} + C_{66}\gamma_{yz} - \beta_6 T$$

or, in matrix notation,

$$\boldsymbol{\sigma} = \boldsymbol{C}\boldsymbol{\epsilon} - \boldsymbol{\beta}T$$

The 36 constants C_{ij} are referred to as *elastic constants* and the six β_i's as *thermal constants*. Considerations of the energy of the elastic solid can be used to show that the elastic constants must satisfy the equations $C_{ij} = C_{ji}$. Thus considered as an array, $\boldsymbol{C}$ is symmetric and there are only 21 independent elastic constants.

Many engineering materials are essentially isotropic, meaning that the form of the stress-strain relations is independent of the coordinate system used. This is best ex-

1-4 BASIC IDEA III: MATERIAL BEHAVIOR: STRESS-STRAIN RELATIONS

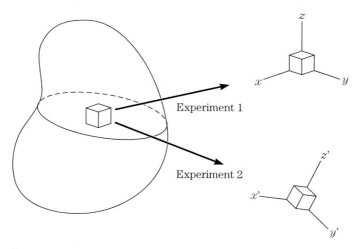

FIGURE 1-28 Material constant experiments.

plained by means of imagining the following set of gedanken experiments. Extract an infinitesimal cube of material from the interior of the solid, as indicated in Figure 1-28. The cube is oriented with respect to an xyz coordinate system as shown.

A complete set of experiments is carried out to determine the elastic constants $\boldsymbol{C}$ in the equation $\boldsymbol{\sigma} = \boldsymbol{C}\boldsymbol{\epsilon}$ for the element oriented along the xyz coordinate system. The cube is reinserted and another cube extracted in the neighborhood of the same point P oriented as per the $x'y'z'$ coordinate system. The equivalent set of experiments is carried out to determine $\boldsymbol{C}'$ in the equation $\boldsymbol{\sigma}' = \boldsymbol{C}'\boldsymbol{\epsilon}'$. If, independent of the orientation of the block, the elastic constants are the same, that is, $\boldsymbol{C} = \boldsymbol{C}'$, then the material is said to be *isotropic*. We will also assume that the values of the coefficients $\boldsymbol{C} = \boldsymbol{C}'$ are not dependent on the point P chosen in the body. Such a body is said to be *homogeneous*. In such a case it turns out that there are only two independent elastic constants and the stress-strain relations can be written as

$$\epsilon_x = \frac{\sigma_x - \nu(\sigma_y + \sigma_z)}{E} + \alpha T \qquad \gamma_{xy} = \frac{\tau_{xy}}{G}$$

$$\epsilon_y = \frac{\sigma_y - \nu(\sigma_z + \sigma_x)}{E} + \alpha T \qquad \gamma_{xz} = \frac{\tau_{xz}}{G} \qquad (1\text{-}12)$$

$$\epsilon_z = \frac{\sigma_z - \nu(\sigma_x + \sigma_y)}{E} + \alpha T \qquad \gamma_{yz} = \frac{\tau_{yz}}{G}$$

where $G = E/(2(1 + \nu))$. E is Young's modulus, ν is Poisson's ratio and G is usually referred to as the shear modulus. The constant α is the isotropic coefficient of thermal expansion. The two independent elastic constants E and ν are usually determined from a standard uniaxial tension test. These six stress-strain relations bring into balance the number of equations (15) and unknowns (15).

It is sometimes necessary to have the stresses expressed in terms of the strains. This is accomplished by solving the stress-strain relations for the stresses as

$$\sigma_x = \frac{E}{(1+\nu)(1-2\nu)}[(1-\nu)\epsilon_x + \nu(\epsilon_y + \epsilon_z)] - \frac{E}{1-2\nu}\alpha T$$

$$\sigma_y = \frac{E}{(1+\nu)(1-2\nu)}[(1-\nu)\epsilon_y + \nu(\epsilon_z + \epsilon_x)] - \frac{E}{1-2\nu}\alpha T \quad (1\text{-}13)$$

$$\sigma_z = \frac{E}{(1+\nu)(1-2\nu)}[(1-\nu)\epsilon_z + \nu(\epsilon_x + \epsilon_y)] - \frac{E}{1-2\nu}\alpha T$$

$$\tau_{xy} = G\gamma_{xy} \qquad \tau_{xz} = G\gamma_{xz} \qquad \tau_{yz} = G\gamma_{yz}$$

For the plane stress case, where σ_z, τ_{xz} and τ_{yz} are taken as zero, the stress-strain relations are

$$\epsilon_x = \frac{\sigma_x - \nu\sigma_y}{E} + \alpha T, \qquad \epsilon_y = \frac{\sigma_y - \nu\sigma_x}{E} + \alpha T, \qquad \gamma_{xy} = \frac{\tau_{xy}}{G} \quad (1\text{-}14)$$

Expressing the stresses in terms of the strains then gives

$$\sigma_x = \frac{E}{1-\nu^2}(\epsilon_x + \nu\epsilon_y) - \frac{E}{1-\nu}\alpha T$$

$$\sigma_y = \frac{E}{1-\nu^2}(\epsilon_y + \nu\epsilon_x) - \frac{E}{1-\nu}\alpha T \quad (1\text{-}15)$$

$$\tau_{xy} = G\gamma_{xy}$$

Note that for the two-dimensional plane stress problem there are two equations of equilibrium, three strain-displacement relations and three stress-strain relations for the determination of the unknowns u, v, σ_x, σ_y, τ_{xy}, ϵ_x, ϵ_y and γ_{xy}.

1-5 Boundary Value Problems: Combination of Basic Ideas

In the previous three sections the equations from considerations of equilibrium, deformation and material behavior were developed for the small deformation plane stress case. These equations are as follows:

Equilibrium:
$$\frac{\partial \sigma_x}{\partial x} + \frac{\partial \tau_{xy}}{\partial y} + B_x = 0$$

$$\frac{\partial \tau_{xy}}{\partial x} + \frac{\partial \sigma_y}{\partial y} + B_y = 0$$

Deformation:
$$\epsilon_x = \frac{\partial u}{\partial x} \qquad \epsilon_y = \frac{\partial v}{\partial y} \quad \text{and} \quad \gamma_{xy} = \frac{\partial u}{\partial y} + \frac{\partial v}{\partial x}$$

Material behavior:
$$\epsilon_x = \frac{\sigma_x - \nu\sigma_y}{E} + \alpha T \qquad \epsilon_y = \frac{\sigma_y - \nu\sigma_x}{E} + \alpha T \quad \text{and} \quad \gamma_{xy} = \frac{\tau_{xy}}{G}$$

1-5 BOUNDARY VALUE PROBLEMS: COMBINATION OF BASIC IDEAS

Note that there are eight unknowns and eight equations. Solving the stress-strain equations for the stresses and eliminating the strains using the strain-displacement equations yields

$$\sigma_x = \frac{E}{1-\nu^2}\left(\frac{\partial u}{\partial x} + \nu\frac{\partial v}{\partial y}\right) - \frac{E}{1-\nu}\alpha T$$

$$\sigma_y = \frac{E}{1-\nu^2}\left(\frac{\partial v}{\partial y} + \nu\frac{\partial u}{\partial x}\right) - \frac{E}{1-\nu}\alpha T$$

$$\tau_{xy} = \frac{E}{2(1+\nu)}\left(\frac{\partial u}{\partial y} + \frac{\partial v}{\partial x}\right)$$

Substituting these equations into the equilibrium equations yields

$$\frac{E}{1-\nu^2}\frac{\partial}{\partial x}\left(\frac{\partial u}{\partial x} + \nu\frac{\partial v}{\partial y}\right) + \frac{E}{2(1+\nu)}\frac{\partial}{\partial y}\left(\frac{\partial u}{\partial y} + \frac{\partial v}{\partial x}\right) - \frac{E\alpha}{1-\nu}\frac{\partial T}{\partial x} + B_x = 0$$

$$\frac{E}{2(1+\nu)}\frac{\partial}{\partial x}\left(\frac{\partial u}{\partial y} + \frac{\partial v}{\partial x}\right) + \frac{E}{1-\nu^2}\frac{\partial}{\partial y}\left(\frac{\partial v}{\partial y} + \nu\frac{\partial u}{\partial x}\right) - \frac{E\alpha}{1-\nu}\frac{\partial T}{\partial y} + B_y = 0$$

(1-16)

a set of coupled second-order linear partial differential equations for the displacements $u(x,y)$ and $v(x,y)$. The task is then to find functions u and v satisfying these equations in the interior of the region D and also appropriate boundary conditions on the boundary B, as shown symbolically in Figure 1-29a.

The boundary B is considered to be partitioned into a portion B_t on which the stresses are known or prescribed, and a portion B_u on which the displacements are prescribed. In equation form this can be expressed as follows.

On B_u: On this portion of the boundary it is necessary that

$$u(x,y) = u_0(B)$$
$$v(x,y) = v_0(B)$$

where $u_0(B)$ and $v_0(B)$ are prescribed functions of position on the boundary. These equations state that at a point on B_u, the functions $u(x,y)$ and $v(x,y)$ must assume

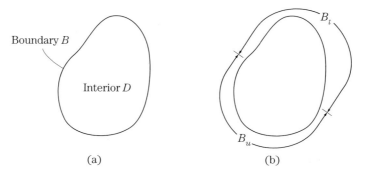

FIGURE 1-29 Stress and displacement portions of the boundary.

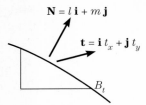

FIGURE 1-30 Stresses prescribed on B_t.

the values of the prescribed functions $u_0(B)$ and $v_0(B)$, respectively. If the boundary is completely fixed, for example, both $u_0(B)$ and $v_0(B)$ would be zero.

On B_t: On this portion of the boundary the stresses, as computed from the displacements in the interior and evaluated on the boundary, would have to be equal to the stresses prescribed on the boundary. In terms of the stress components, these equations are

$$\ell \sigma_x + m \tau_{xy} = t_x(B)$$
$$\ell \tau_{xy} + m \sigma_y = t_y(B)$$

where, as shown in Figure 1-30, $t_x(B)$ and $t_y(B)$ are the prescribed x and y components of the stress vector on B_t.

Expressed in terms of the displacements, these boundary conditions are

$$\ell \frac{E}{1-\nu^2}\left(\frac{\partial u}{\partial x} + \nu \frac{\partial v}{\partial y}\right) + m \frac{E}{2(1+\nu)}\left(\frac{\partial u}{\partial y} + \frac{\partial v}{\partial x}\right) - \ell \frac{E}{1-\nu}\alpha T = t_x(B)$$

$$\ell \frac{E}{2(1+\nu)}\left(\frac{\partial u}{\partial y} + \frac{\partial v}{\partial x}\right) + m \frac{E}{1-\nu^2}\left(\frac{\partial v}{\partial y} + \nu \frac{\partial u}{\partial x}\right) - m \frac{E}{1-\nu}\alpha T = t_y(B)$$

Due to the complexity of the equilibrium equations and boundary conditions, solutions are possible in only a very few instances corresponding to very simple geometries and loadings. In general, numerical solutions of these equations using techniques such as finite-element or boundary-element methods are indicated.

Stress Formulations. In some situations it is desirable to formulate the boundary value problem entirely in terms of the stresses. For a plane stress case with no body forces, the equations of equilibrium are

$$\frac{\partial \sigma_x}{\partial x} + \frac{\partial \tau_{xy}}{\partial y} = 0$$

$$\frac{\partial \tau_{yx}}{\partial x} + \frac{\partial \sigma_y}{\partial y} = 0$$

We attempt to find solutions to this set of equations that also satisfy stress boundary conditions of the form

$$\ell \sigma_x + m \tau_{xy} = t_x$$
$$\ell \tau_{xy} + m \sigma_y = t_y$$

Assuming we are successful, we can then ask what displacements correspond to these stresses. To determine the displacements, we would attempt to integrate the strain-displacement relations

$$\frac{\partial u}{\partial x} = \frac{\sigma_x - \nu\sigma_y}{E} \qquad \frac{\partial v}{\partial y} = \frac{\sigma_y - \nu\sigma_x}{E} \qquad \frac{\partial v}{\partial x} + \frac{\partial u}{\partial y} = \frac{\tau_{xy}}{G}$$

to determine the displacements u and v. It is left for the student to show in the exercises that this is not possible unless the *compatibility* or *integrability* equation

$$\frac{\partial^2 \epsilon_x}{\partial y^2} + \frac{\partial^2 \epsilon_y}{\partial x^2} - \frac{\partial^2 \gamma_{xy}}{\partial x \partial y} = 0$$

is satisfied by the strains. Expressed in terms of the stresses, the compatibility equation for an isotropic solid is

$$\frac{\partial^2 \sigma_x}{\partial y^2} - \nu\frac{\partial^2 \sigma_x}{\partial x^2} + \frac{\partial^2 \sigma_y}{\partial x^2} - \nu\frac{\partial^2 \sigma_y}{\partial y^2} - 2(1+\nu)\frac{\partial^2 \tau_{xy}}{\partial x \partial y} = 0$$

and constitutes an additional requirement on the stresses for a stress formulation. Analogous equations can be developed for the three-dimensional problem.

1-6 Modeling and the Design Process

The dictionary defines an engineer as one who is versed in the design, construction and use of devices that carry out a specific task or a variety of tasks. In the most basic sense, an engineer solves problems. The rational and systematic application of the principles of physics and the tools of mathematics by the engineer to the problem of the development of a device to solve a problem or class of problems is referred to as the *design process*. In this text the design process will be discussed as it relates to devices that will be generally called structures.

Modeling and the design process generally include (a) specification or estimation of the configuration (size, shape, material, etc.) of the structure, (b) identification of the environment(s) in which the structure is to operate, (c) identification of the basic physical principles that govern the problem such as balance of momentum, balance of energy, and so on, (d) development of the mathematical model that is to be used for analyzing the problem, (e) specification of the criteria that must be met in order that the structure function as desired or that the problem can be considered solved, and (f) testing, if necessary, to determine the validity of (a) through (e). The end result of the design process should be a safe, efficient structure that operates as expected or required in response to the inputs or environments for which it is intended.

The first phase, and an integral part of the design process, consists of deciding about the physical and mathematical models. This phase is generally referred to as the *modeling process*. The *physical model* consists of the structure and its surroundings or environment. In the most general design setting, the structure is not even defined at the beginning of the process. A major part of developing the physical model is the visualization of what will be referred to as the *initial configuration*. The *mathematical model* is an equation or set of equations developed as a means for representing and predicting the behavior of the physical model. Two steps are involved in the modeling process:

1. Creation of an idealized physical model by making simplifying assumptions regarding the effects of the environment and the behavior of the structure.
2. Application of the basic principles and laws of physics to the idealized physical model, resulting in the mathematical model.

The mathematical model consists of a set of equations that should contain all the information about the idealized environment and the idealized structure. A schematic outline of the modeling process is depicted in Figure 1-31.

The equations that constitute the mathematical model arise from the application of basic physical principles and laws in mathematical form to the idealized physical model. These equations can be very simple, such as a set of linear algebraic equations, or very complex, as in situations where it is necessary to represent the physics by a set or sets of linear or nonlinear partial differential equations. More complex models are generally better able to accurately describe more of the details of the behavior of the physical model but at the price of substantially increased cost associated with the solution of the equations of the mathematical model. Generally, the complexity of the mathematical model depends on the character of the structure and its environment.

A few general comments regarding the modeling process are in order. Environment refers to the forces that are applied to the structure, the thermal environment to which the structure is subjected, and possible chemical and electrical effects that could influence the behavior of the structure. The collection of phenomena that constitute the environment are sometimes referred to as the *inputs* or *loads*. Structure means the actual pieces or components that occupy space and are composed of material(s). In general, the simplifying assumptions are made with regard to the environment, the structure and their possible interactions.

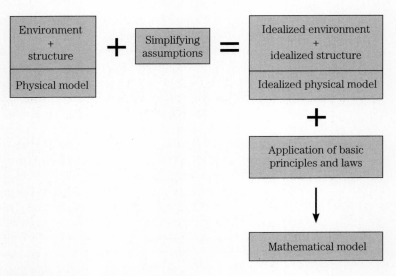

FIGURE 1-31 The modeling process.

The other aspect of the modeling process is the decision as to which basic physical principles and laws should or need to be included. As indicated above, it is absolutely necessary to include Newton's laws of equilibrium or motion in developing the mathematical model in a solid mechanics setting. Additionally, it may be appropriate to enforce thermodynamic principles such as conservation of energy and the second law of thermodynamics. When electrical environments are present, it may also be necessary to consider the laws of Ohm, Ampere and Faraday in generating the mathematical model. Generally, the character of the environment, the structure and the nature of their possible interactions will dictate which basic physical principles and laws are appropriate.

◆ **EXAMPLE 1-4** A tension member of thickness t contains a centrally located hole that must be of diameter d, as shown in Figure 1-32. It is assumed that the load P is specified; it is required to determine suitable values of t and D in order that the member satisfactorily transmit the load P. A mathematical model based on a one-dimensional theory is used to compute the average stress on a section through the hole as

$$\sigma_{avg} = \frac{P}{t(D - d)}$$

Stress levels computed in this fashion are used in connection with the failure criteria discussed in the next section to determine t and D. Tests show, however, that a member designed on the basis of the average stress fails when subjected to a much smaller load than the design load P, that is, the predictions of the mathematical model do not correspond well at all to the behavior of the tension member (that is, the physical model). In the present situation, it is necessary to alter the mathematical model by considering a two-dimensional theory that properly accounts for the effects of the hole. The t and D values determined on the basis of the two-dimensional theory are in fact confirmed by the results of tests. ◆

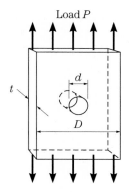

FIGURE 1-32 Tension member containing a hole.

Note carefully that the mathematical model is not unique. Given a single physical model, different assumptions or different degrees of idealization can result in different mathematical models. The success of the mathematical model is ultimately judged on the basis of how well it predicts the behavior of the original physical model. In situations where the agreement between the predictions of the mathematical model and the observed or expected behavior of the physical model is poor, the analyst or designer must have the freedom to reevaluate the mathematical model in order to improve the correspondence between the behavior of the physical model and the predictions of the mathematical model.

One of the most important steps in the design of a structure is the decision as to exactly what is required of the structure. The set of requirements that is imposed upon the structure in an attempt to insure that it can carry out its mission is referred to as the *design criteria* or the *design constraints*. Design constraints generally contain information about the environment and the structure. Factors that are often considered in developing design constraints include strength, stiffness or flexibility, weight, life, corrosion, friction, thermal considerations, cost, size, control, mode of failure and wear.

Once the physical model has been converted into a corresponding mathematical model, and the design constraints developed, the design process is carried out as indicated in Figure 1-33.

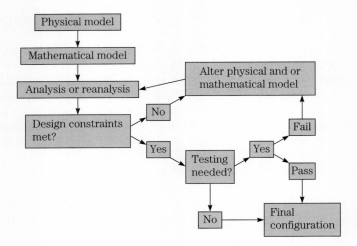

FIGURE 1-33 Partial flowchart depicting the design process.

The flowchart makes clear the fact that in general the design process is iterative, that is, it may be necessary to repeat some or all of the above steps in order to evolve a structure that satisfies all the design constraints. In some situations it may be possible to complete the design process in a relatively small number of iterations spanning perhaps a few hours. For a very complex structure such as a space station, the design process can last several years.

Discussion. The analysis or reanalysis phase of the design process consists of solving some or all of the equations of the mathematical model. At the end of this step the design constraints are reviewed. There may be only a few checks that need to be made to determine if the design constraints have been are met, or there may be many. In a structural setting typical questions that are asked in checking the design constraints include the following:

1. Are critical values of stresses exceeded?
2. Are deformations too large?
3. Is the weight too large?
4. Is the expected life of the structure sufficient?

If all of the design constraints are met and testing is required, another branch in the design process is reached, namely, the testing phase. The test(s) are conducted with the structure either surviving satisfactorily or failing the test. If the test is failed, the physical and or mathematical model must be altered.

It is prudent to distinguish between qualitative and quantitative changes in the mathematical model. A quantitative change in the mathematical model means that details such as size, shape and material properties can change but that the basic form and character of the mathematical model remains unchanged. An example would be a situation where a stress value in a component of the structure was exceeded by comparison with one of the design constraints. In an attempt to reduce the stress, the size of the component might be increased. This would result in a change in the details but not the form of the mathematical model.

A qualitative change in the mathematical model, on the other hand, means a change in the form of the mathematical model. An example would be a situation where it is deemed necessary to change the model from one where the theory is one-dimensional to one where the theory is two-dimensional, resulting in a change in the number and character of the governing equations.

Here is a simple example illustrating these ideas. The mathematical model for a structure might be a second-order differential equation and two boundary conditions of the form

$$a_0(x)u'' + a_1(x)u' + a_2(x)u = a_3(x) \qquad a \leq x \leq b$$
$$u(a) = U_1$$
$$u(b) = U_2$$

Quantitative changes would mean changes in the coefficients a_0, a_1, a_2, a_3 and the constants U_1 and U_2. These would actually result from changes in size, shape, material properties, and so on, of the physical model. Qualitative changes, on the other hand, would be of the sort that would, for instance, result in a change in the order of the differential equations, a change from a linear to a nonlinear differential equation or perhaps a change in the governing differential equation from ordinary to partial.

Note that the mathematical model is possibly changed at two different points in the design process. The first is during the analysis–design constraints check loop of the design process. The mathematical model is solved, yielding displacements, internal forces and stresses that are used directly or indirectly in checking the design constraints. If the design constraints are not satisfied, the mathematical model is changed and the analysis redone. This loop is continued until the design constraints are satisfied. The other point in the process where the mathematical model may have to be altered is if the analysis–design constraints check loop closes but required testing leads to an unsatisfactory result. Here again the mathematical model may be altered and the analysis–design constraints check loop reentered.

The discussion contained in this section is not intended to constitute a complete treatment of mechanical design, as will be apparent to any reader versed in that area. Instead, the intent is to make the student aware of the idea that not all problems in solid mechanics are analysis problems. An analysis problem refers to the sort of problem in which a specific structure—that is, one whose size, shape and materials have already been dictated—is loaded in a prescribed fashion and in which it is required to determine the deformations and stresses on the basis of a particular mathematical model. Alternatively, there is the sort of problem where allowable deformations and stresses are given and the problem is to determine the corresponding allowable loads. However, in a design problem, the task is to determine the size, shape, materials, and so on, in order to accomplish the function required of the structure.

1-7 Failure and Failure Criteria

Any structure or structural component will fail if the loads are sufficiently high. There are many ways that the structure can be deemed as not functioning properly or as having *failed*. These include *excessive deformations*, and *fracture* of one or more components of the structure.

Failure by excessive deformations is generally of two types:

1. *Elastic deformations*. Here the loads on the structure are such that the stresses are within the elastic range but resulting in elastic deformations such that the structure is no longer able to perform its design function. It is also possible for a structure to experience a loss of stability in the elastic range, that is, elastic buckling occurs, again resulting in a situation not within the scope of design of the structure.

2. *Plastic deformations or yielding*. This type of excessive deformation occurs primarily in connection with materials that are classified as *ductile*. As indicated in the uniaxial stress-strain diagram of Figure 1-34a, a ductile material is characterized by significant plastic deformations beyond the elastic region prior to failure. The critical stress is the yield stress σ_Y shown in Figure 1-34a. Criteria for failure in this mode are discussed in the next section.

Failure by fracture can also be grouped into two types:

1. *Fracture due to excessive stress*. Fracture in this manner can occur in materials that are *brittle* such as indicated in Figure 1-34b. When the stress reaches a critical value in a uniaxial member, fracture occurs. For a brittle material the critical stress is denoted by the fracture stress or ultimate stress σ_{ULT}, as shown in Figure 1-34b. There is no apparent evidence of any plastic deformation. Fracture can also occur at the end of a process of plastic deformation, as shown in Figure 1-34a. However, it is usually the case that the extent of the plastic deformation prior to fracture is sufficient for the structure to have been considered as having failed due to excessive plastic deformations.

2. *Fracture due to fatigue*. Fatigue can occur in either ductile or brittle materials in connection with alternating or fluctuating loads. It frequently occurs in connection with a stress concentration associated with the geometry of the structure, that is, sudden changes in cross section, or with stress concentrations associated with microscopic internal flaws. For a given history of mean and alternating stresses, fracture due to fatigue can occur after a certain number of cycles of stress history referred to as *fatigue life*. For a ductile material failure due to fatigue can occur in situations where the stresses are significantly below the levels associated with plastic deformations. For a brittle material failure due to fatigue can occur at stress levels far below those associated with fracture due to excessive stress.

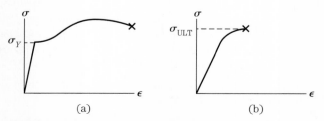

FIGURE 1-34 Typical stress-strain diagrams for (a) ductile and (b) brittle materials.

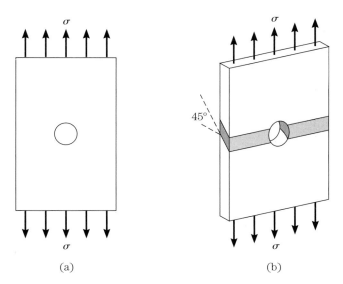

FIGURE 1-35 Tensile test of a ductile material showing Luder bands.

Tests could be performed on each structure to determine the failure loads but these would be prohibitively time-consuming and expensive. In lieu of testing it is necessary to have some sort of rational analytical process to assist in making decisions as whether a structure will fail under the action of a set of loads. Before presenting the theories that have been developed in this regard, this section reviews the results of standard tests.

Consider a tension member containing a central hole, as indicated in Figure 1-35a. The specimen is made from a ductile material and is polished prior to testing. As the load is increased, visible regions called Luder bands will appear on the front and back surfaces, as indicated in Figure 1-35b. These bands, marking the regions over which yielding occurs, make an angle of approximately 45° with the axis of loading, as shown. The shear stress on these 45° planes is at a maximum. Luder bands can also appear when there is no hole. This evidence supports the proposition that for ductile materials, failure occurs when the maximum shear stress reaches a critical value.

If a specimen made from a brittle material is subjected to a uniaxial tension test, as shown in Figure 1-36a, the failure or fracture surface appears as shown in Figure 1-36b. In view of the relationship between the fracture surface and the average normal stress $\sigma_{avg} = P/A$ acting on the fracture surface, it would be suspected that failure is related to the normal stress having exceeded some critical value. The theories of failure for brittle materials are essentially based on this observation.

Consider also the classical experiment consisting of the twisting of a piece of chalk, as shown in Figure 1-37a. Failure occurs as a result of a brittle-type fracture as indicated in Figure 1-37b. The stress resulting from the torque is a shear stress on planes perpendicular to the axis, as indicated in Figure 1-37a. An element of the lateral surface of the piece of chalk is subjected to the shear stresses shown in Figure 1-38a. Isolating an element near the surface of the bar and performing a simple force balance, as indicated in Figure 1-38b, shows that the maximum tensile stress occurs on planes at a 45°, that

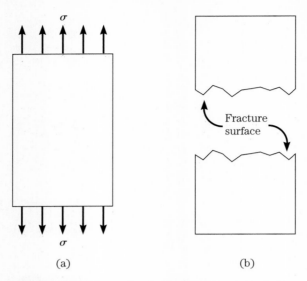

FIGURE 1-36 Failure surface for the uniaxial tension test of a brittle material.

is, precisely the fracture planes. The results of this experiment further support the idea that the failure of a brittle material occurs when the maximum normal stress reaches a critical value.

These test results seem to indicate a connection between failure and stress level. Whether the specimen is loaded in tension or shear (torsion), there is a value of the load and hence a stress or stresses at which failure occurs. The precise nature of the mechanism involved in either a ductile or brittle fracture is actually a complicated function of the microstructure of the material (that is, grain size, grain orientation, and so on). Based on the assumption that we are dealing with a continuum, we will restrict ourselves to making decisions about the mechanism of failure on the basis of observable or

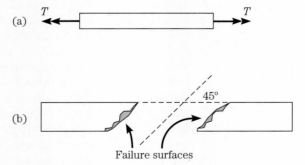

FIGURE 1-37 Failure surface for the torsion test on a piece of chalk.

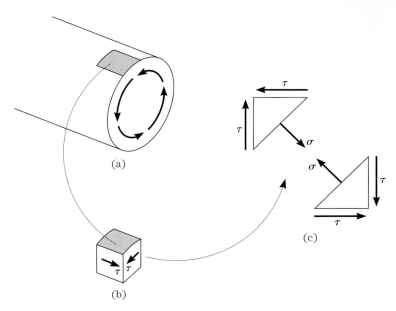

FIGURE 1-38 Shear and normal stress for the torsion test on a piece of chalk.

measurable quantities that appear in the equations of solid mechanics. Stress will be the primary quantity to be used for correlating the experimental data with the predictions of the theories of failure in solid mechanics.

The careful consideration of all the experimental data has resulted in several commonly used theories of failure. These theories are based on the assumption that failure occurs when the maximum shear or normal stress in a structure or structural component reaches a critical value as determined from the results of a uniaxial tension test.

In line with distinguishing between ductile and brittle behavior, it is appropriate to distinguish between failure criteria for materials that fail in a ductile fashion and those that fail in a brittle fashion. For ductile materials (a) the maximum shear stress failure criterion and (b) the von Mises failure criterion will be introduced. For brittle materials (a) the maximum normal-stress failure criterion and (b) the Coulomb-Mohr failure criterion will be introduced. Each of these failure criteria attempts to predict failure in an actual structure based on knowledge of normal stress values on failure surfaces for tests such as the uniaxial tension test and the torsion test described above.

It should be recognized that a structure has a certain awareness in that, if the structure is going to fail, for instance because the maximum normal stress has reached a critical value σ_{cr}, it will do so as soon as the σ_{cr} value is reached *at any point on any plane in the structure*. In other words, the structure, in a sense, "knows" when the critical value is reached irrespective of any analysis that is performed. Thus it is essential that the analysis be able to predict the location and magnitude of the maximum stresses and that these values are the ones compared to the chosen critical values in assessing failure.

1-7-1 Failure Criteria for Ductile Materials

Maximum Shear-Stress Failure Criterion. Charles Coulomb (1773) and Henri Tresca (1868) suggested the maximum shear-stress theory, which states that for all members composed of the same material, failure, as indicated by yielding, occurs when the maximum shear stress in the material reaches the value of the shear stress at yielding as determined by a uniaxial tension (or compression) test.

Consider a uniaxial specimen subjected to a tensile stress σ_0, as indicated in Figure 1-39a. For the coordinate system shown, the stresses are $\sigma_Y = \sigma_0$, with all the other components of stress vanishing. Mohr's circle for stress, shown in Figure 1-39b, clearly shows that for the uniaxial test, the maximum shear stress at yielding is

$$\tau_{max} = \tau_Y = \frac{\sigma_0}{2} \tag{1-17}$$

which is the critical value to be used in the maximum shear-stress criterion of failure. In the above equation for τ_{max}, τ_Y indicates the value of the maximum shear-stress at yield. The maximum shear-stress theory extends this result to the general case involving an arbitrary state of stress by stating that failure occurs when the maximum shear stress on any plane at any point in the structure reaches the value τ_Y. When the three principal stresses $\sigma_1 > \sigma_2 > \sigma_3$ are known at a point, the maximum shear stress is given by

$$\tau_{max} = \frac{\sigma_1 - \sigma_3}{2}$$

and failure occurs when

$$\frac{\sigma_1 - \sigma_3}{2} = \tau_Y = \frac{\sigma_Y}{2}$$

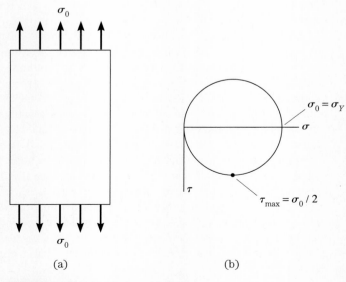

FIGURE 1-39 Uniaxial test specimen.

1-7 FAILURE AND FAILURE CRITERIA

For a plane stress state with $\sigma_z = \tau_{xz} = \tau_{yz} = 0$, one of the principal stresses, say σ_3, is always zero. The other two principal stresses σ_1 and σ_2 will be referred to as in-plane principal stresses. For clarity, the three different combinations of in-plane principal stresses, namely, (a) $\sigma_1 > \sigma_2 > 0$, $\sigma_3 = 0$, (b) $\sigma_2 < \sigma_1 < 0$, $\sigma_3 = 0$ and (c) $\sigma_1 > 0$, $\sigma_2 < 0$, $\sigma_3 = 0$, will be discussed in detail.

Case (a): $\sigma_1 > \sigma_2 > 0$, $\sigma_3 = 0$. Mohr's circle and the block on which the principal stresses σ_1 and σ_2 act are shown in Figure 1-40, parts a and b, respectively. The maximum in-plane shear stress, given by $(\sigma_1 - \sigma_2)/2$ and shown in Figure 1-40a, acts on the plane that bisects the principal planes on which σ_1 and σ_2 act. For this case however, $(\sigma_1 - 0)/2 > (\sigma_1 - \sigma_2)/2$, so that the absolute maximum shear stress is given by

$$\tau_{max} = \frac{\sigma_1}{2}$$

and occurs on the plane bisecting the 1 and 3 principal directions, that is, on the shaded plane indicated in Figure 1-40b.

Case (b): $\sigma_2 < \sigma_1 < 0$, $\sigma_3 = 0$. Mohr's circle and the block on which the principal stresses σ_1 and σ_2 act are shown in Figure 1-41, parts a and b, respectively. The maximum in-plane shear stress, given by $(\sigma_1 - \sigma_2)/2$ and shown in Figure 1-41a, acts on the plane that bisects the principal planes on which σ_1 and σ_2 act. For this case, however, $|\sigma_2 - 0|/2 > (\sigma_1 - \sigma_2)/2$, so that the absolute maximum shear stress is given by

$$\tau_{max} = \frac{|\sigma_2|}{2}$$

and occurs on the plane bisecting the 1 and 3 principal directions, that is, on the shaded plane indicated in Figure 1-41b.

Case (c): $\sigma_1 > 0$, $\sigma_2 < 0$, $\sigma_3 = 0$. For this case the Mohr's circle and the block on which the principal stresses σ_1 and σ_2 act are shown in Figure 1-42, parts a and b, respectively. For this case, $(\sigma_1 - \sigma_2)/2 > \sigma_1/2$ and $(\sigma_1 - \sigma_2)/2 > |\sigma_2|/2$ so that the absolute maximum shear stress is given by

$$\tau_{max} = \frac{\sigma_1 - \sigma_2}{2}$$

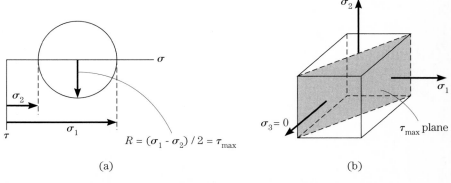

FIGURE 1-40 Maximum shear stresses for $\sigma_1 > \sigma_2 > 0$ and $\sigma_3 = 0$.

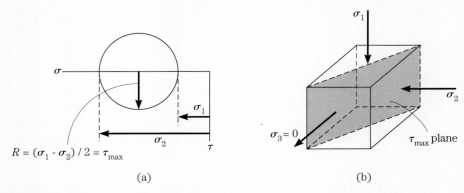

FIGURE 1-41 Maximum shear stresses for $\sigma_2 < \sigma_1 < 0$ and $\sigma_3 = 0$.

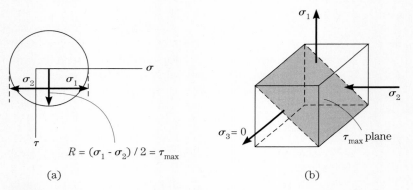

FIGURE 1-42 Maximum shear stresses for $\sigma_1 > 0$, $\sigma_2 < 0$ and $\sigma_3 = 0$.

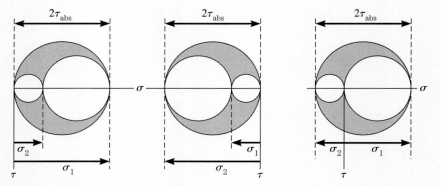

FIGURE 1-43 Mohr's circles for cases *a*, *b* and *c*.

and occurs on the plane bisecting the 1 and 2 principal directions, as indicated by the shaded region in Figure 1-42b.

Thus for a state of plane stress where $\sigma_3 = 0$, the results of cases a, b and c can be summarized by the three corresponding sets of Mohr's circles shown in Figure 1-43, parts a, b, and c respectively. In each case three circles are drawn. One is constructed for the plane associated with the $(\sigma_1, 0)$ principal stresses, one for the $(\sigma_2, 0)$ principal stresses and one for the (σ_1, σ_2) principal stresses. For each case, the absolute maximum shear stress at the point is equal in value to half the diameter of the largest of the three circles. For case (a) this is seen to be the circle corresponding to the $(\sigma_1, 0)$ principal stresses. For case (b) the circle having the largest diameter is the circle corresponding to the $(\sigma_2, 0)$ principal stresses and for case (c) it is the circle corresponding to the (σ_1, σ_2) principal stresses.

Finally, then, for the case where $\sigma_3 = 0$, the process for determining the largest shear stress is to use Mohr's circle to determine the in-plane principal stresses σ_1 and σ_2. The maximum shear stress at the point is then given by

$$\tau_{max} = \max\left(\left|\frac{\sigma_1}{2}\right|, \left|\frac{\sigma_2}{2}\right|, \left|\frac{\sigma_1 - \sigma_2}{2}\right|\right)$$

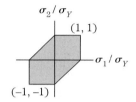

FIGURE 1-44 Maximum shear-stress hexagon.

According to the maximum shear-stress failure criterion, failure will occur if any of the above values exceeds the allowable shear stress given by $\sigma_Y/2$. These collective requirements can be presented very nicely in terms of the maximum shear-stress hexagon depicted in Figure 1-44. Thus, as long as the in-plane principal stresses σ_1 and σ_2 for a biaxial state of stress lie within the shaded region of Figure 1-44, the maximum shear-stress criterion of $\tau_{abs} < \sigma_Y/2$ is satisfied.

EXAMPLE 1-5 As indicated in Figure 1-45a, the stresses at a point P in a body are given by $\sigma_x = \sigma_x$, $\sigma_y = 3$ ksi, $\tau_{xy} = 6$ ksi, with $\sigma_z = \tau_{xz} = \tau_{yz} = 0$. A uniaxial test of the same material indicates yielding at the normal stress value of 16 ksi. What is the allowable value of σ_x?

Solution: Mohr's circle for this state of stress is shown in Figure 1-45b. The maximum shear stress is given by

$$\tau_{max} = \sqrt{\left(\frac{\sigma_x - 3}{2}\right)^2 + 36} \text{ ksi}$$

so that failure occurs when

$$\sqrt{\left(\frac{\sigma_x - 3}{2}\right)^2 + 36} \text{ ksi} = \frac{16}{2} \text{ ksi} = 8 \text{ ksi}$$

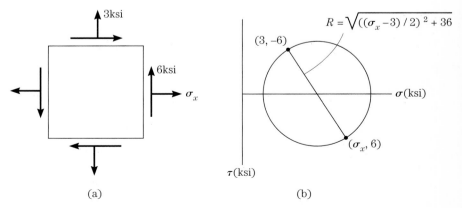

FIGURE 1-45 State of stress at a point in a body.

Solving yields $\sigma_x = 14.58$ ksi and $\sigma_x = -7.58$ ksi; thus, either a tensile stress of 14.58 ksi or a compressive stress of 7.58 ksi will produce a state of stress for which failure is indicated using the maximum shear-stress criterion. ◆

von Mises Failure Criterion. The von Mises failure theory is often referred to as the maximum distortion strain-energy theory. The idea for the theory is motivated as follows: Experiments have shown that homogeneous materials can withstand very high hydrostatic states of stresses ($\sigma_x = \sigma_y = \sigma_z = -\sigma_0, \tau_{xy} = \tau_{xz} = \tau_{xz} = 0$) without yielding. This leads to the thought that failure in a material subjected to a general state of stress may be associated with the difference between the actual state of stress and a hydrostatic state having the value σ equal to the average of the principal stresses at the point, namely, $\sigma = (\sigma_1 + \sigma_2 + \sigma_3)/3$. The von Mises criterion states that failure will occur when the energy associated with the difference in the actual and hydrostatic states is equal to the corresponding energy in a uniaxial tensile test at yielding. The resulting criterion is identical to what is traditionally called the *octahedral shear-stress theory*, which is much easier to develop and understand. For this reason the octahedral shear stress theory will be presented.

Figure 1-46a indicates the block at a point on which the principal stresses act. Figure 1-46b indicates four of the eight directions that make equal angles with adjacent triads of principal directions. Each of these eight directions is perpendicular to what is termed an *octahedral plane*. Using the three-dimensional stress-transformation equations, the normal stress on each of the octahedral planes can be computed as

$$\sigma_{oct} = \frac{(\sigma_1 + \sigma_2 + \sigma_3)}{3}$$

that is, the average of the principal stresses at the point. Also, the shear stress on *each* of the octahedral planes can be computed as

$$\tau_{oct} = \frac{1}{2}\sqrt{(\sigma_1 - \sigma_2)^2 + (\sigma_2 - \sigma_3)^2 + (\sigma_3 - \sigma_1)^2}$$

The octahedral shear-stress theory states that failure occurs when this octahedral shear-stress value is equal to the value of the octahedral stress at yielding in a uniaxial

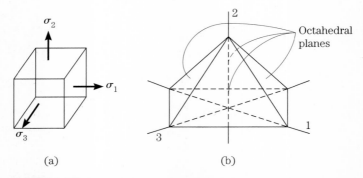

FIGURE 1-46 Principal planes and octahedral planes.

1-7 FAILURE AND FAILURE CRITERIA

tensile test. For the uniaxial tensile test $\sigma_1 = \sigma_Y$ (the yield stress), $\sigma_2 = \sigma_3 = 0$, so that the octahedral shear stress in the uniaxial tensile test is given by

$$\tau_{oct} = \frac{1}{2}\sqrt{(\sigma_Y - 0)^2 + (0 - 0)^2 + (0 - \sigma_Y)^2} = \frac{\sigma_Y}{\sqrt{2}}$$

Equating the two values for τ_{oct} yields

$$\sigma_Y = \frac{1}{\sqrt{2}}\sqrt{(\sigma_1 - \sigma_2)^2 + (\sigma_2 - \sigma_3)^2 + (\sigma_3 - \sigma_1)^2} \quad (1\text{-}18)$$

as the octahedral shear-stress criterion or the von Mises criterion for failure in terms of the principal stresses. For the case where one of the principal stresses, say σ_3, is zero, Equation (1-18) reduces to

$$\sigma_Y = \sqrt{\sigma_1^2 - \sigma_1\sigma_2 + \sigma_2^2}$$

which is frequently written as

$$\left(\frac{\sigma_1}{\sigma_Y}\right)^2 - \left(\frac{\sigma_1}{\sigma_Y}\right)\left(\frac{\sigma_2}{\sigma_Y}\right) + \left(\frac{\sigma_2}{\sigma_Y}\right)^2 = 1$$

This is the equation of the ellipse shown in Figure 1-47.

Failure due to the initiation of yielding is assumed to take place if the state of stress given by the principal stresses σ_1 and σ_2 lies on or outside the ellipse. The von Mises or octahedral shear-stress criterion is generally accepted as the best criterion to use for judging failure, as indicated by initial yielding, for ductile materials. For comparison, the dotted lines in Figure 1-47 indicate the corresponding boundaries for the maximum shear-stress failure criterion.

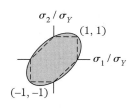

FIGURE 1-47 von Mises ellipse for plane stress.

◆**EXAMPLE 1-6** For a material with a tensile yield of 15 ksi subjected to the stresses $\sigma_x = 9$ ksi, $\sigma_y = -4$ ksi, and $\tau_{xy} = 7$ ksi, assess failure using the von Mises criterion. The principal stresses are determined to be

$$\sigma_{1,2} = \left[\left(\frac{9-4}{2}\right) \pm \sqrt{\left(\frac{9+4}{2}\right)^2 + 7^2}\right] \text{ksi}$$

or $\sigma_1 = 9.55$ ksi and $\sigma_2 = -4.55$ ksi. Substituting these into the equation for the von Mises ellipse yields

$$\left(\frac{9.55}{15}\right)^2 - \left(\frac{9.55}{15}\right)\left(\frac{-4.55}{15}\right) + \left(\frac{-4.55}{15}\right)^2 = 0.69 < 1$$

so failure is *not* indicated using the von Mises criterion. ◆

1-7-2 Failure Criteria for Brittle Materials

Maximum Normal-Stress Failure Criterion. The maximum normal-stress theory or, as it is sometimes called, the maximum principal stress theory, is generally credited to William Rankine. This theory predicts that failure in a brittle material will occur when the maximum principal stress in the material reaches the value of the normal stress

σ_{ULT}, as determined by a uniaxial tension test. For a plane-stress situation, with $\sigma_3 = 0$ and the in-plane principal stresses given by σ_1 and σ_2, failure is predicted when the principal stresses σ_1 and σ_2 lie outside the shaded region of the unit square with axes σ_1/σ_{ULT} and σ_2/σ_{ULT}, as indicated in Figure 1-48. This theory can be useful as long as there is a tensile principal stress against which σ_{ULT} can be compared. Predictions of failure using this theory are not in generally good agreement with experimental results for arbitrary states of stress.

Coulomb-Mohr Failure Criterion. Many materials behaving in a brittle manner exhibit different ultimate strengths in tension and compression. As determined by uniaxial tension and uniaxial compression tests, respectively, denote these ultimate strengths as σ_{UT} and σ_{UC}. When the in-plane principal stresses σ_1 and σ_2 are both positive or both negative, the safe regions in terms of the principal stresses σ_1 and σ_2 are assumed to be indicated by the shaded squares in quadrants I and III in Figure 1-49. When σ_1 and σ_2 are of opposite sign, the theory and experimental results are in good agreement by taking failure to be predicted when the in-plane principal stresses σ_1 and σ_2 lie on or outside of the lines BC and EF in quadrants II and IV, respectively. In quadrant IV, the equation of the line EF is

$$\frac{\sigma_1}{\sigma_{UT}} - \frac{\sigma_2}{\sigma_{UC}} = 1$$

In quadrant II the equation of the line connecting points B and C is given by

$$\frac{\sigma_2}{\sigma_{UT}} - \frac{\sigma_1}{\sigma_{UC}} = 1$$

Failure using the Coulomb-Mohr criterion is predicted when the state of stress given in terms of the in-plane principal stresses σ_1 and σ_2 lies on or outside of the boundaries of the shaded region in Figure 1-49. The Coulomb-Mohr failure criterion is also sometimes referred to as the internal-friction theory.

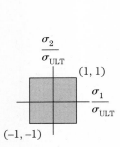

FIGURE 1-48 Failure for the maximum normal-stress theory.

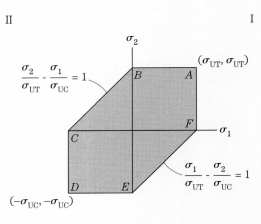

FIGURE 1-49 Safe regions for the Coulomb-Mohr criterion.

1-8 FACTOR OF SAFETY AND WORKING STRESSES

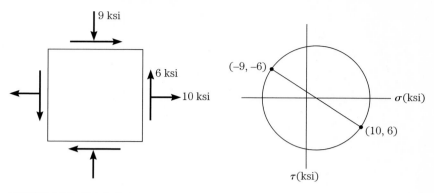

FIGURE 1-50 Mohr's circle.

EXAMPLE 1-7 The stresses at a point P on the surface of a structure made of a brittle material are $\sigma_x = 10$ ksi, $\sigma_y = -9$ ksi and $\tau_{xy} = 6$ ksi. If a uniaxial tension test of the same material indicates failure at 20 ksi, and a uniaxial compression test indicates failure at 33 ksi, will failure occur according to the Coulomb-Mohr criterion?

As indicated in Figure 1-50, the principal stresses using Mohr's circle are

$$\sigma_{1,2} = [0.5 \pm \sqrt{9.5^2 + 6^2}] \text{ ksi}$$

or $\sigma_1 = 11.74$ ksi and $\sigma_2 = -10.74$ ksi. Using the equation in quadrant IV results in

$$\frac{\sigma_1}{\sigma_{UT}} - \frac{\sigma_2}{\sigma_{UC}} = \frac{11.74}{20} - \frac{-10.74}{33} = 0.91 < 1$$

so failure will *not* occur according to the Coulomb-Mohr theory. ◆

It should be emphasized that these failure criteria are only applicable to isotropic materials in that each criterion predicts failure when the state of stress at a point reaches a critical level independent of the actual orientation of the stresses at the point.

In summary, the maximum shear-stress and von Mises failure criterion are used for ductile materials, with the von Mises criterion generally considered to more accurately predict results consistent with experiment. For brittle materials, the Coulomb-Mohr criterion generally shows the best agreement with experiment.

1-8 Factor of Safety and Working Stresses

In the analysis and design of a structure there are many aspects about which there is incomplete or imprecise data. These include the actual magnitudes of the loads, values of the elastic constants, possible residual stresses due to machining and/or heat treating, deviations from intended geometry, stress or load levels that will actually cause failure, and so on.

If the probability of failure is to be reduced to an acceptable level, the loads that a structure can theoretically sustain must be in excess of those which are assumed to actually be present. It is customary to define a *factor of safety (FS)* given by

$$FS = \frac{\text{Theoretical load}}{\text{Actual load}}$$

or

$$FS = \frac{\text{Theoretical load}}{\text{Allowable load}}$$

Clearly, the factor of safety must be greater than 1. Values of the factor of safety starting at 1.5 are common, with much larger values being appropriate in situations where there are many uncertainties in the model.

In terms of stresses the factor of safety is usually expressed as

$$FS = \frac{\text{Ultimate stress}}{\text{Allowable stress}}$$

where for brittle materials the ultimate stress is taken to be the fracture stress and for ductile materials the ultimate stress usually corresponds to the yield stress. Given a factor of safety and the ultimate stress, the equation for the factor of safety serves to determine the allowable stress. It is also customary to refer to the allowable stress as the *working stress* σ_W. Thus, it is possible to write

$$\sigma_W = \frac{\text{Ultimate stress}}{\text{Factor of safety}}$$

For ductile materials, the ultimate stress is taken to be the yield stress σ_Y with the working or allowable stress given by

$$\sigma_W = \frac{\sigma_Y}{FS}$$

with either the maximum shear-stress theory or the von Mises theory used to assure that the working stress is not exceeded. For brittle materials, the ultimate stress is taken to be the fracture stress σ_F with the working or allowable stress given by

$$\sigma_W = \frac{\sigma_F}{FS}$$

Here the Coulomb-Mohr theory would probably be used for assessing the working stress.

When a factor of safety is used in connection with the maximum shear-stress criterion, the diagram appearing in Figure 1-51 applies. The dotted lines in the figure correspond to a factor of safety of unity.

For the von Mises criterion with one zero principal stress, the diagram appearing in Figure 1-52 applies. Again, the dotted line in the figure corresponds to a unit factor of safety.

Finally, for a brittle material with the same factor of safety used for the tensile ultimate σ_{UT} and the compression ultimate σ_{UC}, the diagram for the Coulomb-Mohr

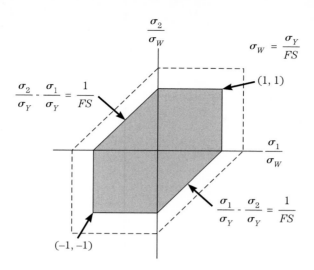

FIGURE 1-51 Maximum shear-stress criterion diagram with a factor of safety.

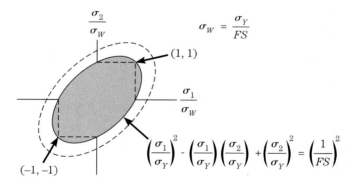

FIGURE 1-52 von Mises criterion diagram with a factor of safety.

failure criterion appearing in Figure 1-53 applies. Again, the dotted lines correspond to $FS = 1$.

There are many considerations that influence the decision as to the value to be used for the factor of safety. These include anticipated mode(s) of failure, the degree of confidence in the mathematical model used in the design and the many aspects involved in the manufacturing process that produces the structure or its components.

The mode of failure for a particular component of the structure may be sudden (no advance warning) or gradual. If a component of a structure fails it may be possible for a redistribution of the loads to take place internally and for the structure to continue to function. On the other hand, the failure of a single component may result in failure of the entire structure. In the first scenario, the factor of safety could be lower than in the second.

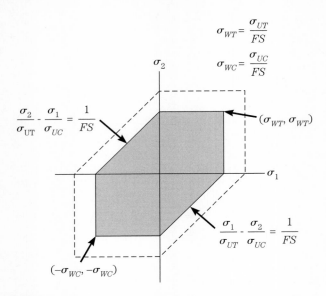

FIGURE 1-53 Diagram for the Coulomb-Mohr criterion with a factor of safety.

There are many unknowns involved in the mathematical model. The inputs or loads may be static or dynamic in nature. Dynamic loads are frequently less well defined than static loads. The actual nature of the materials used is such that their flexibility and strength properties must be considered somewhat variable. In simple models there is generally a larger degree of approximation involved in the process of converting the physical model into the mathematical model than for more complex models. On the other hand, a complex mathematical model may have to be analyzed approximately.

To actually create the structure or device in a form suitable for testing or for actual use, the results of the analysis and design must be converted from the drawing board to the actual configuration by some sequence of manufacturing/assembly processes. The manufacturing/assembly process generally involves many steps, each of which can result in a slight deviation of the structure from what is intended in the mathematical model. These deviations can result from a variety of sources, including misinterpretation of the drawings, poor workmanship, failure to produce components with the specified geometric tolerances, and incorrect processing of materials. All of these deviations essentially introduce further unknowns into the analysis and design process.

In the final analysis it is often the case that the factor of safety is chosen by an experienced engineer or group of engineers after considerable evaluation of all the available information about the physical and mathematical models, as well as all previous experience and knowledge in the area.

1-9 Approximate Formulations: Mechanics of Materials

The potential difficulties in solving the general equations of solid mechanics developed in the above sections suggest that simpler theories and sets of equations that pertain to less general classes of problems need to be developed. This step is part of the modeling process discussed in section 1-6. These simpler theories fit generally into the realm referred to as *mechanics of materials* or *strength of materials*. As will be seen throughout much of the remainder of the text, these theories have the following characteristics in common:

1. They are developed for a specific geometry, that is, the boundary of the region to which the theory applies is predefined as part of the theory.
2. The type of loading allowed is not completely general, consisting of special surface loadings, body forces and force resultants.
3. Force resultants, consistent with the assumptions mentioned in item 2, rather than stresses, are frequently used as the primary kinetic variables.
4. Specific simplifying assumptions are made regarding the character of the displacements, essentially reducing the number of independent variables and hence the complexity of the governing equations.

The intent and effect of these assumptions is generally to reduce partial differential equations to one or more ordinary differential equations or to a substantially less complicated set of partial differential equations, that is, to a simpler mathematical model. The resulting theories are very useful and quite accurate as long as all the assumptions that go into making up the theory are satisfied. The rest of this section will demonstrate these ideas as applied to the classical axial deformation problem.

EXAMPLE 1-8 *Axial deformation of a prismatic bar.* There are many applications where the primary function of a structural member is to transmit a single force resultant along its length. The elementary theory for such a problem is referred to as the *axial problem*. The theory is developed as follows.

Geometry: The first consideration in formulating the axial problem in solid mechanics is a careful definition of the region. The region is taken to be a bar with a cross section unchanging in shape but with the possibility of a gradual change in area, as indicated in Figure 1-54. It is also necessary to state that the orientation of the cross section is not a function of position along the axis of the bar, that is, there is no twist.

External loading: The external loads are assumed to consist of distributed loadings $q(x)$ that can arise from a combination of body forces in the interior, from stresses on the lateral surface of the bar, and from concentrated loads such as P_0 applied at the

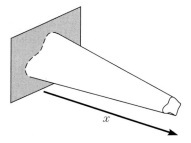

FIGURE 1-54 Definition of the region for an axial deformation problem.

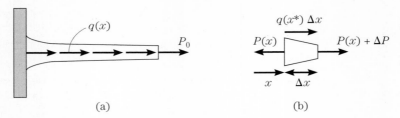

FIGURE 1-55 Differential element for the axial deformation problem.

boundary, as shown in Figure 1-55a. It will be shown later in this section that the external loads must be such that the only force resultant transmitted at any location along the axis of the bar is an axial load $P(x)$ whose line of action is the centroid of the cross section.

Equilibrium: The bar or any portion of the bar must be in equilibrium, or must be such that the appropriate equation of motion is satisfied. For a one-dimensional problem the free body diagram is taken to be a portion Δx of the bar, as shown in Figure 1-55b. The equilibrium equation is obtained from

$$\sum F_x = 0 = -P(x) + \int_x^{x+\Delta x} q(x)\, dx + P(x) + \Delta P(x)$$

Assume that $q(x)$ is a continuous function so that a mean value theorem can be used to replace the integral by $q(x^*)\,\Delta x$, where x^* is a suitably chosen value within the differential element. The equilibrium equation can then be written as $\Delta P/\Delta x + q(x^*) = 0$ or, after passing to the limit as $\Delta x \to 0$,

$$P' + q(x) = 0 \tag{1-19}$$

which is the differential equation of equilibrium. Note that this equation relates the *internal force P* to the *external loading q*.

The other equation that arises from consideration of equilibrium relates the resultant force P to the normal stress σ_x in the bar. Consulting Figure 1-56 it is seen that the resultant axial force P is given by

$$P = \int_{Area} \sigma_x\, dA \tag{1-20}$$

and that as a result of the assumption regarding the loading

$$M_z = -\int_{Area} y\sigma_x\, dA = 0 \quad \text{and} \quad M_y = \int_{Area} z\sigma_x\, dA = 0 \tag{1-21}$$

Deformation: Each point in a particular cross section located a distance x from the left end of the bar is assumed to *translate* to a new position $x + u(x)$, as shown in Figure 1-57. This assumption is classically referred to as "plane sections remain plane" in the context of the axial deformation problem. The assumption of "plane sections remain plane" will be discussed in much more detail in Chapter 4. Note at this juncture, however, that the deformation is assumed to consist of only one displacement, as op-

1-9 APPROXIMATE FORMULATIONS: MECHANICS OF MATERIALS

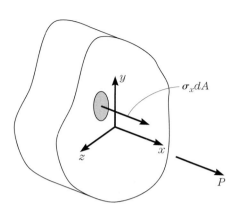

FIGURE 1-56 Relation between internal force resultants and stress.

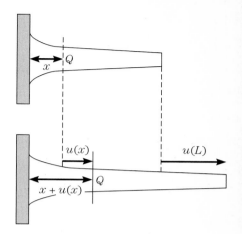

FIGURE 1-57 Axial translational displacement.

posed to three for the general case, with that one displacement assumed to depend on only one of the spatial variables, namely, x.

The extensional strain, defined as the change in length per unit of length, can then be shown to be

$$\varepsilon_x = \frac{du}{dx} = u' \qquad (1\text{-}22)$$

Material behavior: At this point, on the basis of a consideration of equilibrium and deformation, the following five equations have been developed:

$$P' + q(x) = 0 \qquad (1\text{-}19)$$

$$P = \int_{\text{Area}} \sigma_x \, dA \qquad (1\text{-}20)$$

$$M_z = -\int_{\text{Area}} y \sigma_x \, dA = 0 \quad \text{and} \quad M_y = \int_{\text{Area}} z \sigma_x \, dA = 0 \qquad (1\text{-}21)$$

$$\varepsilon_x = \frac{du}{dx} = u' \qquad (1\text{-}22)$$

These five equations involve the four specific unknowns P, σ_x, ε_x and u, plus the unknown location from which y and z are measured within the cross section. It might be possible to integrate Equation (1-19) to determine the force variable P and then to estimate the stress σ_x from Equation (1-20). At that point there would be no way to relate this information about the force variables to the deformation variables ε_x and u contained in Equation (1-22). Also notice that there is currently no information contained in the problem formulation regarding what type of material constitutes the bar. The additional information and the bridge that connects the force variables P and σ_x to the displacement variables u and ε_x are contained in an equation pertaining to the material behavior. The simplest relationship known, and fortunately one of the most useful in terms of modeling many engineering solids, is one that relates the stress σ_x to the strain ε_x in a linear fashion, namely,

$$\sigma_x = E\varepsilon_x \qquad (1\text{-}23)$$

usually termed a *stress-strain relationship*. E is an experimentally determined material constant known as Young's modulus or the modulus of elasticity. There are now four equations for the four unknowns P, σ_x, ε_x and u.

Combination: First substitute Equation (1-22) into Equation (1-23) to obtain

$$\sigma_x = Eu' = \sigma_x(x) \qquad (1\text{-}24)$$

Then substitute Equation (1-24) into Equation (1-20) to obtain

$$\begin{aligned} P(x) &= \int_{\text{Area}} Eu'\, dA \\ &= AEu' \end{aligned} \qquad (1\text{-}25)$$

which is known as the *force-displacement relation*. Note that once the force P has been determined Equation (1-25) can be integrated to determine the displacement u. Also note from Equation (1-25) that

$$\frac{P(x)}{A(x)} = Eu' = \sigma_x(x)$$

showing that *at any particular location along the axis of the bar* the stress σ_x is constant over the area of the cross section. This equation is frequently written simply as

$$\sigma_x = \frac{P}{A} \qquad (1\text{-}26)$$

Returning to Equations (1-21) and using the fact that σ_x is constant over the area of the cross section and can be passed through the integral signs, there results

$$\sigma_x \int_{\text{Area}} y\, dA = 0 \quad \text{and} \quad \sigma_x \int_{\text{Area}} z\, dA = 0$$

or

$$\int_{\text{Area}} y\, dA = 0 \quad \text{and} \quad \int_{\text{Area}} z\, dA = 0$$

showing that y and z must be measured from the centroid of the cross section, that is, that the line of action of the resultant P must pass the centroid. Finally, substituting Equation (1-25) into Equation (1-19) yields

$$(AEu')' + q(x) = 0 \qquad (1\text{-}27)$$

as the governing equation of equilibrium expressed in terms of the displacement u.

Equation (1-27) is a linear second-order ordinary differential equation that must be integrated subject to two boundary conditions, *one at each end*. The two most frequent boundary conditions are

1. u is prescribed at a boundary.
2. $P = AEu'$ is prescribed at a boundary.

At a boundary *either* the force P or the displacement u, but not both, can be prescribed. Thus an elementary mathematical model for the axial deformation problem is the equilibrium equation

$$(AEu')' + q(x) = 0$$

together with two boundary conditions, one at each end. Solving the differential equation and boundary conditions yields the displacement variable $u(x)$ with the force variable $P(x)$ then determined in terms of the derivative $u'(x)$ according to $P = AEu'$. Thus the goals of the formulation of a problem in solid mechanics—namely, determining the deformation $u(x)$ and the internal force distribution $P(x)$—are met.

It is frequently possible to obtain exact solutions to the displacement boundary value problem $(AEu')' + q = 0$ together with two appropriate boundary conditions. However, it must be kept in mind that in general, when considered within the context of the three-dimensional character of the bar, equilibrium, for example, is not satisfied. A simple example is the problem of a bar of varying cross section, as shown in Figure 1-58a. Figure 1-58b indicates that in order for the shaded piece to be in equilibrium, it is necessary that shear stresses τ_{xy} and normal stresses σ_y must be present, although in the axial deformation theory it is assumed that only a normal stress σ_x is present. As long as the cross section changes gradually (one of the assumptions of the elementary axial deformation theory) the stresses σ_y and τ_{xy} are quite small with the primary stress σ_x and the axial displacement u predicted with good accuracy by the approximate theory.

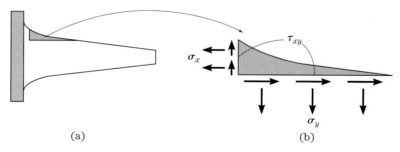

FIGURE 1-58 Approximate character of the axial deformation theory.

Any of the elementary theories that will be developed throughout this text will represent different mathematical models than those associated with the general development contained in sections 1-2 through 1-5. As mentioned at the beginning of this section, they are intended generally to apply to specific geometries and loadings. If all the assumptions are satisfied these theories can result in mathematical models that can be a valid part of the design process. If not, it may be necessary to employ a more general model.

1-10 Summary

The goals of solid mechanics are the determination of the deformations and the internal force distributions, usually meaning the stresses. Generally, the derivation of the governing equations for any class of problems in solid mechanics involves the careful consideration of several aspects of the problem, namely:

1. The geometry of the region.
2. Permissible loadings.
3. The requirements of equilibrium.
4. Deformation.
5. Material behavior.

For a displacement formulation, these considerations are combined to yield a set of equations in terms of the displacements. These equations then must be solved subject to suitable boundary conditions on displacements and/or stresses. There is no concern regarding compatibility since the displacements are the dependent variables.

For a stress formulation, the equations of equilibrium expressed in terms of the stresses are considered to be the governing equations, with the deformation and material behavior combined to produce additional equations (that is, the compatibility equations) that the stresses must satisfy to insure that there is a suitable set of displacements that correspond to the stresses.

The thrust in the remainder of the text will be primarily in the direction of approximate formulations that are appropriate for special classes of problems where the regions are considered to be one- or two-dimensional in character. In these settings the five ideas spelled out above in connection with the formulation and solution of the different classes of problems will be considered.

REFERENCES

1. Timoshenko, S. P., and Goodier, J. N. *Theory of Elasticity*, 2nd ed. McGraw-Hill, New York, 1951.
2. Boresi, A. P., and Lynn, P. P. *Elasticity in Engineering Mechanics*. Prentice-Hall, Englewood Cliffs, N.J., 1974.
3. Fung, Y. C. *Foundations of Solid Mechanics*. Prentice-Hall, Englewood Cliffs, N.J., 1965.
4. Chou, P. C., and Pagano, N. J. *Elasticity*. Van Nostrand, New York, 1967.
5. Reismann, H., and Pawlik, P. S. *Elasticity: Theory and Applications*. Wiley, New York, 1980.
6. Bickford, W. B. *Mechanics of Solids: Concepts and Applications*. Irwin, Burr Ridge, IL, 1993.

EXERCISES

Section 1-2

For each of the following determine
  a. The stress vector on the plane indicated.
  b. The normal stress on the plane indicated.
  c. The shear stress and its direction on the plane indicated.

EXERCISES

1. $\sigma_x = 10, \sigma_y = -7, \sigma_z = 5, \tau_{xy} = -3, \tau_{xz} = 0, \tau_{yz} = 2,$
   $\ell = 0.3579, m = 0.6437 \text{ and } n = 0.6764$
2. $\sigma_x = 8, \sigma_y = 4, \sigma_z = -2, \tau_{xy} = 4, \tau_{xz} = 4, \tau_{yz} = -3,$
   $\ell = 0.5478, m = -0.4367 \text{ and } n = -0.7136$
3. $\sigma_x = 10, \sigma_y = -7, \sigma_z = 5, \tau_{xy} = -3, \tau_{xz} = 0, \tau_{yz} = 2,$
   $\ell = -0.7539, m = 0.5364 \text{ and } n = 0.3794$
4. $\sigma_x = 9, \sigma_y = -7, \sigma_z = 5, \tau_{xy} = -6, \tau_{xz} = 0, \tau_{yz} = 6,$
   $\ell = 0.8, m = 0.4 \text{ and } n = 0.4472$
5. $\sigma_x = -5, \sigma_y = -7, \sigma_z = 5, \tau_{xy} = -4, \tau_{xz} = 0, \tau_{yz} = -5,$
   $\ell = 0.5774, m = 0.5774 \text{ and } n = 0.5774$
6. $\sigma_x = 70, \sigma_y = 40, \sigma_z = 50, \tau_{xy} = -30, \tau_{xz} = 0, \tau_{yz} = 0,$
   $\ell = 0.8954, m = 0.2315 \text{ and } n = -0.3804$
7. $\sigma_x = 0, \sigma_y = -17, \sigma_z = 25, \tau_{xy} = -23, \tau_{xz} = 10, \tau_{yz} = -19$
   $\ell = 0.0, m = 0.6 \text{ and } n = 0.8$
8. $\sigma_x = 15, \sigma_y = -7, \sigma_z = 5, \tau_{xy} = -3, \tau_{xz} = 0, \tau_{yz} = -4,$
   $\ell = 0.7071, m = 0.0 \text{ and } n = 0.7071$
9. $\sigma_x = 20, \sigma_y = -17, \sigma_z = 15, \tau_{xy} = -45, \tau_{xz} = 0, \tau_{yz} = 17,$
   $\ell = 0.350, m = 0.640 \text{ and } n = 0.684$

In each of the following determine
   a. The principal stresses.
   b. The principal directions.
   c. The absolute maximum shear stress and the planes on which it acts.

10. $\sigma_x = 10, \sigma_y = -7, \sigma_z = 5, \tau_{xy} = -3, \tau_{xz} = 0, \tau_{yz} = 2$
11. $\sigma_x = 20, \sigma_y = -17, \sigma_z = 15, \tau_{xy} = -12, \tau_{xz} = 0, \tau_{yz} = 22$
12. $\sigma_x = 8, \sigma_y = 4, \sigma_z = -2, \tau_{xy} = 4, \tau_{xz} = 4, \tau_{yz} = -3$
13. $\sigma_x = 10, \sigma_y = -7, \sigma_z = 5, \tau_{xy} = -3, \tau_{xz} = 0, \tau_{yz} = 2$
14. $\sigma_x = 9, \sigma_y = -7, \sigma_z = 5, \tau_{xy} = -6, \tau_{xz} = 0, \tau_{yz} = 6$
15. $\sigma_x = -5, \sigma_y = -7, \sigma_z = 5, \tau_{xy} = -4, \tau_{xz} = 0, \tau_{yz} = -5$
16. $\sigma_x = 70, \sigma_y = 40, \sigma_z = 50, \tau_{xy} = -30, \tau_{xz} = 0, \tau_{yz} = 0$
17. $\sigma_x = 0, \sigma_y = -17, \sigma_z = 25, \tau_{xy} = -23, \tau_{xz} = 10, \tau_{yz} = -19$
18. $\sigma_x = 15, \sigma_y = -7, \sigma_z = 5, \tau_{xy} = -3, \tau_{xz} = 0, \tau_{yz} = -4$
19. $\sigma_x = 20, \sigma_y = -17, \sigma_z = 15, \tau_{xy} = -45, \tau_{xz} = 0, \tau_{yz} = 17$

In each of the following use a 2×2 eigenvalue problem formulation to determine
   a. The principal stresses.
   b. The principal directions.
   c. The absolute maximum shear stress and the planes on which it acts.

20. $\sigma_x = 10, \sigma_y = -7, \tau_{xy} = -3; \sigma_z = \tau_{xz} = \tau_{yz} = 0$
21. $\sigma_x = 40, \sigma_y = -30, \tau_{xy} = 20; \sigma_z = \tau_{xz} = \tau_{yz} = 0$
22. $\sigma_x = 100, \sigma_y = -70, \tau_{xy} = 60; \sigma_z = \tau_{xz} = \tau_{yz} = 0$
23. $\sigma_x = -10, \sigma_y = 17, \tau_{xy} = 13; \sigma_z = \tau_{xz} = \tau_{yz} = 0$
24. $\sigma_z = 12, \sigma_y = 8, \tau_{yz} = -3; \sigma_x = \tau_{xz} = \tau_{xy} = 0$

25. $\sigma_x = 10, \sigma_z = 13, \tau_{xz} = -8; \sigma_y = \tau_{xy} = \tau_{yz} = 0$
26. Repeat problem 20 using Mohr's circle.
27. Repeat problem 21 using Mohr's circle.
28. Repeat problem 22 using Mohr's circle.
29. Repeat problem 23 using Mohr's circle.
30. Repeat problem 24 using Mohr's circle.
31. Repeat problem 25 using Mohr's circle.
32. By requiring equilibrium for an element $tr\Delta r\Delta\theta$, show that the two-dimensional differential equations of equilibrium for cylindrical coordinates are

$$\frac{\partial \sigma_r}{\partial r} + \frac{\sigma_r - \sigma_\theta}{r} + \frac{1}{r}\frac{\partial \tau_{r\theta}}{\partial \theta} + R = 0$$

and

$$\frac{1}{r}\frac{\partial \sigma_\theta}{\partial \theta} + \frac{\partial \tau_{r\theta}}{\partial r} + \frac{2\tau_{r\theta}}{r} + \Theta = 0$$

where R and Θ are the body forces in the radial and circumferential directions, respectively.

33. By requiring equilibrium for an element $r\Delta r\Delta\theta\Delta z$, show that the differential equations of equilibrium for *axisymmetric* problems are

$$\frac{\partial \sigma_r}{\partial r} + \frac{\sigma_r - \sigma_\theta}{r} + \frac{\tau_{rz}}{\partial z} + R = 0$$

and

$$\frac{\partial \tau_{rz}}{\partial r} + \frac{\partial \sigma_z}{\partial z} + \frac{\tau_{rz}}{r} + Z = 0$$

where R and Z are the body forces in the radial and axial directions, respectively.

34. By requiring equilibrium for an element $r\Delta r\Delta\theta\Delta z$, show that the differential equations of equilibrium for the three-dimensional problem in cylindrical coordinates are

$$\frac{\partial \sigma_r}{\partial r} + \frac{1}{r}\frac{\partial \tau_{r\theta}}{\partial \theta} + \frac{\tau_{rz}}{\partial z} + \frac{\sigma_r - \sigma_\theta}{r} + R = 0$$

$$\frac{\partial \tau_{rz}}{\partial r} + \frac{1}{r}\frac{\partial \tau_{\theta z}}{\partial \theta} + \frac{\partial \sigma_z}{\partial z} + \frac{\tau_{rz}}{r} + Z = 0$$

$$\frac{\partial \tau_{r\theta}}{\partial r} + \frac{1}{r}\frac{\partial \sigma_\theta}{\partial \theta} + \frac{\partial \tau_{\theta z}}{\partial z} + \frac{2\tau_{r\theta}}{r} + \Theta = 0$$

where R, Θ and Z are the body forces in the radial, circumferential and axial directions, respectively. Observe that by appropriate specialization these can be reduced to each of the results obtained in problems 32 and 33.

Section 1-3

35. Integrate to determine the most general solution of the equations

$$\epsilon_x = \frac{\partial u}{\partial x} = 0, \quad \epsilon_y = \frac{\partial v}{\partial y} = 0, \quad \gamma_{xy} = \frac{\partial u}{\partial y} + \frac{\partial v}{\partial x} = 0$$

that is, the most general displacements that correspond to zero linear strains in xy coordinates. Interpret your results.

36. Proceeding along lines similar to those used in section 1-3, take the radial, circumferential and axial components of displacements as $u(r, \theta, z)$, $v(r, \theta, z)$ and $w(r, \theta, z)$ and show that the nonlinear strain displacement relations for cylindrical coordinates are

$$\epsilon_r = \frac{\partial u}{\partial r} + \frac{1}{2}\left[\left(\frac{\partial u}{\partial r}\right)^2 + \left(\frac{\partial v}{\partial r}\right)^2 + \left(\frac{\partial w}{\partial r}\right)^2\right]$$

$$\epsilon_\theta = \frac{1}{r}\frac{\partial v}{\partial \theta} + \frac{u}{r} + \frac{1}{2r^2}\left[\left(\frac{\partial u}{\partial \theta} - v\right)^2 + \left(u + \frac{\partial v}{\partial \theta}\right)^2 + \left(\frac{\partial w}{\partial \theta}\right)^2\right]$$

$$\epsilon_z = \frac{\partial w}{\partial z} + \frac{1}{2}\left[\left(\frac{\partial u}{\partial z}\right)^2 + \left(\frac{\partial v}{\partial z}\right)^2 + \left(\frac{\partial w}{\partial z}\right)^2\right]$$

$$\gamma_{r\theta} = \frac{1}{r}\frac{\partial u}{\partial \theta} + \frac{\partial v}{\partial r} - \frac{v}{r} + \frac{1}{r}\left[\frac{\partial u}{\partial r}\left(\frac{\partial u}{\partial \theta} - v\right) + \frac{\partial v}{\partial r}\left(u + \frac{\partial v}{\partial \theta}\right) + \frac{\partial w}{\partial r}\frac{\partial w}{\partial \theta}\right]$$

$$\gamma_{\theta z} = \frac{\partial v}{\partial z} + \frac{1}{r}\frac{\partial w}{\partial \theta} + \frac{1}{r}\left[\left(\frac{\partial u}{\partial \theta} - v\right)\frac{\partial u}{\partial z} + \left(u + \frac{\partial v}{\partial \theta}\right)\frac{\partial v}{\partial z} + \frac{\partial w}{\partial z}\frac{\partial w}{\partial \theta}\right]$$

$$\gamma_{rz} = \frac{\partial u}{\partial z} + \frac{\partial w}{\partial r} + \left[\frac{\partial u}{\partial r}\frac{\partial u}{\partial z} + \frac{\partial v}{\partial r}\frac{\partial v}{\partial z} + \frac{\partial w}{\partial r}\frac{\partial w}{\partial z}\right]$$

37. Integrate to determine the most general solution of the equations

$$\epsilon_r = \frac{\partial u}{\partial r} = 0, \qquad \epsilon_\theta = \frac{1}{r}\frac{\partial v}{\partial \theta} + \frac{u}{r} = 0, \qquad \gamma_{r\theta} = \frac{1}{r}\frac{\partial u}{\partial \theta} + \frac{\partial v}{\partial r} - \frac{v}{r} = 0$$

that is, the most general displacements that correspond to zero linear strains in $r\theta$ coordinates. Interpret your results.

Sections 1-4 – 1-5

In each of the following problems the student is asked to formulate problems in terms of the stresses. When this approach is taken it is necessary for the stresses to satisfy certain compatibility conditions. However, when the stresses are constants these compatibility conditions are satisfied identically, guaranteeing that the strain displacement relations can in fact be integrated.

38. A rectangular plate is subjected to constant shear and normal stresses on the edges, as shown. Show that all the equilibrium equations and the compatibility equation is satisfied by taking the stresses in the interior of the plate to be equal to the corresponding values prescribed on the boundary.

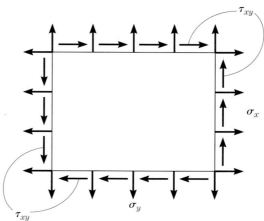

39. The rectangular block with lengths shown is subjected to a uniform normal stress $\sigma_x = -\sigma_0$.

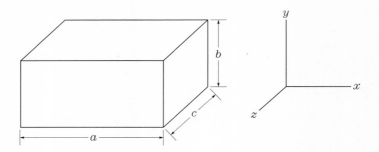

Assuming the stresses to be $\sigma_x = -\sigma_0$, $\sigma_y = \sigma_z = \tau_{xy} = \tau_{xz} = \tau_{yz} = 0$ throughout the block, determine the changes in the lengths a, b and c. Also determine the change in the volume.

40. The rectangular block of problem 39 is subjected to a uniform normal stress $\sigma_x = -\sigma_0$ and a uniform normal stress $\sigma_y = \sigma^*$. Assuming the stresses to be $\sigma_x = -\sigma_0$, $\sigma_y = \sigma^*$, $\sigma_z = \tau_{xy} = \tau_{xz} = \tau_{yz} = 0$ throughout the block, determine σ^* so that the length b does not change. Also determine the changes in the lengths of a and c.

41. The rectangular block of problem 39 is subjected to a uniform normal stress $\sigma_x = -\sigma_0$ and to uniform normal stresses $\sigma_y = \sigma_1$ and $\sigma_z = \sigma_2$. Assuming the stresses to be $\sigma_x = -\sigma_0$, $\sigma_y = \sigma_1$, $\sigma_z = \sigma_2$, $\tau_{xy} = \tau_{xz} = \tau_{yz} = 0$ throughout the block, determine σ_1 and σ_2 in terms of σ_0 such that the lengths b and c do not change. Determine the change in the length of a.

42. A steel plate with dimensions as shown is subjected to stresses on the edges of $\sigma_x = 110$ MPa and $\sigma_y = -70$ MPa. Determine (a) the change in the length of the 2.0 m side, (b) the change in length of the 1.0 m side, (c) the change in length of a diagonal and (d) the change in the volume. Take the thickness to be 0.01 m, $E = 200$ GPa and $\nu = 0.3$.

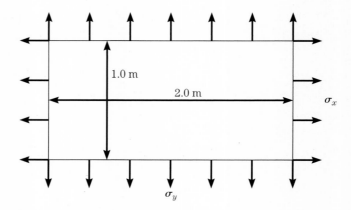

43. A plate in the shape of a rhombus is subjected to stresses on the edges as shown. Determine (a) the stresses σ_x, σ_y and τ_{xy} and (b) the change in the length of the long diagonal. Take $a = 100$ in, $b = 50$ in, $\sigma_1 = 20$ ksi, $\sigma_2 = 10$ ksi, $E = 10^7$ psi and $\nu = 0.3$.

EXERCISES

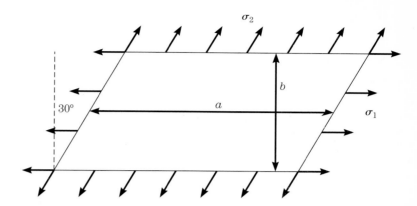

44. The rectangular block of problem 38 is subjected to a uniform temperature increase T_0. Assuming the block is completely free to expand, determine the changes in the lengths a, b and c.

45. The rectangular block of problem 39 is subjected to a uniform temperature increase T_0. Assuming the block is completely free to expand in the y and z directions but is constrained at the ends against motion in the x direction, determine the changes in the lengths b and c and the stress in the x direction.

46. The rectangular block of problem 39 is subjected to a uniform temperature increase T_0. Assuming the block is completely free to expand in the z direction but is constrained at the ends against motion in the x and y directions, determine the change in the length c and the stresses in the x and y directions.

Section 1-7

47. A ductile steel specimen is subjected to a pure shear test and observed to yield when the shear stress reaches 80 MPa. If the specimen were subjected to a uniaxial tension test, at what extensional stress would you expect yielding to occur using (a) the maximum shear-stress criterion and (b) the von Mises criterion?

48. A brittle material is subjected to uniaxial tension and compression tests resulting in failure strengths of $\sigma_{UT} = 4$ ksi and $\sigma_{UC} = 17$ ksi. For a member of the same material subjected to a state of plane stress, the stresses at a point are $\sigma_x = -5$ ksi, $\sigma_y = 0$ and $\tau_{xy} = 15$ ksi. Evaluate failure on the basis of the Coulomb-Mohr theory.

49. A plane stress state at a point in a body consists of the stresses $\sigma_x = 60$ MPa, $\sigma_y = -30$ MPa and $\tau_{xy} = 18$ MPa. For a yield stress in tension of 200 MPa, determine whether yielding is indicated on the basis of (a) the maximum shear-stress criterion and (b) the von Mises criterion.

50. Uniaxial tension and compression tests of a brittle isotropic material show failure strengths of $\sigma_{UT} = 6$ ksi and $\sigma_{UC} = 27$ ksi. At a point in a member subjected to a state of plane stress the stresses are $\sigma_x = \sigma_0$ ksi, $\sigma_y = 0$ and $\tau_{xy} = 5$ ksi. Determine the permissible range for σ_0 in order that the Coulomb-Mohr criterion be satisfied.

51. A steel with a tensile yield stress of 385 MPa is subjected to a state of plane stress with $\sigma_y = -160$ MPa and $\tau_{xy} = 100$ MPa. Using the von Mises criterion determine whether yielding occurs when (a) $\sigma_x = 150$ MPa, (b) $\sigma_x = 190$ MPa and (c) $\sigma_x = 230$ MPa.

52. Repeat problem 51 using the maximum shear-stress criterion.

53. A structural component with a tensile yield of 40 ksi is subjected to a state of plane stress with $\sigma_x = 30$ ksi, $\sigma_y = 20$ ksi. Determine whether yielding occurs using the maximum shear-stress criterion for (a) $\tau_{xy} = 9$ ksi, (b) 18 ksi and (c) 27 ksi.
54. Repeat problem 53 using the von Mises criterion.
55. For the state of stress $\sigma_x = 20$ ksi, $\sigma_y = 15$ ksi, $\sigma_z = 10$ ksi, $\tau_{xy} = 12$ ksi, $\tau_{xz} = \tau_{yz} = 0$. If $\sigma_Y = 35$ ksi determine whether yielding occurs on the basis of (a) the maximum shear-stress criterion and (b) the von Mises criterion.
56. For the state of stress $\sigma_x = 180$ Mpa, $\sigma_y = 100$ Mpa, $\sigma_z = -60$ MPa, $\tau_{xy} = 100$ MPa, $\tau_{xz} = \tau_{yz} = 0$. If $\sigma_Y = 240$ MPa determine whether yielding occurs on the basis of (a) the maximum shear-stress criterion and (b) the von Mises criterion.
57. A ductile material is subjected to the state of stress given by $\sigma_x = 6$ ksi, $\sigma_y = -8$ ksi, $\tau_{xy} = 10$ ksi, $\tau_{xz} = \tau_{yz} = 0$. For $\sigma_Y = 30$ ksi determine the permissible range for σ_z based on (a) the maximum shear-stress criterion and (b) the von Mises criterion.
58. A ductile material is subjected to the state of stress given by $\sigma_x = 65$ MPa, $\sigma_y = -55$ MPa, $\tau_{xy} = 70$ MPa, $\tau_{xz} = \tau_{yz} = 0$. For $\sigma_Y = 200$ MPa determine the permissible range for σ_z based on (a) the maximum shear-stress criterion and (b) the von Mises criterion.
59. A ductile material is subjected to the state of stress given by $\sigma_x = 19$ ksi, $\sigma_y = 16$ ksi, $\sigma_z = 10$ ksi, $\tau_{xz} = \tau_{yz} = 0$. For $\sigma_Y = 36$ ksi determine the permissible range for τ_{xy} based on the maximum shear-stress criterion.
60. A ductile material is subjected to the state of stress given by $\sigma_x = 180$ MPa, $\sigma_y = 100$ MPa, $\sigma_z = -50$ MPa, $\tau_{xz} = \tau_{yz} = 0$. For $\sigma_Y = 240$ MPa determine the permissible range for τ_{xy} based on the maximum shear-stress criterion.
61. Repeat problem 59 using the von Mises criterion.
62. Repeat problem 60 using the von Mises criterion.

Section 1-9

For each of the following, state and solve the appropriate boundary value problem for the elementary axial problem indicated. Determine both $u(x)$ and $P(x)$.

63.
64.
65.
66.

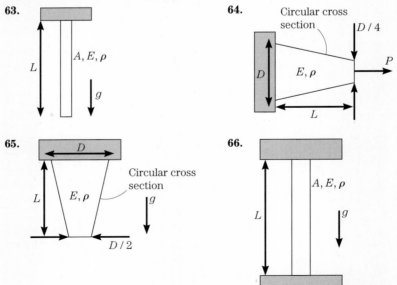

EXERCISES

67.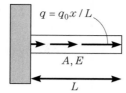

68.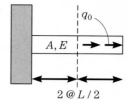

69.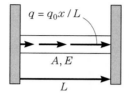

70.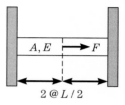

71.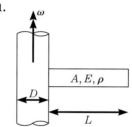

72.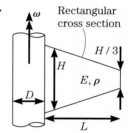

2 Variational and Approximate Methods

2-1 Introduction
2-2 Work and Potential Energy
2-3 Strain Energy and Stiffness Formulations
    2-3-1 Strain Energies for Structural Members
    2-3-2 Stationary Potential Energy
    2-3-3 Structural Analysis: Castigliano's First Theorem
2-4 Approximate Methods
    2-4-1 Method of Ritz
    2-4-2 Finite-Element Methods
2-5 Complementary Energy Ideas
    2-5-1 Complementary Energies for Structural Members
    2-5-2 Dummy Loads
    2-5-3 Statically Indeterminate Systems: Principle of Least Work
2-6 Summary
    References
    Exercises

2-1 Introduction

In Chapter 1 the governing equations of solid mechanics were developed using what is referred to as the *vectorial approach*, wherein Newton's second law was used directly in terms of drawing free body diagrams and requiring equilibrium by summing forces and moments. There are a number of alternative formulations that are frequently very valuable, referred to as *variational approaches*. These involve the use of integrated statements that, for many classes of problems, are equivalent to the vectorial approach. The perspective of the vectorial approach is analogous to the direct application of $\mathbf{F} = m\mathbf{a}$ in a first course in dynamics, whereas the variational approach is reminiscent of work energy for a particle or rigid body, stated briefly as $WD = \Delta T$ (work done equals change in kinetic energy) and obtained by integrating $\mathbf{F} = m\mathbf{a}$ with respect to position.

For a deformable body the work done consists of two parts. The first part accounts for the work done by external forces and is conceptually the same as for a particle or rigid body. The second part accounts for the deformations associated with internal forces and is essentially the generalization of the familiar $W = -ke^2/2$ term for the work done by a linear spring. For a deformable body, the negative of the internal work done is usually referred to as a *strain energy*.

In this chapter strain energy and its uses in formulating and analyzing structural models will be discussed using the displacements as the independent variables. Such formulations lead to *stiffness formulations*. The concept of complementary energy will be defined and used in connection with situations where the force resultants are taken as the independent variables. These are the so-called *flexibility formulations*. Additionally, it will be shown how the strain energy formulations lead naturally to the very powerful and useful displacement finite-element method.

2-2 Work and Potential Energy

Underlying the idea of the energy of a mechanical system is the concept of work. The work of a force is defined as a product of the force and the displacement. More specifically, when a force **F** moves along a path, such as indicated in Figure 2-1, the work done by **F** in passing from A to B is

$$W(A, B) = \int_A^B \mathbf{F} \cdot \mathbf{dr} = \int_A^B F \cos \alpha \, ds$$

where the integral is a line integral evaluated along the specified curve from A to B. In general, the value of W is path dependent, that is, when different paths are used in passing from A to B, different values of W may result.

There are many practical technical situations where the same amount of work is done by the force in passing from A to B regardless of which path is chosen, that is, *the work done is independent of the path*. Such a force is called a *conservative force* with the *potential energy* at the configuration **x** associated with the force **F** defined to be

$$V(\mathbf{x}) - V(\mathbf{x_0}) = -\int_{\mathbf{x_0}}^{\mathbf{x}} \mathbf{F} \cdot \mathbf{dr} \tag{2-1}$$

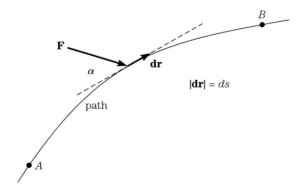

FIGURE 2-1 Work done by a force.

where $V(\mathbf{x_0})$ is the potential energy at some standard configuration $\mathbf{x} = \mathbf{x_0}$. $V(\mathbf{x_0})$ is a constant that can usually be taken as zero. Equation (2-1) clearly states that aside from an arbitrary constant associated with a standard configuration, the potential energy associated with a general configuration $\mathbf{x}$ is the *negative* of the work done in passage from the standard configuration $\mathbf{x_0}$ to the general configuration $\mathbf{x}$.

◀ **EXAMPLE 2-1** Consider the earth's gravitational force acting on a mass near the surface of the earth. The work done by the gravitational force of the earth is

$$W(A, B) = -\int_A^B \mathbf{F} \cdot \mathbf{dr} = \int_A^B (-mg\mathbf{j}) \cdot (\mathbf{i}\, dx + \mathbf{j}\, dy)$$
$$= -mg \int_{h_1}^{h_2} dy = -mg(h_2 - h_1) = -mgy$$

where, as indicated in Figure 2-2, y is a coordinate measured from the initial configuration. With the value of the integral clearly independent of the path (the x component of a displacement $\mathbf{dr}$ along the path AB results in no work being done), it follows that the potential energy associated with the gravitational force between the earth and a mass m is given by

$$V(y) = -W(A, B) = mgy$$

In a completely similar manner it follows that the potential energy of a constant force F acting through a distance X, as shown in Figure 2-3a, is $V(X) = -FX$ and that the potential energy of a constant moment acting through an angle Θ is $V(\Theta) = -M\Theta$, as shown in Figure 2-3b. ◆

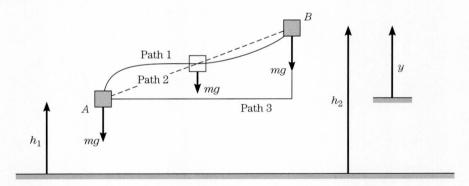

FIGURE 2-2 Potential energy of a mass near the earth.

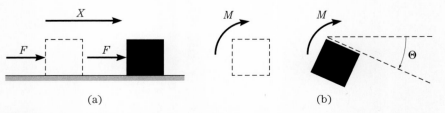

FIGURE 2-3 Potential energies of constant forces and moments.

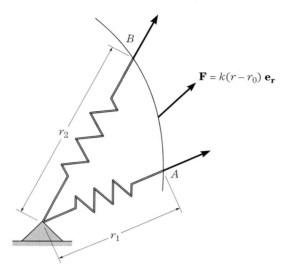

FIGURE 2-4 Potential energy of a linear spring.

EXAMPLE 2-2 Consider the potential energy of the force in a linear spring, as shown in Figure 2-4. Take the unstretched length of the spring to be r_0. Note carefully that the force shown in the figure is opposite to the force that is applied by the spring, that is, $\mathbf{F}_{sp} = -k(r - r_0)\mathbf{e_r}$, where $\mathbf{e_r}$ is the unit vector in the radial direction. The potential energy associated with the force in the spring is then (potential energy = negative the work done)

$$V = -\int_A^B -k(r - r_0)\mathbf{e_r} \cdot \mathbf{dr} = \int_A^B k(r - r_0)\mathbf{e_r} \cdot (\mathbf{e_r}\, dr + \mathbf{e_\theta} r\, d\theta)$$

$$= \int_{r_1}^{r_2} k(r - r_0)\, dr = k\frac{(r_2 - r_0)^2 - (r_1 - r_0)^2}{2}$$

$$= k\frac{e_2^2 - e_1^2}{2}$$

where e_1 and e_2 are the initial and final elongations of the spring. Here again the work done is independent of the path with only the component of displacement along the direction of the spring resulting in any work. For the case where the spring is originally unstretched and the final elongation is denoted by e, $V = ke^2/2$, a familiar result. It is very common to simply refer to the potential energy of a linear spring rather than to the potential energy of the force associated with a linear spring. As will be seen presently, this particular potential energy is typical of the potential energy for a linearly elastic structure.

When all the forces associated with a mechanical system are conservative, the mechanical system is called a *conservative mechanical system*. In such a case the equilibrium equations for the mechanical system can be derived using a variational principle called the *principle of stationary potential energy*.

Suppose the potential energy of a conservative structure is expressed in terms of the degrees of freedom of the system, $x_1, x_2, \ldots, x_n$. The number of degrees of freedom is the minimum number of variables necessary to completely specify the configuration of the mechanical system. The principle of stationary potential energy states that

> with respect to variations in the degrees of freedom the potential energy $V(x_1, x_2, \ldots, x_n)$ of a conservative mechanical system is stationary at an equilibrium configuration,

that is,

$$\frac{\partial V}{\partial x_k} = 0 \quad k = 1, 2, \ldots, n$$

These equations are equivalent, and in some cases identical, to the equations developed on the basis of the vectorial approach of drawing FBDs and summing forces and/or moments.

EXAMPLE 2-3 Consider the system of initially unstretched linear springs, as shown in Figure 2-5. The loads P_1 and P_2 are applied to the two masses whose displacements u_1 and u_2 are taken to be the degrees of freedom. The total potential energy of the system consists of the sum of the potential energies of the springs and the potential energies of the external loads P_1 and P_2. Each of the springs has a potential energy of the form $V = ke^2/2$ so that with the elongations given by $e_1 = u_1$, $e_2 = u_2 - u_1$ and $e_3 = -u_2$, the total potential energy of the system can be expressed as

$$V(u_1, u_2) = \frac{1}{2}[k_1 u_1^2 + k_2(u_2 - u_1)^2 + k_3(-u_2)^2] - P_1 u_1 - P_2 u_2$$

The principle of stationary potential energy then yields

$$\frac{\partial V}{\partial u_1} = k_1 u_1 - k_2(u_2 - u_1) - P_1 = 0$$

$$\frac{\partial V}{\partial u_2} = k_2(u_2 - u_1) + k_3 u_2 - P_2 = 0$$

The reader should show that the vectorial approach yields the same results. To continue the solution of this problem, let $k_1 = k_2 = k_3 = k$, after which the equations

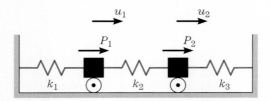

FIGURE 2-5 System of linear springs.

can be written in augmented form as

$$2u_1 - u_2 = \frac{P_1}{k}$$

$$-u_1 + 2u_2 = \frac{P_2}{k}$$

with solution $u_1 = (2P_1 + P_2)/3k$ and $u_2 = (P_1 + 2P_2)/3k$. The internal forces in the springs can then be computed according to

$$f_1 = k_1 e_1 = k u_1 = \frac{2P_1 + P_2}{3}$$

$$f_2 = k_2 e_2 = k(u_2 - u_1) = \frac{P_2 - P_1}{3}$$

$$f_3 = k_3 e_3 = k(-u_2) = -\frac{P_1 + 2P_2}{3}$$

The reader should show that these internal forces are in balance with the loads by drawing the proper FBDs of each of the masses. This method of formulation and solution is typical of the stiffness method, covered in detail later in the chapter. ◆

2-3 Strain Energy and Stiffness Formulations

The discussions and examples in the previous section dealt with discrete systems. In this section analogous ideas for continuous systems will be developed. It is still frequently possible to form the total potential energy in terms of what will be termed strain energy plus the potential energy of the external loads, and then to obtain equilibrium equations by requiring that the total potential energy be stationary. The strain energy for continuous systems is analogous to terms of the form $k\Delta^2/2$ for discrete systems with the potential energy of the external loads being analogous to terms of the form $-P\Delta$. The finite-element method, which is in many instances based on these ideas, is discussed later in this chapter.

To begin discussions of strain energy, consider the elementary problem of the axial deformation of a uniform elastic bar, as indicated in Figure 2-6. Assume that the bar is in equilibrium under the action of the distributed loading $q(x)$ and the force P_0 at the end of the bar. The principle of virtual work states that during a virtual displacement

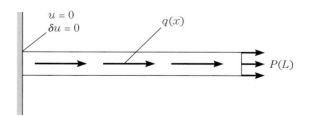

FIGURE 2-6 Axial deformation problem.

$\delta u(x)$ consistent with the constraints, the sum of δW_e, the virtual work done by the external forces, and δW_i, the virtual work done by the internal forces, is zero, that is,

$$\delta W_e + \delta W_i = 0$$

From the figure it is easily seen that

$$\delta W_e = \int_0^L q\ \delta u\ dx + P_0\ \delta u(L)$$

where δu is an arbitrary continuous function vanishing at the support, that is, $\delta u(0) = 0$. Since the bar is in equilibrium, the equilibrium equation $P' + q = 0$ can be used to write

$$\delta W_e = -\int_0^L P'\ \delta u\ dx + P_0\ \delta u(L)$$

or, upon integration by parts,

$$\delta W_e = -P(L)\ \delta u(L) + P(0)\ \delta u(0) + \int_0^L P\ \delta u'\ dx + P_0\ \delta u(L)$$

or, since $P(L) = P_0$ and $\delta u(0) = 0$,

$$\delta W_e = \int_0^L P\ \delta u'\ dx = \int_0^L P\ \delta \epsilon\ dx$$

Then, with $A\ dx = dvol$,

$$\delta W_e = \int_{vol} \sigma\ \delta \epsilon\ dvol$$

In general, the first law of thermodynamics states that

$$\delta W_e + \delta Q = \delta U + \delta T \qquad (2\text{-}2)$$

where δW_e is the work done by the external forces, δQ is the heat added, δU is the internal energy and δT is the change in kinetic energy. Restricting consideration to adiabatic situations with no appreciable change in kinetic energy, the first law reduces to

$$\delta W_e = \delta U = \int_{vol} \sigma\ \delta \epsilon\ dvol$$

In the context of solid mechanics, the internal energy U is usually called the *strain energy*. Assume the existence of a strain energy density function U_0 depending only on the strains and perhaps the temperature such that

$$\delta U = \int_{vol} \delta U_0\ dvol$$

from which it follows that $\delta U_0 = \sigma\ d\epsilon$. Note that this development also contains the result that $\delta U = -\delta W_i$ stating that the change in internal strain energy is negative the work done by the internal forces.

In connection with Figure 2-7 it is thus seen that U_0 is the area under the stress-strain curve for the material in question, that is,

$$U_0 = \int \sigma\ d\epsilon$$

2-3 STRAIN ENERGY AND STIFFNESS FORMULATIONS

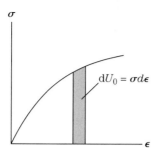

FIGURE 2-7 Strain energy density.

For the case where the one-dimensional body is composed of a linearly elastic material characterized by $\sigma = E\epsilon$, it is seen that

$$U_0 = \int E\epsilon \, d\epsilon = \frac{E\epsilon^2}{2}$$

When temperature changes are present, the stress-strain relation is $\sigma = E(\epsilon - \alpha T)$, resulting in

$$U_0 = \int E(\epsilon - \alpha T) \, d\epsilon = E\left(\frac{\epsilon^2}{2} - \epsilon\alpha T\right)$$

The specific case of the axial deformation of the bar will be considered in the next section.

For situations involving shear stresses and shear strains, the strain energy density is

$$U_0 = \int \tau \, d\gamma$$

which, for linearly elastic bodies with $\tau = G\gamma$, becomes

$$U_0 = \int G\gamma \, d\gamma = \frac{G\gamma^2}{2}$$

In any event, the total strain energy is obtained by integrating the strain energy density over the volume of the body in question.

In a three-dimensional situation, the strain energy density is a function of all the strains with

$$\delta U = \int_D [\sigma_x \delta\epsilon_x + \sigma_y \delta\epsilon_y + \sigma_z \delta\epsilon_z + \tau_{xy} \delta\gamma_{xy} + \tau_{xz} \delta\gamma_{xz} + \tau_{yz} \delta\gamma_{yz}] \, dvol$$

$$= \int_D \left[\frac{\partial U_0}{\partial \epsilon_x} \delta\epsilon_x + \frac{\partial U_0}{\partial \epsilon_y} \delta\epsilon_y + \frac{\partial U_0}{\partial \epsilon_z} \delta\epsilon_z + \frac{\partial U_0}{\partial \gamma_{xy}} \delta\gamma_{xy} + \frac{\partial U_0}{\partial \gamma_{yz}} \delta\gamma_{yz} + \frac{\partial U_0}{\partial \gamma_{xz}} \delta\gamma_{xz}\right] dvol$$

so that

$$\sigma_x = \frac{\partial U_0}{\partial \epsilon_x}, \quad \sigma_y = \frac{\partial U_0}{\partial \epsilon_y}, \quad \sigma_z = \frac{\partial U_0}{\partial \epsilon_z}$$

and

$$\tau_{xy} = \frac{\partial U_0}{\partial \gamma_{xy}}, \quad \tau_{xz} = \frac{\partial U_0}{\partial \gamma_{xz}}, \quad \tau_{yz} = \frac{\partial U_0}{\partial \gamma_{yz}} \quad (2\text{-}3)$$

relating the stresses to the strain energy density. Given the relations between the stresses and the strains, these equations can be integrated to determine the strain energy density. Specifics for this case are left to the exercises.

2-3-1 Strain Energies for Structural Members Many structural analyses involve representing portions of the structure as bars or beams, that is, as idealized structural components. As seen in the first course in solids mechanics, these components are assumed to transmit axial loads, torsional loads and shear and bending loads. In this section the expressions for the potential energies for such idealized members will be developed.

Axial Loading. As discussed above, the axial problem is characterized as shown in Figure 2-8 where q represents distributed loadings, P_0 an axial load and $u(x)$ the axial component of displacement. Assuming that the bar is composed of a linearly elastic material, the resulting strain energy density is then taken to be $U_0 = E\epsilon_x^2/2$ with the total strain energy

$$U = \int_{vol} \frac{E\epsilon_x^2}{2}\, dvol = \int_{vol} \frac{Eu'^2}{2}\, dvol$$

which, with $dvol = A\, dx$, can be expressed as

$$U = \int_0^L \frac{AEu'^2}{2}\, dx \quad (2\text{-}4)$$

When temperature is included, it is left for the student to verify that the strain energy can be written as

$$U = \int_0^L \left(\frac{AEu'^2}{2} - AE\alpha T u' \right) dx$$

Torsional Loading. The torsional problem for a circular cross section bar is characterized as shown in Figure 2-9, where $t(x)$ represents distributed external torques, T_0

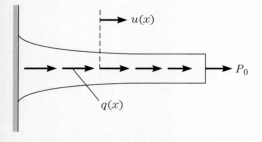

FIGURE 2-8 Axial loading of a bar.

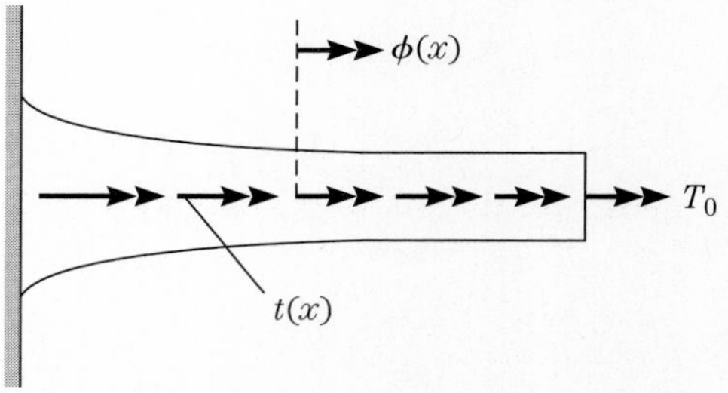

FIGURE 2-9 Torsional loading of a bar.

2-3 STRAIN ENERGY AND STIFFNESS FORMULATIONS

an applied torque and $\phi(x)$ the angle of twist at the cross section in question. As developed in the first course in solid mechanics, the shear strain can be represented as $\gamma = r\phi'$, where r is the radial position of the point in question. The resulting strain energy density is then $U_0 = G\gamma^2/2$ with the total strain energy

$$U = \int_{vol} \frac{G\gamma^2}{2} \, dvol = \int_{vol} \frac{Gr^2\phi'^2}{2} \, dvol$$

which, with $dvol = A \, dx$ and $\int r^2 \, dA = J$ as the polar moment of the area, can be expressed as

$$U = \int_0^L \frac{JG\phi'^2}{2} \, dx \tag{2-5}$$

Bending Deformations. The bending problem for a beam having a cross section symmetrical with respect to the xz plane is shown in Figure 2-10, where $q(x)$ represents distributed transverse loading, with P a transverse load, M an applied moment and $w(x)$ the transverse displacement. Considering only the energy associated with the bending deformation, the bending strain is given by $\epsilon_x = -zw''(x)$ where z is measured from the centroid of the cross section. The resulting strain energy density for bending is then $U_0 = E\epsilon_x^2/2$ with the total strain energy

$$U = \int_{vol} \frac{E\epsilon_x^2}{2} \, dvol = \int_{vol} \frac{Ez^2 w''^2}{2} \, dvol$$

which, with $dvol = A \, dx$ and $\int z^2 \, dA = I$ as the second moment of the area about the y axis, can be expressed as

$$U = \int_0^L \frac{EI w''^2}{2} \, dx \tag{2-6}$$

When temperature is included

$$U = \int_0^L \left(\frac{EI w''^2}{2} + E\alpha w'' m_T \right) dx$$

where $m_T = \int zT \, dA$.

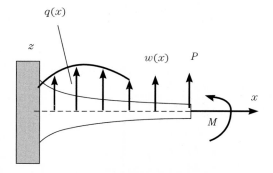

FIGURE 2-10 Transverse loading of a bar.

In the next section it will be indicated how these strain energies can be used in connection with the principle of stationary potential energy to derive equilibrium equations and boundary conditions.

2-3-2 Stationary Potential Energy

For a body subjected to surface tractions and body forces, the total potential energy can be written as

$$V = \int_D U_0 \, dvol - \int_S \mathbf{t_n} \cdot \mathbf{u} \, dS - \int_D \mathbf{F} \cdot \mathbf{u} \, dSvol$$
$$= U + \Omega$$

where

$$U = \int_D U_0 \, dvol$$

is the internal potential or strain energy and

$$\Omega = -W_e = -\int_S \mathbf{t_n} \cdot \mathbf{u} \, dS - \int_D \mathbf{F} \cdot \mathbf{u} \, dvol$$

represents the potential energy of the surface tractions $\mathbf{t_n}$ and the body forces $\mathbf{F}$, respectively, that is, the potential energy of the external forces. As mentioned above, the terms U & Ω are analogous to the quantities $(1/2)kx^2$ and $-Fx$, respectively, for linear discrete problems. In general the strains are represented in terms of the displacements using the strain-displacement relations. The total potential energy is then a function of the displacements and their derivatives. Requiring the total potential energy to be stationary results in differential equations of equilibrium. In what follows, the use of stationary potential energy will be demonstrated in connection with several elementary problems and then its general significance discussed.

EXAMPLE 2-4 For an example of the application of the principle of stationary potential energy to an elementary problem, consider the one-dimensional axial deformation problem discussed in section 1-6 and shown here in Figure 2-11. As outlined in section 2-3-1, the strain energy can be represented as

$$U = \int_D \frac{E\epsilon_x^2}{2} \, dvol = \int_D \frac{Eu'^2}{2} \, dvol = \int_0^L \frac{AEu'^2}{2} \, dx$$

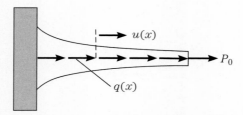

FIGURE 2-11 Axial loading of a bar.

2-3 STRAIN ENERGY AND STIFFNESS FORMULATIONS

For the problem shown in Figure 2-11, the potential energy of the external loads q and P_0 is

$$\Omega = -\int_0^L qu \, dx - P_0 u(L)$$

so that the total potential energy is

$$V = \int_0^L \frac{AEu'^2}{2} dx - \int_0^L qu \, dx - P_0 u(L) \tag{2-7}$$

The principle of stationary potential energy was demonstrated for discrete systems in the previous section. Requiring that the value of the potential energy given by Equation (2-7) be stationary should give equations that represent equilibrium. Understanding that this is in fact true requires a knowledge of the calculus of variations, discussed in appendix A, that is, the calculus of variations is the tool that is used to investigate the stationary value(s) of the potential energy given as an integral such as in Equation (2-7).

Applying the results of the calculus of variations to Equation (2-7) leads to

$$0 = \Delta V = V(u + \delta u) - V(u)$$

$$= \int_0^L (AEu' \, \delta u' - q \, \delta u) \, dx - P_0 \, \delta u(L) + \frac{1}{2} \int_0^L AE(\delta u')^2 \, dx$$

$$= \delta V + \frac{1}{2} \delta^2 V$$

where δu is an arbitrary continuous function of x vanishing at $x = 0$. Requiring δV to vanish gives

$$0 = \int_0^L (AEu' \, \delta u' - q \, \delta u) \, dx - P_0 \, \delta u(L)$$

or, upon integration by parts and remembering that $\delta u(0) = 0$,

$$0 = (AEu'(L) - P_0) \, \delta u(L) - \int_0^L [(AEu')' + q] \, \delta u \, dx$$

Thus, on the basis of the arbitrary character of $\delta u(x)$,

$$(AEu')' + q(x) = 0$$
$$AEu'(L) = P_0$$

which are the equilibrium equation and the nonessential or natural boundary condition. This shows the equivalence of the vectorial and variational formulations for the elementary axial deformation problem. In the expression for ΔV above, the term $\delta^2 V$ is called the second variation and is used to assess stability of an equilibrium configuration. ◆

◆**EXAMPLE 2-5**   As shown in Figure 2-12 a beam is subjected to a distributed loading $q(x)$ as well as a shear load Q and a moment M_0 at the boundary $x = L$. Use stationary potential energy to determine the equation of equilibrium and the mechanical boundary conditions.

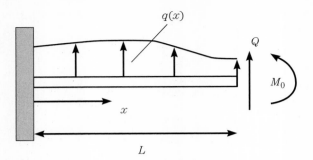

FIGURE 2-12 Transverse deformations of a beam.

Solution: As developed in the previous section, the strain energy of bending for a uniform cross-section linearly elastic beam is given by

$$U = \int_0^L \frac{EI(w'')^2}{2}\, dx$$

The potential energy associated with the external loading can be expressed as

$$\Omega = -\int_0^L qw\, dx - Qw(L) - M_0 w'(L)$$

so that

$$V = \int_0^L \frac{EI(w'')^2}{2}\, dx - \int_0^L qw\, dx - Qw(L) - M_0 w'(L)$$

The essential boundary conditions are $w(0) = 0$ and $w'(0) = 0$ so that $\delta w(0) = 0$ and $\delta w'(0) = 0$ during the process of requiring that V be stationary, leading to

$$0 = \int_0^L [EIw''\, \delta w'' - q\, \delta w]\, dx - Q\, \delta w(L) - M_0\, \delta w'(L)$$

Integrating the first integral by parts twice and collecting terms then yields

$$0 = \int_0^L [EIw'''' - q]\, \delta w\, dx + (EIw''(L) - M_0)\, \delta w'(L) - (EIw'''(L) + Q)\, \delta w(L)$$

Based on the arbitrariness of δw, it follows that

$$EIw'''' - q = 0 \qquad 0 \leq x \leq L$$

for the differential equation of equilibrium that must be satisfied by the displacement $w(x)$ with

$$EIw''(L) = M_0$$

and

$$EIw'''(L) - Q$$

as the mechanical boundary conditions that must be satisfied at $x = L$. Again it is seen that the principle of stationary potential energy generates the equilibrium equations and mechanical boundary conditions and serves as an approach for deriving the governing equations. ◆

In this more general setting, where the total potential energy is expressed in terms of the displacements, the principle of stationary potential energy is stated as follows:

> Of all displacements satisfying continuity and the displacement boundary conditions, those that also satisfy equilibrium and the stress boundary conditions correspond to a stationary value of the total potential energy.

Not only does this approach provide an alternate route for the derivation of the equations of equilibrium of a conservative mechanical system, but it also indicates the appropriate mechanical boundary conditions that must be enforced.

This approach—namely, to express the total potential energy in terms of the displacements and invoke stationary potential energy—will be used to derive the equations of equilibrium and boundary conditions for various applications throughout the remainder of the text.

2-3-3 Structural Analysis: Castigliano's First Theorem

Consider an elastic body acted upon by external forces and moments and sufficiently supported against rigid body motion, as shown in Figure 2-13. Assume that by using the strain-displacement relations, the strain energy has been expressed in terms of a set of displacement degrees of freedom $\Delta_i, i = 1, 2, \ldots, N$, according to

$$U = U(\Delta_1, \Delta_2, \ldots, \Delta_N)$$

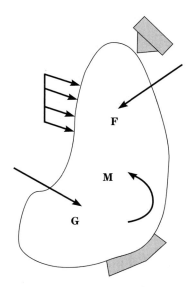

FIGURE 2-13 Elastic body loaded by forces and couples.

and that the external potential energy Ω has been represented as

$$\Omega = -\sum P_i \Delta_i$$

resulting in a so-called discrete problem. Stationary potential energy is then used to write

$$\frac{\partial V}{\partial \Delta_k} = \frac{\partial U}{\partial \Delta_k} - P_k = 0$$

leading to

$$\frac{\partial U}{\partial \Delta_k} = P_k \qquad (2\text{-}8a)$$

usually referred to as *Castigliano's first theorem* [1]. In case the degree of freedom in question is a rotation θ, there results the corollary

$$\frac{\partial U}{\partial \theta_k} = M_k \qquad (2\text{-}8b)$$

where M_k is the moment corresponding to θ_k.

In what follows several situations will be indicated where it is possible to reduce a continuous system to an equivalent discrete system. Castigliano's first theorem can then be used to generate equations of equilibrium that are algebraic. When possible, this approach is preferable from the standpoint that algebraic equations are generally simpler to solve than differential equations.

EXAMPLE 2-6 An axial member consisting of constant cross-section segments is loaded by forces, as shown in Figure 2-14. Determine the displacements at the loads and the internal forces in each of the members.

Solution: With the displacements u_1 and u_2 as shown in the figure, the strains in each of the three members are $\varepsilon_1 = u_1/L_1$, $\varepsilon_2 = (u_2 - u_1)/L_2$ and $\varepsilon_3 = -u_2/L_3$. It follows that the total strain energy—namely, $U = U_1 + U_2 + U_3$ where U_1, U_2 and U_3 are the potential energies of the three bars—can be expressed as

$$U = \int_{vol_1} E_1 \left(\frac{u_1}{L_1}\right)^2 dvol_1 + \int_{vol_2} E_2 \left(\frac{u_2 - u_1}{L_2}\right)^2 dvol_2 + \int_{vol_3} E_3 \left(\frac{u_2}{L_3}\right)^2 dvol_3$$

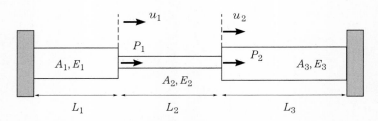

FIGURE 2-14 Axial loading of a segmented bar.

2-3 STRAIN ENERGY AND STIFFNESS FORMULATIONS

or, with $dvol_k = A_k\,dx$, as

$$U = \frac{A_1 E_1 u_1^2}{2L_1} + \frac{A_2 E_2 (u_2 - u_1)^2}{2L_2} + \frac{A_3 E_3 u_2^2}{2L_3}$$

$$= \frac{k_1 u_1^2}{2} + k_2 \frac{(u_2 - u_1)^2}{2} + \frac{k_3 u_2^2}{2}$$

where $k_i = A_i E_i / L_i$. Note carefully that the observation that the strain in each of the members is constant allows us to express the strains and hence the potential energies in terms of displacements at the ends of the bars, that is, to consider the problem discrete in nature. Using Castigliano's first theorem leads to

$$\frac{\partial U}{\partial u_1} = k_1 u_1 + k_2(u_1 - u_2) = P_1$$

and

$$\frac{\partial U}{\partial u_2} = k_2(u_2 - u_1) + k_3 u_2 = P_2$$

which are the equilibrium equations. Take $k_1 = 3k$, $k_2 = k$, $k_3 = 2k$, $P_1 = P$ and $P_2 = 2P$, leading to

$$4u_1 - u_2 = \frac{P}{k}$$

and

$$-u_1 + 3u_2 = \frac{2P}{k}$$

Solving yields $u_1 = 5P/11k$ and $u_2 = 9P/11k$. The internal forces p_i can then be computed according to $p_i = A_i \sigma_i = A_i E_i \varepsilon_i$, leading to

$$p_1 = \frac{A_1 E_1 u_1}{L_1} = k_1 u_1 = \frac{15P}{11}$$

$$p_2 = \frac{A_2 E_2 (u_2 - u_1)}{L_2} = k_2(u_2 - u_1) = \frac{4P}{11}$$

$$p_3 = \frac{A_3 E_3 (-u_2)}{L_3} = -k_3 u_2 = -\frac{18P}{11}$$

The reader should show that these internal forces satisfy equilibrium. ◆

EXAMPLE 2-7 A beam is loaded by an end force and moment as shown in Figure 2-15a. Express the potential energy in terms of the displacement and slope at the loaded end of the beam and then use Castigliano's first theorem to derive the equations of equilibrium.

Solution: In the exercises the student is asked to show that for a beam with degrees of freedom $w(0) = w_1$, $w'(0) = \theta_1$, $w(L) = w_2$ and $w'(L) = \theta_2$ the strain energy can

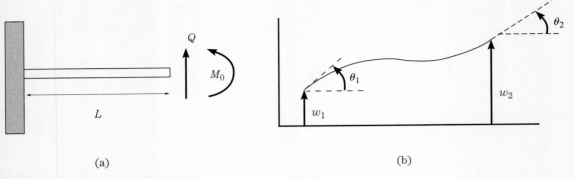

FIGURE 2-15 End-loaded cantilever beam.

be expressed as

$$U = \frac{2EI}{L}\left[\theta_1^2 + \theta_1\theta_2 + \theta_2^2 - 3\left(\frac{w_2 - w_1}{L}\right)(\theta_1 + \theta_2) + 3\left(\frac{w_2 - w_1}{L}\right)^2\right]$$

where w_1, θ_1, w_2 and θ_2 are shown in Figure 2-15b. For the beam shown, $w_1 = \theta_1 = 0$ due to the fixity at $x = 0$. Castigliano's first theorem then gives

$$\frac{\partial U}{\partial w_2} = Q = \frac{2EI}{L}\left[\frac{6w_2}{L^2} - \frac{3\theta_2}{L}\right]$$

and

$$\frac{\partial U}{\partial \theta_2} = M_0 = \frac{2EI}{L}\left[2\theta_2 - \frac{3w_2}{L}\right]$$

Solving for w_2 and θ_2 gives

$$w_2 = \frac{QL^3}{3EI} + \frac{M_0 L^2}{2EI}$$

and

$$\theta_2 = \frac{QL^2}{2EI} + \frac{M_0 L}{EI}$$

Using other results from the exercises, it can then be shown that the moment M and shear V in the beam can be expressed as

$$M = EIw''(x) = Q(L - x) + M_0$$

and

$$V = -EIw''' = Q_0$$

which, by drawing a free body diagram, can easily be verified as being correct.

In general, application of Castigliano's first theorem to a structure leads to a system of equilibrium equations expressed in terms of the displacement degrees of freedom. For a linearly elastic structure undergoing small strains and displacements, the internal energy is a quadratic function of the displacement degrees of freedom and can be expressed as

$$\begin{aligned} U &= k_{11}x_1^2 + 2k_{12}x_1x_2 + 2k_{13}x_1x_3 + \ldots \\ &\quad + k_{22}x_2^2 + 2k_{23}x_2x_3 + \ldots \\ &\quad + \ldots \\ &\quad + 2k_{N,N-1}x_Nx_{N-1} + k_{N,N}x_N^2 \\ &= \tfrac{1}{2}\mathbf{x}^T\mathbf{k}\mathbf{x} \end{aligned}$$

where $\mathbf{k}$ is referred to as the *stiffness matrix*. Computing

$$\frac{\partial U}{\partial x_i} = P_i$$

according to Castigliano's first theorem leads to a system of linear algebraic equations that can be expressed as

$$\mathbf{k}\boldsymbol{x} = \boldsymbol{P}$$

These equations are solved for the unknowns $\boldsymbol{x}$ after which the desired internal forces can be computed. A formulation that considers the displacements as the independent variables is generally referred to as a *stiffness formulation* or the *stiffness method*. This is the case whether or not the structure is statically determinate or statically indeterminate. When using the displacement method, there is generally no additional effort required to solve for the displacements and internal forces for a statically indeterminate system than for a statically determinate one.

2-4 Approximate Methods

In terms of the goals of solid mechanics—namely, to determine the displacements and stresses—the total potential energy for a conservative mechanical system is only the starting point for deriving the governing equations of equilibrium using the principle of stationary potential energy. It is the solution of the governing equations of equilibrium and associated boundary conditions that yields the desired displacements and stresses. Frequently, however, the complexity of the boundary value problem—that is, the equations of equilibrium and boundary conditions—makes it practically impossible to actually solve the boundary value problem and thus generate a classical solution. It turns out, however, that the potential energy can be used to construct approximate solutions using the celebrated method of Ritz [2], which we now discuss.

2-4-1 Method of Ritz
In this section elementary examples of applications of the method of Ritz will be presented. The method will also be described in general.

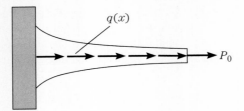

FIGURE 2-16 Typical axial problem.

For a typical application of the Ritz method, consider the axial problem shown in Figure 2-16. The total potential energy can be expressed as

$$V(u) = \int_0^L \left(\frac{AEu'^2}{2} - qu \right) dx - P_0 u(L)$$

with $u(0) = 0$ as the displacement boundary condition. For this one-dimensional axial problem, the method of Ritz is carried out by assuming that an approximate solution $u(x) = U(x)$ can be represented according to

$$U(x) = \sum_{n=1}^{N} a_n \varphi_n(x)$$

where the a_n are unknown constants to be determined by the Ritz method and *each* of N functions $\varphi_n(x)$ must be chosen so as to satisfy the displacement boundary condition $u(0) = 0$. $U(x)$ is frequently referred to as an *N term Ritz approximation*. Substituting the approximate solution U into the potential energy functional V yields

$$V = \int_0^L \left\{ \frac{AE}{2} \left(\sum_{n=1}^{N} a_n \varphi'_n(x) \right)^2 - q(x) \sum_{n=1}^{N} a_n \varphi_n(x) \right\} dx - P \sum_{n=1}^{N} a_n \varphi_n(L)$$

that is, a *function* of the form $V = V(a_1, a_2, \ldots, a_N)$. Thus, this initial step of the Ritz method has converted the functional $V(u)$, corresponding to a continuous problem with an infinite number of degrees of freedom, into a function $V = V(a_1, a_2, \ldots, a_N)$ with a finite number (N, in fact) of degrees of freedom.

The stationary value of the *function* $V(a_1, a_2, \ldots, a_N)$ is determined in the usual manner according to

$$\frac{\partial V}{\partial a_i} = 0 \quad i = 1, 2, \ldots, N$$

that is, N equations for the N unknown a_i. The reader should show that these equations are linear algebraic equations having the form **ku** = **f** where

$$k_{ij} = \int_0^L AE\varphi'_i \varphi'_j \, dx$$

and

$$f_j = \int_0^L q\varphi_j \, dx - P_0 \varphi_j(L)$$

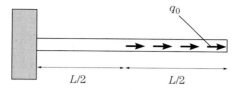

FIGURE 2-17 Example axial problem.

After solving $\mathbf{ku} = \mathbf{f}$ for the a_i, the approximate displacement $U = \Sigma\, a_n \varphi_n(x)$ and the approximate internal force $P = AE\, \Sigma\, a_n \varphi_n'(x)$ can be determined.

EXAMPLE 2-8 As a specific example, consider the problem of a bar loaded by a distributed load q_0 over half of its length, as shown in Figure 2-17. The solution is taken in the form

$$U = \sum_{1}^{N} a_n x^n$$

where each of the $\varphi_n(x) = x^n$ clearly satisfies the displacement or essential boundary condition $U(0) = 0$. Evaluating the coefficients k_{ij} and f_j yields

$$k_{ij} = \frac{ij AE L^{i+j-1}}{i+j-1}$$

$$f_j = \frac{q_0 L^{j+1}\left[1 - \left(\frac{1}{2}\right)^{j+1}\right]}{j+1}$$

Consider specifically the case $N = 2$. The augmented set of equations becomes

$$\begin{bmatrix} AEL & AEL^2 & \bigg| & \dfrac{3q_0 L^2}{8} \\ AEL^2 & \dfrac{4AEL^3}{3} & \bigg| & \dfrac{7q_0 L^3}{24} \end{bmatrix}$$

Define $\alpha_1 = a_1$, $\alpha_2 = L a_2$ and $\Phi = q_0 L / AE$, leading to

$$\begin{bmatrix} 1 & \dfrac{1}{2} & \bigg| & \dfrac{3\Phi}{8} \\ \dfrac{1}{2} & \dfrac{1}{3} & \bigg| & \dfrac{7\Phi}{24} \end{bmatrix}$$

with the solution in terms of the a's as $a_1 = 5 q_0 L / 8 AE$ and $a_2 = -2 q_0 / 8 AE$. The displacement and internal force transmitted are then easily determined to be

$$U_2(x) = \frac{q_0 L^2}{8 AE}\left[5\frac{x}{L} - 2\left(\frac{x}{L}\right)^2\right]$$

and

$$P_2(x) = AE U_2' = \frac{q_0 L}{8}\left[5 - 4\frac{x}{L}\right]$$

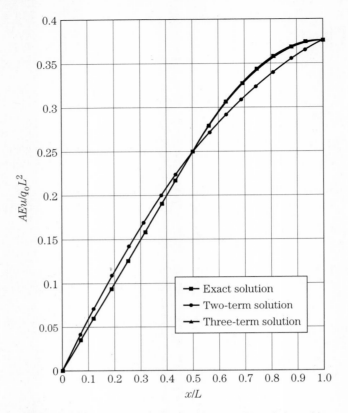

FIGURE 2-18 Comparison of two- and three-term Ritz solutions with the exact solution for the displacement.

which are shown in Figures 2-18 and 2-19. The student is asked to show in the exercises that the $N = 3$ term solution is

$$U_3(x) = \frac{q_0 L^2}{32AE}\left[15\frac{x}{L} + 7\left(\frac{x}{L}\right)^2 - 10\left(\frac{x}{L}\right)^3\right]$$

with

$$P_3(x) = AEU_3' = \frac{q_0 L}{32}\left[15 + 14\frac{x}{L} - 30\left(\frac{x}{L}\right)^2\right]$$

which are also shown in Figures 2-18 and 2-19.

As can be seen from Figure 2-18, the two- and three-term solutions for the displacement both compare reasonably well with the exact solution. As can be seen from Figure 2-19, the two-term solution $P_2(x)$ does not approximate the internal force well, whereas the three-term solution provides a more accurate picture of the internal force distribution. In addition, observe that the value of the force transmitted at $x = 0$, namely, $P(0) = AEu'(0)$, seems to be approaching the correct value of $q_0 L/2$. Similarly,

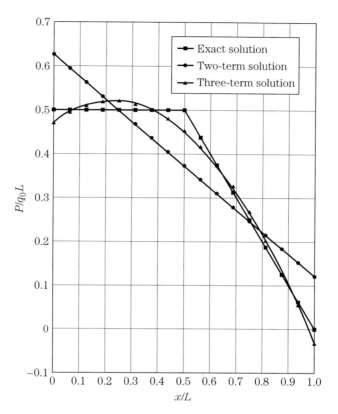

FIGURE 2-19 Comparison of two- and three-term Ritz solutions with the exact solution for the internal force.

the satisfaction of the force or nonessential boundary condition $P(L) = AEu'(L) = 0$ is improved from the two- to the three-term Ritz solution.

The character of the solution $u(x)$ and the derived variable $P = AEu'$ for the example are typical of approximate solutions obtained using the Ritz method. It is usually the case that a relatively few number of terms are necessary in order to adequately approximate the displacements, that is, convergence is fairly rapid to the displacements that are the dependent variables. The internal forces, which are generally obtained by taking derivatives of the displacements, are not as well approximated. In situations where the internal forces are of primary importance, the number of terms necessary should be assessed on the basis of the convergence of the derived variable, $P = AEu'$ in the present instance, rather than on the dependent variable u. The natural boundary condition, $P(L) = AEu'(L) = 0$, will be satisfied in the limit as the number of terms in the Ritz solution is increased. ◆

In using the Ritz method, displacement boundary conditions must be satisfied by each of the admissible functions. When the admissible functions satisfy only the displacement boundary conditions, the force boundary conditions will in general be satisfied only in the limit as the number of terms taken in the Ritz solution is increased. It is sometimes possible to choose a set of admissible functions that satisfies not only the displacement boundary conditions but also the force boundary conditions, resulting in faster convergence. Generally, the additional effort required to produce a set of admissible functions that satisfies both the displacement boundary conditions and the force boundary conditions is not warranted.

Note that a very important property of the Ritz method is that if, for some choice of the a_n, the exact solution is contained within the assumed approximate solution $U(x)$, the Ritz method will in fact choose those values for the a_n. Of course, if this were not the case the method would not be useful at all.

In general for a solid mechanics problem that is modeled as linear, the total potential energy turns out to be a functional $V(\mathbf{u})$ that can be represented as

$$V(\mathbf{u}) = \int_D Q(\mathbf{u})\, dvol$$

where $Q(\mathbf{u})$ is a quadratic function of the displacements and displacement derivatives. Invoking stationary potential energy leads to a boundary value problem consisting of linear differential equations and boundary conditions. For the Ritz method, assume that the displacements can be represented according to

$$u(x, y, z) = \sum_{i=1}^{N_1} a_i \phi_i(x, y, z)$$

$$v(x, y, z) = \sum_{j=1}^{N_2} b_j \psi_j(x, y, z)$$

$$w(x, y, z) = \sum_{k=1}^{N_3} c_k \chi_k(x, y, z)$$

where *each* of the *admissible* functions ϕ_n, ψ_n and χ_n is chosen so as to satisfy the displacement boundary conditions on B_u. Substitution of these *approximate* displacements into the functional $V(\mathbf{u})$ and subsequent evaluation of all the integrals yields $V = V(\mathbf{a}, \mathbf{b}, \mathbf{c})$, that is, a function of the unknown coefficients $\mathbf{a}$, $\mathbf{b}$ and $\mathbf{c}$. The Ritz method effectively converts a continuous problem given by $V(\mathbf{u})$ into a discrete problem given by $V(\mathbf{a}, \mathbf{b}, \mathbf{c})$, an approximate total potential energy. The *principle* of stationary potential energy then dictates that the function $V(\mathbf{a}, \mathbf{b}, \mathbf{c})$ be stationary so that

$$\frac{\partial V(\mathbf{a}, \mathbf{b}, \mathbf{c})}{\partial a_i} = 0 \quad i = 1, 2, \ldots, N_1$$

$$\frac{\partial V(\mathbf{a}, \mathbf{b}, \mathbf{c})}{\partial b_j} = 0 \quad j = 1, 2, \ldots, N_2$$

$$\frac{\partial V(\mathbf{a}, \mathbf{b}, \mathbf{c})}{\partial c_k} = 0 \quad k = 1, 2, \ldots, N_3$$

which is a set of linear algebraic equations of the form $\mathbf{Kx} = \mathbf{F}$, where $\mathbf{x} = [\mathbf{a}\ \mathbf{b}\ \mathbf{c}]^T$. The solution can then be used to determine $u(x, y, z)$, $v(x, y, z)$ and $w(x, y, z)$ with the stresses computed using the stress-strain relations and the strain-displacement relations. The displacements are often approximated quite accurately for relatively small values of N_1, N_2 and N_3, whereas the approximation of the derived variables—namely, the stresses—is generally less accurate. This is typical of any numerical method wherein the process of computing derivatives results in a degradation of the accuracy.

2-4-2 Finite Element Methods

The finite element method is a general technique that can be used to obtain accurate approximate solutions to a variety of different types of problems in engineering and applied mathematics. For many problems in solid mechanics, the finite-element method can be considered an application of the method of Ritz. The first exposure to the finite-element method will be in terms of its use in connection with the classical axial deformation problem.

The first step in generating a finite-element model is termed *discretization*. In this step the region—the bar in this case—is subdivided into a number of subregions called *elements*. The points that define the extents of the element are referred to as *nodes*. These ideas are illustrated in Figure 2-20.

The next step involves making an assumption as to the behavior of the dependent variable $u(x)$ within an element. To see how this is carried out, it is desirable to consider the plot of a typical solution shown as the solid line in Figure 2-21. Also shown in Figure 2-21 as a dotted line is an approximate representation of the solution $u(x)$ in terms of piecewise linear functions connecting the nodal values u_i. This process is referred to generally as *interpolation*. In the present example, linear interpolation has clearly been used to represent $u(x)$ within an element. Representing the displacement on a typical element as $u_e(x) = \alpha + \beta x$ and requiring that $u_e(x_i) = u_i$ and $u_e(x_{i+1}) = u_{i+1}$ yields

$$u_e(x) = \frac{x_{i+1} - x}{x_{i+1} - x_i} u_i + \frac{x - x_i}{x_{i+1} - x_i} u_{i+1} = u_i N_i(x) + u_{i+1} N_{i+1}(x)$$

where $N_i(x)$ and $N_{i+1}(x)$ are referred to as linear interpolation functions. They along with their derivatives $N_i'(x)$ and $N_{i+1}'(x)$ are shown in Figures 2-22a and 2-22b, respectively, where $\ell_e = x_{i+1} - x_i$ is the length of the element. The character of $N_i'(x)$ and

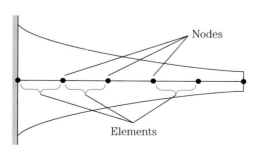

FIGURE 2-20 Nodes and elements.

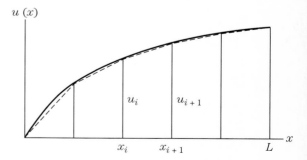

FIGURE 2-21 Depiction of linear interpolation.

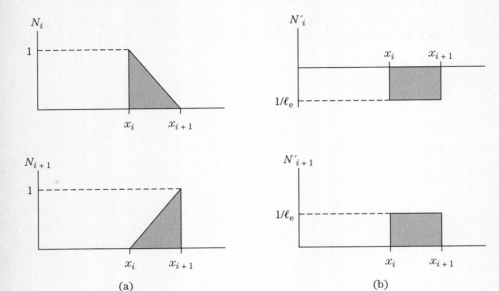

FIGURE 2-22 Interpolation functions and their derivatives.

$N'_{i+1}(x)$—namely, that they are constants—guarantees that the derivative of the solution $u'_e(x)$ is also a constant with the elemental value

$$u'_e(x) = \frac{u_{i+1} - u_i}{\ell_e}$$

For the developments that follow, it is convenient to write the elemental solution and its derivative as

$$u_e(x) = \mathbf{u_e^T N} = \mathbf{N^T u_e} \tag{2-9a}$$

and

$$u'_e(x) = \mathbf{u_e^T N'} = \mathbf{N'^T u_e} \tag{2-9b}$$

where

$$\mathbf{u_e^T} = [u_i \ u_{i+1}]$$

and

$$\mathbf{N^T} = [N_i(x) \ N_{i+1}(x)]$$

are the elemental displacement and interpolation vectors, respectively.

Consider then the typical axial problem, as shown in Figure 2-23, where the total potential energy is expressed as

$$V(u) = \int_0^L \left(\frac{AEu'^2}{2} - qu\right) dx - P_0 u(L)$$

2-4 APPROXIMATE METHODS

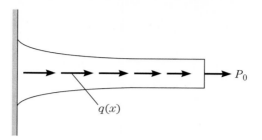

FIGURE 2-23 Typical axial problem.

with $u(0) = 0$ as the displacement boundary condition. Consistent with the discretization, represent this as

$$V = \sum_e \int_{x_i}^{x_{i+1}} \left(\frac{AEu'^2}{2} - qu \right) dx - P_0 u(L)$$

$$\approx \sum_e \int_{x_i}^{x_{i+1}} \left(\frac{AEu'_e(x)^2}{2} - q u_e(x) \right) dx - P_0 u(L)$$

Consider one of the integrals

$$\int_{x_i}^{x_{i+1}} \frac{AEu'_e(x)^2}{2} dx = \int_{x_i}^{x_{i+1}} \frac{AE\mathbf{u}_e^T \mathbf{N}' \mathbf{N}'^T \mathbf{u}_e}{2} dx$$

$$= \frac{1}{2}\mathbf{u}_e^T \int_{x_i}^{x_{i+1}} \mathbf{N}' AE \mathbf{N}'^T dx \, \mathbf{u}_e = \frac{1}{2}\mathbf{u}_e^T \mathbf{k}_e \mathbf{u}_e$$

where

$$\mathbf{k}_e = \int_{x_i}^{x_{i+1}} \mathbf{N}' AE \mathbf{N}'^T dx \tag{2-10}$$

is the *elemental stiffness matrix* for the axial deformation problem. The reader should show that upon substitution of the expressions for the derivatives of the interpolation functions,

$$\mathbf{k}_e = \begin{bmatrix} 1 & -1 \\ -1 & 1 \end{bmatrix} \int_{x_i}^{x_{i+1}} \frac{AE \, dx}{\ell_e^2}$$

Assuming that E is constant and that A is a reasonably smooth function of x, replace the integral by $EA_{\text{avg}} \ell_e \approx E(A_i + A_{i+1})/2 \ell_e$ so that

$$\mathbf{k}_e = \frac{E(A_i + A_{i+1})}{2\ell_e} \begin{bmatrix} 1 & -1 \\ -1 & 1 \end{bmatrix} = k_i \begin{bmatrix} 1 & -1 \\ -1 & 1 \end{bmatrix} \tag{2-11}$$

In an application, this amounts to replacing the original bar by a group of connected constant cross-section bars with areas $(A_i + A_{i+1})/2$, as indicated in Figure 2-24. Each of the constant cross-section bars acts as a linear spring with stiffness $k_i = E(A_i + A_{i+1})/2\ell_e$.

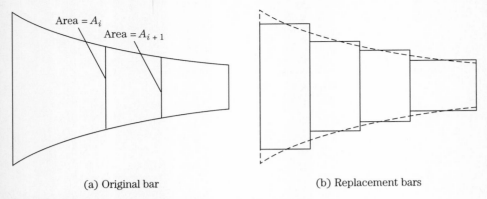

(a) Original bar (b) Replacement bars

FIGURE 2-24 Interpretation of the elemental stiffness matrices.

Consider next one of the integrals

$$\int_{x_i}^{x_{i+1}} q(x)u_e(x)\ dx = \int_{x_i}^{x_{i+1}} q(x)\mathbf{u_e^T N}\ dx$$
$$= \mathbf{u_e^T} \int_{x_i}^{x_{i+1}} q(x)\mathbf{N}\ dx = \mathbf{u_e^T q_e}$$

where

$$\mathbf{q_e} = \int_{x_i}^{x_{i+1}} q(x)\mathbf{N}\ dx \qquad (2\text{-}12)$$

is the *elemental load matrix*, as indicated in Figure 2-25. Assume that $q(x)$ is smooth enough that it can be approximated on an element as

$$q(x) = q_i N_i(x) + q_{i+1} N_{i+1}(x)$$

The reader should show that the elemental load matrix becomes

$$\mathbf{q_e} \approx \frac{\ell_e}{6} \begin{bmatrix} 2q_i + q_{i+1} \\ q_i + 2q_{i+1} \end{bmatrix} \qquad (2\text{-}13)$$

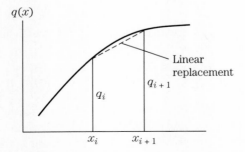

FIGURE 2-25 Linear representation of loading $q(x)$.

Some insight into this expression can be gained by taking the loading to be constant—say, $q = q_0$—in which case

$$\mathbf{q_e} = \begin{bmatrix} q_0 \dfrac{\ell_e}{2} \\ q_0 \dfrac{\ell_e}{2} \end{bmatrix} \quad (2\text{-}14)$$

which clearly indicates that half of the total load $q_0 \ell_e$ should be applied to each of the nodes at the ends of the element. This is essentially the lumping procedure usually used in a situation with constant distributed loading. For linear interpolation, Equation (2-13) can be thought of as a way of generalizing and applying the usual lumping process to situations where the loading is not constant.

For concentrated loads the nodes should be chosen so that a node coincides with the location of each of the concentrated loads. The potential energy of a concentrated load P is then given very simply by

$$\Omega = -Pu(x_i) = -Pu_i$$

where u_i is the nodal displacement at the node in question.

The representation of the original integral form of the potential energy resulting in the elemental matrices $\mathbf{k_e}$ and $\mathbf{q_e}$ is referred to as the *elemental formulation*. It is through this development that the physics of the problem is essentially built into the mathematical model.

In terms of the elemental matrices developed above, the total potential energy for the typical axial problem shown in Figure 2-20 can then be represented as

$$V(u_1, u_2, \ldots, u_{N+1}) = \frac{1}{2} \sum_e \mathbf{u_e^T k_e u_e} - \sum_e \mathbf{u_e^T q_e} - P_0 u_{N+1}$$

which is clearly a function of the nodal displacements u_i. In the above it is assumed that there are N elements and $N+1$ nodes.

To see more clearly the essential nature of each of the terms in the potential energy, consider the term corresponding to the first element,

$$\mathbf{u_{e1}^T k_{e1} u_{e1}} = [u_1 \ u_2] \begin{bmatrix} (k_1)_{11} & (k_1)_{12} \\ (k_1)_{21} & (k_1)_{22} \end{bmatrix} \begin{bmatrix} u_1 \\ u_2 \end{bmatrix}$$

Thinking in terms of the total number of degrees of freedom, write this as

$$\mathbf{u_{e1}^T k_{e1} u_{e1}} = [u_1 \ u_2 \ u_3 \ u_4 \ \ldots \ u_{N+1}] \begin{bmatrix} (k_1)_{11} & (k_1)_{12} & 0 & 0 & \ldots \\ (k_1)_{21} & (k_1)_{22} & 0 & 0 & \ldots \\ 0 & 0 & 0 & 0 & \ldots \\ 0 & 0 & 0 & 0 & \ldots \end{bmatrix} \begin{bmatrix} u_1 \\ u_2 \\ u_3 \\ u_4 \\ \vdots \\ u_{N+1} \end{bmatrix}$$

$$= \mathbf{u_G^T k_{G1} u_G}$$

where $\mathbf{u_G} = [u_1 \ u_2 \ u_3 \ u_4 \ \ldots \ u_{N+1}]^T$ is referred to as the *global displacement vector*.

Similarly, the second term can be written as

$$\mathbf{u}_{e2}^T \mathbf{k}_{e2} \mathbf{u}_{e2} = [u_2 \; u_3] \begin{bmatrix} (k_2)_{11} & (k_2)_{12} \\ (k_2)_{21} & (k_2)_{22} \end{bmatrix} \begin{bmatrix} u_2 \\ u_3 \end{bmatrix}$$

$$= [u_1 \; u_2 \; u_3 \; u_4 \; \ldots \; u_{N+1}] \begin{bmatrix} 0 & 0 & 0 & 0 & 0 & 0 \\ 0 & (k_2)_{11} & (k_2)_{12} & 0 & . & 0 \\ 0 & (k_2)_{21} & (k_2)_{22} & 0 & 0 & . \\ 0 & 0 & 0 & 0 & & . \\ 0 & 0 & 0 & 0 & & \end{bmatrix} \begin{bmatrix} u_1 \\ u_2 \\ u_3 \\ u_4 \\ \vdots \\ u_{N+1} \end{bmatrix}$$

$$= \mathbf{u}_G^T \mathbf{k}_{G2} \mathbf{u}_G$$

Summing the internal potential energies over all of the element gives

$$U = \frac{1}{2}[\mathbf{u}_G^T \mathbf{k}_{G1} \mathbf{u}_G + \mathbf{u}_G^T \mathbf{k}_{G2} \mathbf{u}_G + \ldots + \mathbf{u}_G^T \mathbf{k}_{GN} \mathbf{u}_G]$$

$$= \frac{1}{2} \mathbf{u}_G^T \mathbf{K}_G \mathbf{u}_G$$

where

$$\mathbf{K}_G = \sum_e \mathbf{k}_{Gi}$$

is the *global stiffness matrix*.

Consider also a typical loading contribution $\mathbf{u}_e^T \mathbf{q}_e$. For the first element this can be written as

$$\mathbf{u}_{e1}^T \mathbf{q}_{e1} = [u_1 \; u_2] \begin{bmatrix} (q_1)_1 \\ (q_1)_2 \end{bmatrix} = [u_1 \; u_2 \; u_3 \; u_4 \; \ldots \; u_{N+1}] \begin{bmatrix} (q_1)_1 \\ (q_1)_2 \\ 0 \\ 0 \\ \vdots \end{bmatrix}$$

where $(q_1)_1$ and $(q_1)_2$ indicate the first and second components of the elementary load vector for element 1. Similarly,

$$\mathbf{u}_{e2}^T \mathbf{q}_{e2} = [u_2 \; u_3] \begin{bmatrix} (q_2)_1 \\ (q_2)_2 \end{bmatrix} = [u_1 \; u_2 \; u_3 \; u_4 \; \ldots \; u_{N+1}] \begin{bmatrix} 0 \\ (q_2)_1 \\ (q_2)_2 \\ 0 \\ 0 \\ \vdots \end{bmatrix}$$

When all these terms are added there results

$$\sum_e \mathbf{u}_e^T \mathbf{q}_e = [u_1 \; u_2 \; u_3 \; u_4 \; \ldots \; u_{N+1}] \begin{bmatrix} (q_1)_1 \\ (q_1)_2 + (q_2)_1 \\ (q_2)_2 + (q_3)_1 \\ 0 \\ 0 \\ \cdot \\ (q_{N-1})_2 + (q_N)_1 \\ (q_N)_2 \end{bmatrix} = \mathbf{u}_G^T \mathbf{Q}_G$$

where $\mathbf{Q_G}$ is the portion of the global load vector arising from the distributed loading. The process of expanding each of the elemental potential energies and loads into the corresponding global arrays is referred to as *assembly*.

As mentioned above, concentrated loads must be added to $\mathbf{Q_G}$ in the location corresponding to the node at which the load is applied. Denoting these loads as $\mathbf{P_G}$, and the total of the external loads as $\mathbf{F_G} = \mathbf{Q_G} + \mathbf{P_G}$, the total potential energy can be expressed as

$$V(\mathbf{u_G}) = \frac{1}{2}\mathbf{u_G^T K_G u_G} - \mathbf{u_G^T F_G}$$

and as indicated is, as a result of modeling the bar using the finite-element method, a function of the nodal displacements $\mathbf{u_G}$. The student should show that

$$\frac{\partial V}{\partial \mathbf{u_G}} = \frac{1}{2}(\mathbf{K_G} + \mathbf{K_G^T})\mathbf{u_G} - \mathbf{F_G} = 0$$

or, since the global stiffness matrix is symmetric,

$$\mathbf{K_G u_G} = \mathbf{F_G}$$

The equations $\mathbf{K_G u_G} = \mathbf{F_G}$ represent equilibrium equations in terms of the displacements $\mathbf{u_G}$. Before these equations can be solved, it is generally necessary to satisfy the boundary conditions of the original continuous problem by applying constraints to the equilibrium equations. For the case of the typical axial deformation problem shown in Figure 2-17, the boundary condition is clearly $u(0) = 0$ with the corresponding constraint $u_1 = 0$. This condition can be enforced by replacing the first of the equilibrium equations by the equation $u_1 = 0$. In augmented form, this would appear as

$$\begin{bmatrix} 1 & 0 & 0 & 0 & . & . & | & 0 \\ K_{21} & K_{22} & K_{23} & 0 & . & . & | & F_1 \\ 0 & K_{32} & K_{33} & K_{34} & . & . & | & F_2 \\ . & . & . & . & . & . & | & . \\ . & . & . & . & . & . & | & . \end{bmatrix}$$

where the first row clearly states that $u_1 = 0$. For the sake of ecomony of solution, the elementary row operation(s) necessary to restore symmetry are carried out, leading to

$$\begin{bmatrix} 1 & 0 & 0 & 0 & . & . & | & 0 \\ 0 & K_{22} & K_{23} & 0 & 0 & . & | & F_2 \\ 0 & K_{32} & K_{33} & K_{34} & 0 & . & | & F_3 \\ . & . & . & . & . & . & | & . \\ . & . & . & . & . & . & | & . \end{bmatrix}$$

The process of enforcing the boundary conditions is referred to as *constraints*.

The equations are then solved, preferably using software that takes account of the symmetry of the coefficient matrix and of the banded nature of the equations, when present. This step is logically referred to as *solution*. The components of the resulting displacement vector $\mathbf{u_G}$ generally represent accurate approximations to the corresponding exact values of the solution $u(x_i)$ to the problem.

As stated at the outset, one of the primary goals of solid mechanics is to determine the stresses or distribution of internal forces. For the axial problem, the internal force resultant is the axial load P. P is related to the displacement u by the force displacement relation $P = AEu'(x)$. Within an element $u'(x)$ is given by

$$u'_e(x) = \mathbf{u_e^T N'} = \mathbf{N'^T u_e} = \frac{u_{i+1} - u_i}{\ell_e}$$

so that

$$P(x) = \frac{A(x)E(u_{i+1} - u_i)}{\ell_e}$$

Moan has shown that the optimum point in the interval at which to evaluate this term is the midpoint [3]. We will assume that $A(x)$ is smooth enough so that we can approximate this as

$$P\left(\frac{x_{i+1} + x_i}{2}\right) = A\left(\frac{x_{i+1} + x_i}{2}\right)\frac{E(u_{i+1} - u_i)}{\ell_e}$$

$$\approx \frac{A_{i+1} + A_i}{2} E \frac{u_{i+1} - u_i}{\ell_e} = \frac{E(A_i + A_{i+1})}{2\ell_e}(u_{i+1} - u_i)$$

$$= k_i(u_{i+1} - u_i)$$

Thus, in order to compute the internal forces, it is only necessary to multiply the stiffness of the element by $e_i = u_{i+1} - u_i$, the elongation of the element. Note that the stiffness k_i has already been computed at the elemental formulation stage and can be reused for computing the internal force. Note also that, when computed in this manner, the internal force in each element is a constant with the result that an internal force plot will appear as a piecewise constant function, as shown in Figure 2-26. This result is clearly a consequence of having used linear interpolation for the finite-element model. Also shown in Figure 2-26 is a dotted line connecting the midpoints of the piecewise constant segments representing $P(x)$. The logic behind this representation follows from Moan stating that the derived variable is most accurately represented at the midpoint of the interval.

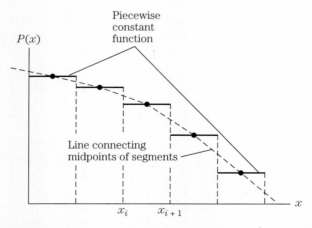

FIGURE 2-26 Piecewise constant character of P and line connecting midpoints.

It is easily seen why a finite-element model generated in this fashion is referred to as the Ritz finite-element model in that the original continuous problem is converted into a discrete problem with the nodal displacements as the degrees of freedom. For further information on the finite-element method, the reader is referred to Bickford [4].

◆ **EXAMPLE 2-9** For a specific example, consider the problem of Example 2-8, which was solved using the classical Ritz Method. For the first finite-element model, take two equal-length elements, as shown in Figure 2-27. The student should show that each of the elemental stiffness matrices is given by

$$\mathbf{k_e} = k \begin{bmatrix} 1 & -1 \\ -1 & 1 \end{bmatrix}$$

where $k = 2AE/L$ and that

$$\mathbf{K_G} = k \begin{bmatrix} 1 & -1 & 0 \\ -1 & 2 & -1 \\ 0 & -1 & 1 \end{bmatrix}$$

Similarly, $\mathbf{q_{e1}} = \mathbf{0}$ and

$$\mathbf{q_{e2}} = \frac{q_0 L}{4} \begin{bmatrix} 1 \\ 1 \end{bmatrix}$$

so that the final assembled equations can be represented in augmented form as

$$\begin{bmatrix} 1 & -1 & 0 & | & 0 \\ -1 & 2 & -1 & | & \phi \\ 0 & -1 & 1 & | & \phi \end{bmatrix}$$

where $\phi = q_0 L^2 / 8AE$. From the boundary condition $u(0) = 0$, the constraint is $u_1 = 0$. Enforcing the constraint and restoring symmetry leads to

$$\begin{bmatrix} 1 & 0 & 0 & | & 0 \\ 0 & 2 & -1 & | & \phi \\ 0 & -1 & 1 & | & \phi \end{bmatrix}$$

with solution $u_2 = 2q_0 L^2/8AE$ and $u_3 = 3q_0 L^2/8AE$, shown in Figure 2-28. The internal forces transmitted are

$$P_1 = k_1(u_2 - 0) = \frac{q_0 L}{2}$$

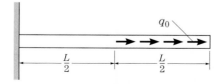

FIGURE 2-27 Example axial problem.

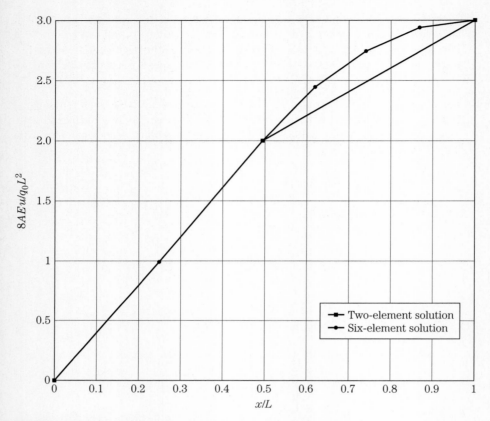

FIGURE 2-28 Displacements for the two- and six-element models.

and

$$P_2 = k_2(u_3 - u_2) = \frac{q_0 L}{4}$$

as shown in Figure 2-29.

The reader should verify that taking two elements in the unloaded portion and four elements in the loaded portion of the bar results in the augmented constrained equations

$$\begin{bmatrix} 1 & 0 & 0 & & & & & | & 0 \\ 0 & 2 & -1 & & & & & | & 0 \\ 0 & -1 & 3 & -2 & & & & | & \phi \\ & & -2 & 4 & -2 & & & | & 2\phi \\ & & & -2 & 4 & -2 & & | & 2\phi \\ & & & & -2 & 4 & -2 & | & 2\phi \\ & & & & & -2 & 2 & | & \phi \end{bmatrix}$$

2-4 APPROXIMATE METHODS

where $\phi = q_0L^2/64AE$. The solution for the displacement vector is

$$\mathbf{u_G^T} = [0.0\ \ 8.0\ \ 16.0\ \ 19.5\ \ 22.0\ \ 23.5\ \ 24.0]\frac{q_0L^2}{64AE}$$

and for the internal forces transmitted is

$$\mathbf{P^T} = [8\ \ 8\ \ 7\ \ 5\ \ 3\ \ 1]\frac{q_0L}{16}$$

These are shown in Figures 2-28 and 2-29, respectively.

The nodal values for the displacements coincide with the exact values for both the two- and four-element models, a result that can be shown to be true in general when the exact solution is quadratic. The internal forces are constant within each element, an

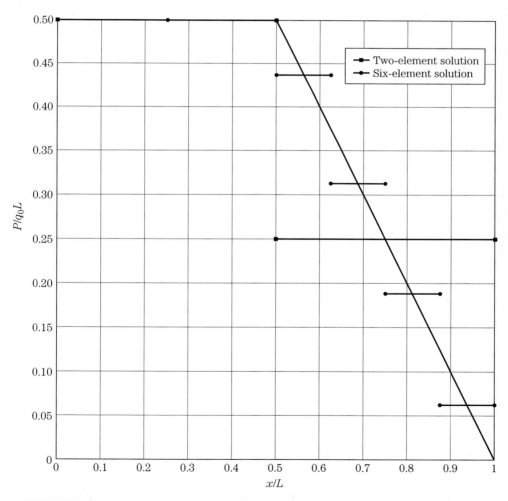

FIGURE 2-29 Internal forces for the two- and six-element models.

inescapable consequence of having taken the displacements to be linear within the element. Note, however, from Figure 2-29, that if the midpoints of the lines representing the elemental forces are connected, the straight line that results coincides with the corresponding exact solution. This procedure for estimating the internal force doesn't give the exact solution in general but does produce results that accurately represent the exact solution for the derived variable as the number of elements is increased.

The results obtained by application of the finite-element method to the axial deformation problem are typical of situations where a one-dimensional problem is being analyzed. The basic steps outlined, however, are independent of the particular application and of the number of spatial dimensions involved, that is, with appropriate discretization and interpolation, the finite-element method can be applied to a wide variety of problems in solid mechanics. Through its application to many classes of different problems throughout the remainder of the text, the power and versatility of the finite-element method will be demonstrated. ◆

2-5 Complementary Energy Ideas

In the principles discussed so far in this chapter, the independent variables were displacement variables. There is a dual set of ideas, referred to generally as complementary ideas, in which the independent variables are force variables. These ideas play an important part in the analysis of structures and will be treated in detail in what follows.

The essentials of the complementary approach can be discerned from a simple uniaxial stress-strain curve, as shown in Figure 2-30. As discussed above, the strain energy density U_0 is defined as

$$U_0 = \int \sigma(\epsilon)\, d\epsilon$$

and represents the area under the stress-strain curve. The total strain energy associated with a body is obtained by integrating the strain energy density over the volume according to

$$U = \int_{vol} U_0\, dvol$$

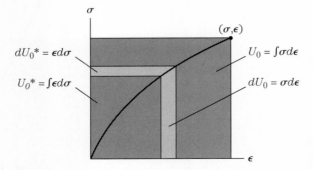

FIGURE 2-30 Stress-strain diagram and energy densities for a one-dimensional conservative elastic solid.

2-5 COMPLEMENTARY ENERGY IDEAS

In situations where stationary potential energy or Castigliano's first theorem is used to derive the governing equations, the stress is represented in terms of the strain for evaluating the strain energy density, after which the strains are expressed in terms of the displacement degrees of freedom leading to the equations of equilibrium in terms of the displacements, that is, a stiffness formulation.

Note that the relations among the strain energy density, the stress and the strain can be expressed as

$$\frac{dU_0}{d\epsilon} = \sigma$$

whereas according to Castigliano's first theorem, the relations among the strain energy, the load and the displacement can be expressed as

$$\frac{dU}{d\Delta} = P$$

The parallel is obvious.

In connection with Figure 2-30 for a one-dimensional situation, there is an alternate point of view leading to formulations where the force variables are the independent variables for the problem. Considering U_0^*, the area above the stress-strain curve, it is apparent that

$$dU_0^* = \epsilon\, d\sigma$$

or $dU_0^*/d\sigma = \epsilon$, and that

$$U_0^* = \int \epsilon(\sigma)\, d\sigma$$

U_0^* is called the *complementary energy density*. The use of the word complementary refers to the fact that with respect to the area represented by the product $\sigma \cdot \epsilon$, U_0 and U_0^* are complements in the sense that

$$\sigma \cdot \epsilon = U_0 + U_0^*$$

When the relation between stress and strain is linear—say, $\sigma = E\epsilon$—the complementary energy density for a uniaxial state is expressed as

$$U_0^* = \int \frac{\sigma}{E}\, d\sigma = \frac{\sigma^2}{2E}$$

For a simple shear situation characterized by $\tau = G\gamma$, the complementary energy density is represented as

$$U_0^* = \int \gamma\, d\tau = \int \frac{\tau}{G}\, d\tau = \frac{\tau^2}{2G}$$

In any event, the total internal complementary energy is obtained by integrating the complementary energy density over the volume, namely,

$$U^* = \int_{vol} U_0^*\, dvol$$

Note carefully that the complementary energy density U_0^* is a function of the stresses.

2-5-1 Complementary Energies for Structural Members

Many structures are analyzed on the basis of simplifying assumptions that result in the structure being modeled with idealized structural elements such as straight bars transmitting axial, torsional and bending force resultants. In this section expressions for the corresponding complementary energies will be developed and used in analyzing typical structures.

Axial Loading. Consider the general problem of a member transmitting an internal axial force resultant arising from concentrated and distributed loads, as shown in Figure 2-31. Assume that the axial stress in the bar can be computed by dividing the internal force $P(x)$ by the local area $A(x)$, that is, $\sigma = P(x)/A(x)$, so that $U_0^* = \sigma^2/2E = P^2/2A^2E$ with the complementary energy as

$$U^* = \int_{vol} \frac{P^2}{2EA^2} \, dvol$$

or, with $dvol = A(x)\,dx$,

$$U^* = \int_0^L \frac{P^2 \, dx}{2AE}$$

If AE is a constant—that is, a homogeneous uniform cross-section bar—and if the only load is the force F at the end of the bar, it follows that $P = F$ and that

$$U^* = \frac{F^2 L}{2AE}$$

Note in passing that

$$\frac{dU^*}{dF} = \frac{FL}{AE} = \Delta$$

which is the correct expression for the displacement of the bar at the point of application of the load F.

The parallel between this equation and the equation relating the complementary energy density, the stress and the strain, namely,

$$\frac{dU_0^*}{d\sigma} = \epsilon \qquad (2\text{-}15)$$

is obvious.

Torsional Loading. Consider the general problem of a member with a circular cross-section transmitting an internal torque arising from concentrated and distributed tor-

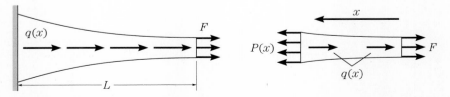

FIGURE 2-31 Typical axial member.

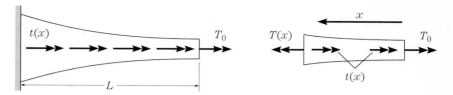

FIGURE 2-32 Typical torsion problem.

ques, as shown in Figure 2-32. Assume that the shear stress in the bar can be computed according to the classical torsional formula $\tau = Tr/J$, where $T = T(x)$ is the torque transmitted, r is the distance from the axis of the circular bar and $J = \pi a^4/2$ is the polar moment of the area. The complementary energy density is taken to be $U_0^* = \tau^2/2G = (Tr/J)^2/2G$, with the complementary energy then represented as

$$U^* = \int_{vol} \frac{1}{2G}\left(\frac{Tr}{J}\right)^2 dvol$$

or, with $dvol = dA\, dx$,

$$U^* = \int_0^L \left(\int_{Area} \frac{T^2 r^2}{2J^2 G} dA\right) dx = \int_0^L \frac{T^2\, dx}{2J^2 G} dx \left(\int_{Area} r^2\, dA\right)$$

$$= \int_0^L \frac{T^2}{2JG} dx$$

If a is a constant—that is, a uniform circular cross-section bar—and if the only load is the torque T_0 applied at the end of the bar, it follows that $T = T_0$ and that

$$U^* = \frac{T_0^2 L}{2JG}$$

and that

$$\frac{dU^*}{dT_0} = \frac{T_0 L}{JG} = \phi$$

which is the correct expression for the rotation of the bar at the point of application of the torque T_0.

Bending and Shear Loading. The general problem of a straight bar transmitting shear and bending loads is shown in Figure 2-33. According to the elementary theory,

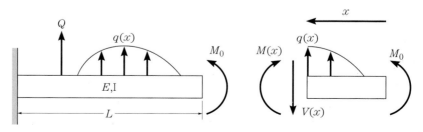

FIGURE 2-33 Typical transverse bending problem.

the bending moment M results in bending stresses according to $\sigma = -My/I$ with the shear force V giving rise to shear stresses $\tau = VQ/It$. To account for both normal and shear contributions, the complementary energy density is taken to be $U_0^* = \sigma^2/2E + \tau^2/2G$, resulting in

$$U^* = \int_{vol} \left[\frac{1}{2E}\left(\frac{My}{I}\right)^2 + \frac{1}{2G}\left(\frac{VQ}{It}\right)^2 \right] dvol$$

Then, with $dvol = dA\, dx$, and

$$\int_{Area} y^2\, dA = I,$$

$$U^* = \int_0^L \frac{M^2\, dx}{2EI} + \int_0^L \frac{V^2\, dx}{2kAG}$$

The dimensionless constant k is given by

$$k^{-1} = \int_{Area} \frac{\left(\dfrac{Q}{t}\right)^2 dA}{\dfrac{I^2}{A}}$$

and depends on the shape of the cross section. Its value is of the order of or slightly higher than unity for most cross sections.

For the specific problem of the cantilever beam loaded as shown in Figure 2-34 with x measured from the loaded end, the moment and shear are $M = M_0 + Px$ and $V = P$, leading to

$$U^* = \frac{\left(M_0^2 L + M_0 P L^2 + \dfrac{P^2 L^3}{3}\right)}{2EI} + \frac{P^2 L}{2kAG}$$

for the complementary energy. It is then possible to compute the partial derivatives

$$\frac{\partial U^*}{\partial P} = \frac{PL^3}{3EI} + \frac{PL}{kAG} + \frac{M_0 L^2}{2EI}$$

$$\frac{\partial U^*}{\partial M_0} = \frac{PL^2}{2EI} + \frac{M_0 L}{EI}$$

which are the correct expressions for the transverse displacement and rotation, respectively, at the point of application of the load P and M_0. The terms involving EI represent

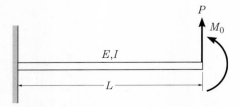

FIGURE 2-34 Transverse bending of a cantilever.

2-5 COMPLEMENTARY ENERGY IDEAS

the contributions due to bending, and the term containing the AG represents the contribution due to shear. It is left to the homework problems to investigate the relative magnitudes of the bending and shear terms.

In a more general situation where a straight untwisted bar is simultaneously transmitting all six force and moment resultants, as shown in Figure 2-35, the complementary energy can be expressed as

$$U^* = \int_0^L \frac{P^2\, dx}{2AE} + \int_0^L \frac{T^2\, dx}{2J^*G} + \int_0^L \frac{V_z^2\, dx}{2k_z AG}$$
$$+ \int_0^L \frac{V_y^2\, dx}{2k_y AG} + \int_0^L \frac{M_y^2\, dx}{2EI_{yy}} + \int_0^L \frac{M_z^2\, dx}{2EI_{zz}} \quad (2\text{-}16)$$

where J^* is a suitable torsional constant and y and z are principal axes of the cross section. The shear constants k_y and k_z depend on the shape of the cross section.

Engesser's First Theorem. Consider then a structure acted upon by external forces and/or couples and distributed loadings, as shown in Figure 2-36. For a statically determinate system, the equilibrium equations can be used to express the stresses in the interior of the body in terms of the external loading. The complementary energy U^* is then expressed in terms of the external forces and moments, that is, $U^* = U^*(F_1, F_2, \ldots, M_1, M_2)$.

Engesser's first theorem [5] states that

If a sufficiently supported elastic system undergoing small displacements is acted upon by forces F_i and moments M_j the displacement Δ_i in the direction of F_i at the point of application of F_i is given by

$$\frac{\partial U^*}{\partial F_i} = \Delta_i$$

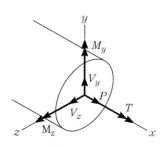

FIGURE 2-35 Bar transmitting force and moment resultants.

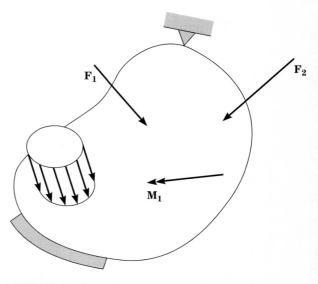

FIGURE 2-36 Generally loaded structure.

Similarly the rotation θ_j about an axis parallel to M_j at the point of application of M_j is given by

$$\frac{\partial U^*}{\partial M_j} = \theta_j$$

Small displacements mean that the initial and final configurations of the body do not differ appreciably, so the equilibrium equations can be written with respect to the undeformed structure. Note that although Engesser's first theorem is restricted to elastic bodies undergoing small displacements, it is not restricted to linearly elastic materials, that is, Engesser's first theorem can be used in situations where the stress-strain relations are nonlinear. The cases above for the axial, torsional and bending loadings are examples of the use of Engesser's first theorem. The student is asked to investigate other situations of this type in the exercises.

Thus, for an idealized statically determinate structure for which the complementary energy can be represented as in Equation (2-16), the displacement Δ_1 in the direction of F_1 at the point of application of F_1, for example, is given by

$$\Delta_1 = \frac{\partial U^*}{\partial F_1} = \int_0^L \frac{P \frac{\partial P}{\partial F_1} dx}{AE} + \int_0^L \frac{T \frac{\partial T}{\partial F_1} dx}{J^*G} + \int_0^L \frac{V_z \frac{\partial V_z}{\partial F_1} dx}{k_z AG}$$

$$+ \int_0^L \frac{V_y \frac{\partial V_y}{\partial F_1} dx}{k_y AG} + \int_0^L \frac{M_y \frac{\partial M_y}{\partial F_1} dx}{EI_{yy}} + \int_0^L \frac{M_z \frac{\partial M_z}{\partial F_1} dx}{EI_{zz}}$$

When the structure is composed of linearly elastic materials, the areas above and below the stress-strain curve of Figure 2-24 are equal, that is, $U^* = U$, and the first of the two equations that comprise Engesser's first theorem becomes

$$\frac{\partial U}{\partial F_i} = \Delta_i$$

which is known as *Castigliano's second theorem*. In what follows U^* will be used exclusively in connection with complementary energy and U in connection with strain energy irrespective of whether the material is linear, that is, $\partial U/\partial \Delta = P$ will be used in connection with stiffness formulations and $\partial U^*/\partial P = \Delta$ in connection with flexibility formulations.

EXAMPLE 2-10 A structure consisting of three linearly elastic bars is loaded as shown in Figure 2-37. The internal forces in the bars are easily determined to be $P_1 + P_2 + P_3$, $P_2 + P_3$ and

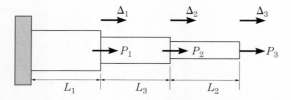

FIGURE 2-37 Axial loading of segmented bar.

2-5 COMPLEMENTARY ENERGY IDEAS

P_3, respectively, so that with $U^* = U_1^* + U_2^* + U_3^*$

$$U^* = \frac{(P_1 + P_2 + P_3)^2 L_1}{2A_1 E_1} + \frac{(P_2 + P_3)^2 L_2}{2A_2 E_2} + \frac{P_3^2 L_3}{2A_3 E_3}$$

Engesser's first theorem then states that

$$\Delta_1 = \frac{\partial U^*}{\partial P_1} = \frac{(P_1 + P_2 + P_3) L_1}{A_1 E_1} = e_1$$

$$\Delta_2 = \frac{\partial U^*}{\partial P_2} = \frac{(P_1 + P_2 + P_3) L_1}{A_1 E_1} + \frac{(P_2 + P_3) L_2}{A_2 E_2} = e_1 + e_2$$

$$\Delta_3 = \frac{\partial U^*}{\partial P_3} = \frac{(P_1 + P_2 + P_3) L_1}{A_1 E_1} + \frac{(P_2 + P_3) L_2}{A_2 E_2} + \frac{P_3 L_3}{A_3 E_3} = e_1 + e_2 + e_3$$

where e_1, e_2 and e_3 are the elongations of the three bars, respectively. Denoting $k_i = A_i E_i / L_i$, that is, the stiffnesses, these equations can be written as

$$\Delta_1 = \frac{P_1}{k_1} + \frac{P_2}{k_1} + \frac{P_3}{k_1}$$

$$\Delta_2 = \frac{P_1}{k_1} + P_2\left(\frac{1}{k_1} + \frac{1}{k_2}\right) + P_3\left(\frac{1}{k_1} + \frac{1}{k_2}\right)$$

$$\Delta_3 = \frac{P_1}{k_1} + P_2\left(\frac{1}{k_1} + \frac{1}{k_2}\right) + P_3\left(\frac{1}{k_1} + \frac{1}{k_2} + \frac{1}{k_3}\right)$$

or $\Delta = \mathbf{S}\mathbf{P}$, where $\mathbf{S}$ is referred to as a flexibility matrix. The reader is asked in the exercises to show that $\mathbf{SK} = \mathbf{I}$, where $\mathbf{K}$ is the corresponding stiffness matrix for the structure.

2-5-2 Dummy Loads It frequently turns out that displacements and/or rotations are desired at points in the structure at which there is no load or moment applied. In such a case a fictitious load or moment, called a dummy load or dummy moment, is used to determine the corresponding displacement or rotation. This idea is illustrated in the example below.

EXAMPLE 2-11 The structure of Example 2-10 is shown in Figure 2-38, where only the load P_3 is applied. Determine the displacements Δ_1, Δ_2 and Δ_3.

Solution: Dummy loads p_1 and p_2 are applied at the locations of the desired displacements Δ_1 and Δ_2. To indicate that p_1 and p_2 are dummy loads, we use dotted arrows

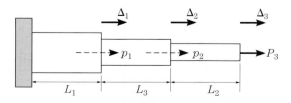

FIGURE 2-38 Dummy loads on a segmented bar.

and lowercase letters. In terms of the actual load P_3 and the dummy loads p_1 and p_2, the total complementary energy can be expressed as

$$U^* = \frac{(p_1 + p_2 + P_3)^2 L_1}{2A_1 E_1} + \frac{(p_2 + P_3)^2 L_2}{2A_2 E_2} + \frac{P_3^2 L_3}{2A_3 E_3}$$

Temporarily considering p_1 and p_2 to be actual loads, Engesser's first theorem then states that

$$\Delta_1 = \frac{\partial U^*}{\partial p_1} = \frac{(p_1 + p_2 + P_3)L_1}{A_1 E_1}$$

$$\Delta_2 = \frac{\partial U^*}{\partial p_2} = \frac{(p_1 + p_2 + P_3)L_1}{A_1 E_1} + \frac{(p_2 + P_3)L_2}{A_2 E_2}$$

$$\Delta_3 = \frac{\partial U^*}{\partial P_3} = \frac{(p_1 + p_2 + P_3)L_1}{A_1 E_1} + \frac{(p_2 + P_3)L_2}{A_2 E_2} + \frac{P_3 L_3}{A_3 E_3}$$

If p_1 and p_2 are then set equal to zero, there results

$$\Delta_1 = \frac{\partial U^*}{\partial p_1}\bigg|_{\substack{p_1=0 \\ p_2=0}} = \frac{P_3 L_1}{A_1 E_1}$$

$$\Delta_2 = \frac{\partial U^*}{\partial p_2}\bigg|_{\substack{p_1=0 \\ p_2=0}} = \frac{P_3 L_1}{A_1 E_1} + \frac{P_3 L_2}{A_2 E_2}$$

$$\Delta_3 = \frac{\partial U^*}{\partial P_3}\bigg|_{\substack{p_1=0 \\ p_2=0}} = \frac{P_3 L_1}{A_1 E_1} + \frac{P_3 L_2}{A_2 E_2} + \frac{P_3 L_3}{A_3 E_3}$$

demonstrating the concept and use of the dummy load. ◆

EXAMPLE 2-12 A right-angle bent is loaded as shown in Figure 2-39. Each portion of the bent has a circular cross section. Determine the z-component of displacement and the rotation about the y axis at the point of application of the load P.

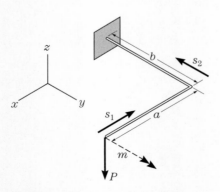

FIGURE 2-39 Dummy loads on a right-angle bent.

Solution: For the purpose of determining the required rotation, a dummy moment m is applied as shown. Accounting for only the bending and torsional energy, the total complementary energy can be expressed as

$$U^* = \int_0^a \frac{(m + Ps_1)^2 \, ds_1}{2EI} + \int_0^b \frac{(m + Pa)^2 \, ds_2}{2JG} + \int_0^b \frac{(Ps_2)^2 \, ds_2}{2EI}$$

Using Engesser's first theorem results in

$$\Delta = \left. \frac{\partial U^*}{\partial P} \right|_{m=0} = \int_0^a \frac{(Ps_1)(s_1) \, ds_1}{EI} + \int_0^b \frac{(Pa)(a) \, ds_2}{JG} + \int_0^b \frac{(Ps_2)(s_2) \, ds_2}{EI}$$

$$= \frac{Pa^3}{3EI} + \frac{Pa^2 b}{JG} + \frac{Pb^3}{3EI}$$

and

$$\theta = \left. \frac{\partial U^*}{\partial m} \right|_{m=0} = \int_0^a \frac{(Ps_1)(1) \, ds_1}{EI} + \int_0^b \frac{(Pa)(1) \, ds_2}{JG} = \frac{Pa^2}{2EI} + \frac{Pab}{JG} \quad \blacklozenge$$

When using a dummy load, it must be remembered that the dummy loads are set to zero only *after* taking any necessary derivatives.

2-5-3 Statically Indeterminate Systems: Principle of Least Work

For statically indeterminate systems it is not possible to determine all the reactions and/or internal forces in the structure solely on the basis of statics in that the number of unknowns exceeds the number of independent equations of equilibrium available. The number of additional unknowns are referred to as *redundants* in that they are not necessary in order that the structure be able to support external loads. A very simple example is shown in Figure 2-40a, where it is clear that either support could be removed without altering the ability of the structure to transmit the axial load, that is, there is one redun-

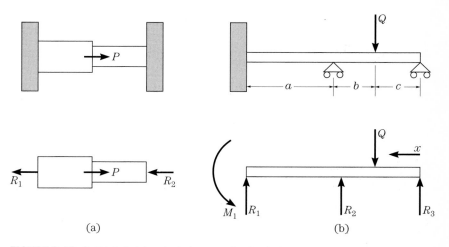

FIGURE 2-40 Statically indeterminate forces and redundants.

dant. For the beam of Figure 2-40b, the pairs of reactions M_1 and R_2, or R_2 and R_3 could be discarded, that is, there are two redundants.

Each of these examples involves reactions at the boundary of the structure as redundants. Once the redundants are known, the internal force results are easily determined. In the case of the axial problem, the single equation of equilibrium is

$$\sum F = P - R_1 - R_2 = 0$$

or $R_1 + R_2 = P$. The internal force in the member on the right is $P_2 = -R_2$ and in the member on the left, $P_1 = R_1$. Using the equilibrium equation to eliminate R_1, it is possible to express the forces in both members in terms of R_2 according to

$$P_1 = P - R_2 \quad \text{and} \quad P_2 = -R_2$$

With $k_1 = A_1 E_1/L_1$ and $k_2 = A_2 E_2/L_2$ as the stiffnesses of the two bars, the total complementary energy can then be expressed as

$$U^* = \frac{(P - R_2)^2}{2k_1} + \frac{(-R_2)^2}{2k_2}$$

In this instance the displacement at the point of application of the reaction R_2 is known to be zero so that as a direct consequence of Engesser's first theorem we can state that

$$\frac{\partial U^*}{\partial R_2} = \Delta_2 = 0 = -\frac{P - R_2}{k_1} + \frac{R_2}{k_2} \tag{2-17}$$

a single equation for the unknown R_2. The solution is $R_2 = k_2 P/(k_1 + k_2)$, after which the equilibrium equation is used to solve for the other reaction as $R_1 = k_1 P/(k_1 + k_2)$.

Note that $(P - R_2)/k_1 = P_1/k_1 = e_1$ and $R_2/k_2 = -P_2/k_2 = -e_2$ so that Equation (2-17) states that

$$e_1 + e_2 = 0$$

that is, that the sum of the elongations of the members is zero as required by the constraints at the ends. Further interpretation will be considered later in this section.

For the beam shown in Figure 2-40b the student should show that the equilibrium equations are

$$R_1 + R_2 + R_3 - Q = 0$$

and

$$M_1 + R_2 a - Q(a + b) + R_3(a + b + c) = 0$$

By selecting R_2 and R_3 as the redundants and measuring x as shown in Figure 2-40b, it is possible to express the moment M_1 and hence the complementary energy in terms of R_2, R_3 and Q, that is, $U^* = U^*(R_2, R_3, Q)$. Engesser's first theorem is then used to generate the two equations

$$\frac{\partial U^*}{\partial R_2} = \Delta_2 = 0$$

2-5 COMPLEMENTARY ENERGY IDEAS

and

$$\frac{\partial U^*}{\partial R_3} = \Delta_3 = 0$$

which will turn out to be two linear algebraic equations for the two unknowns R_2 and R_3. Once R_2 and R_3 have been determined, R_1 and M_1 can be determined from the equilibrium equations.

Contrast these examples with that shown in Figure 2-41a, where the support reactions R_2 and R_3 are easily determined but where the internal forces remain unknown. The unknown internal forces are exposed by deleting the redundant lower horizontal member of the structure, as shown in Figure 2-41b. In this case the two unknowns H and M can be taken to be redundants. Denoting the upper portion of the structure by subscript 1 and the redundant horizontal member by subscript 2, the total complementary energy can be expressed in terms of the redundants H and M:

$$U^*(H, M) = U_1^*(H, M) + U_2^*(H, M)$$

Compute

$$\frac{\partial U^*}{\partial H} = \frac{\partial U_1^*}{\partial H} + \frac{\partial U_2^*}{\partial H}$$

The term $\partial U_1^*/\partial H = \Delta_{AB}$ represents the decrease in distance between points A and B in the upper structure, whereas $\partial U_2^*/\partial H = \delta_{AB}$ represents the increase in distance be-

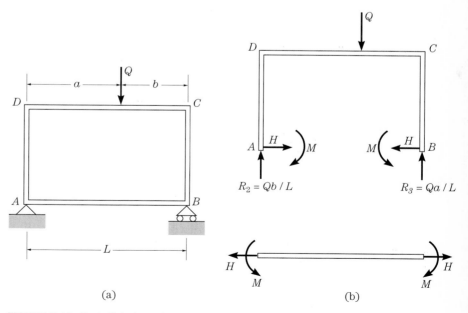

FIGURE 2-41 Statically indeterminate structure with internal redundants.

tween points A and B in the lower structure. In order for the structure to fit together it must be the case that $\Delta_{AB} = -\delta_{AB}$. Hence,

$$\frac{\partial U^*}{\partial H} = \Delta_{AB} + \delta_{AB} = 0$$

Similar arguments exist for the redundant M so that

$$\frac{\partial U^*}{\partial M} = 0$$

must also be satisfied.

These are examples of the application of Engesser's second theorem [6] or the principle of least work, stated as follows:

> Of all possible values of the redundants that satisfy equilibrium for a statically indeterminate structure, the correct values, which make the complementary energy stationary, also satisfy compatibility.

For the general statically indeterminate problem, suppose that there are M independent equations of equilibrium with $N > M$ unknowns so that there are $NR = N - M$ redundants. With $\mathbf{X}$ as the redundants and $\mathbf{p}$ as the known external forces, write the equilibrium equations in partitioned form as

$$[\mathbf{A}\!:\!\mathbf{B}] \begin{bmatrix} \mathbf{a} \\ \mathbf{X} \end{bmatrix} = \mathbf{p} \qquad (2\text{-}18)$$

where $\mathbf{A}$ is M by M and **nonsingular**, $\mathbf{B}$ is M by NR, $\mathbf{a}$ is M by 1 and $\mathbf{X}$ is NR by 1. This set of equations is solved for the M unknowns $\mathbf{a}$ in terms of the redundants $\mathbf{X}$. It is then possible to express the complementary energy in terms of the redundants $\mathbf{X}$ and the external loads $\mathbf{p}$, that is, $U^* = U^*(\mathbf{X}\!:\!\mathbf{p})$. The principle of least work is then used to generate the NR additional equations

$$\frac{\partial U^*}{\partial X_i} = 0 \qquad i = 1, 2, \ldots, NR$$

When the structure is composed of linearly elastic materials, this NR by NR set of equations is linear and can be represented as

$$\mathbf{S}\mathbf{X} = \mathbf{Q}(\mathbf{p})$$

where $\mathbf{Q}(\mathbf{p})$ indicates that the right-hand side depends on the external loads $\mathbf{p}$. After solving for $\mathbf{X}$ the remaining unknowns $\mathbf{a}$ can be determined from Equation (2-18). Engesser's first theorem can then be used in the usual manner to determine displacements if desired.

◆ **EXAMPLE 2-13** Consider the two-dimensional frame problem shown in Figure 2-42. The three equations of equilibrium are

$$\sum F_H = H_1 - H_2 = 0$$
$$\sum F_V = V_1 + V_2 - Q = 0$$
$$\sum M = M_2 + 2V_2 L - QL - H_2 L = 0$$

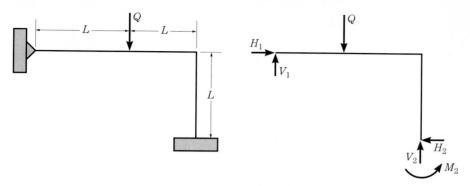

FIGURE 2-42 Two-dimensional frame.

There are five unknowns and three equations, hence two redundants. In principle any two of the unknowns can be taken as the redundants as long as the resulting partitioned set of equations can be solved for the remaining unknowns. In this instance it is clear that both V_1 and V_2 could not be taken as redundants since it would be impossible for the structure to support the load Q in that case. The reader should verify that this choice for the redundants leads to a singular matrix **A** in terms of the discussion above, that is, it would be impossible to solve for H_1, H_2 and M_2 with this choice for the redundants. A permissible choice for the redundants is H_1 and V_1, resulting in the equations

$$H_2 = H_1$$
$$V_2 = Q - V_1$$
$$-H_2 L + 2V_2 L + M_2 = QL$$

clearly solvable for H_2, V_2 and M_2. However, for this problem it is not actually necessary to solve these equations for H_2, V_2 and M_2 if we work from the left end in drawing FBDs and determining internal force resultants since only H_1 and V_1, the redundants, will be involved in the equations of equilibrium and internal force resultants. The student should show that proceeding in this manner leads to

$$U^* = \int_0^L \frac{(V_1 x)^2}{2EI} dx + \int_L^{2L} \frac{(V_1 x - Q(x-L))^2}{2EI} dx + \int_0^L \frac{(V_1 2L - QL + H_1 x)^2}{2EI} dx$$

where only the bending energy has been included. The principle of least work is then used according to

$$\frac{\partial U^*}{\partial V_1} = 0$$

$$= \int_0^L \frac{(V_1 x)(x)}{EI} dx + \int_L^{2L} \frac{(V_1 x - Q(x-L))(x)}{EI} dx$$
$$+ \int_0^L \frac{(V_1 2L - QL + H_1 x)(2L)}{EI} dx$$

and

$$\frac{\partial U^*}{\partial H_1} = 0 = \int_0^L \frac{(V_1 2L - QL + H_1 x)(x)}{EI} dx$$

Performing the integrations and simplifying leads to

$$20V_1 + 3H_1 = 17Q/2$$
$$3V_1 + H_1 = 3Q/2$$

from which $V_1 = 4Q/11$ and $H_1 = 9Q/22$. The equilibrium equations then yield $H_2 = 9Q/22$, $V_2 = 7Q/11$ and $M_2 = 3QL/22$. With the reactions and internal forces entirely known, Engesser's first theorem can be used to determine displacements. ◆

2-6 Summary

In this chapter it has been shown that it is possible to formulate and solve problems in solid mechanics using variational or energy methods. These energy principles for solid mechanics provide alternatives to the vectorial approaches for determining the relations between forces and displacements. Depending on what information is desired, the energy methods may involve less computational effort than the corresponding vectorial approach.

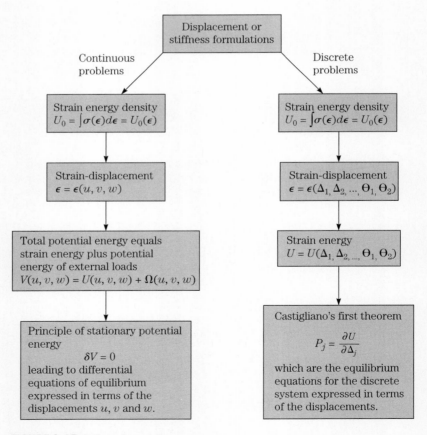

FIGURE 2-43 Stiffness formulations.

2-6 SUMMARY

In one approach the displacements are taken as the independent variables. Generally, the strain energy density U_0 is expressed in terms of the strains. The strain displacement relations are then used to express the strains in terms of the displacements Δ_i with the strain energy U expressed in terms of the displacements. Stationary potential energy or Castigliano's first theorem is then used to generate the equilibrium equations in terms of the displacements according to $P_i = \partial U/\partial \Delta_i$. These ideas are summarized in Figure 2-43.

In the other basic approach, forces are taken to be the independent variables. The complementary energy density U_0^* is expressed in terms of the stresses. The equilibrium equations are used to express the stresses in terms of the loads, after which it is possible to express the complementary energy U^* in terms of the loads, that is, $U^* = U^*(P_i)$. For statically determinate problems Engesser's first theorem is then used to generate compatiblility equations in terms of the forces according to $\Delta_i = \partial U^*/\partial P_i$. For statically indeterminate problems the principle of least work is first used to determine the redundants, after which Engesser's first theorem is used to generate compatiblility equations in terms of the forces. These ideas are summarized in Figure 2-44.

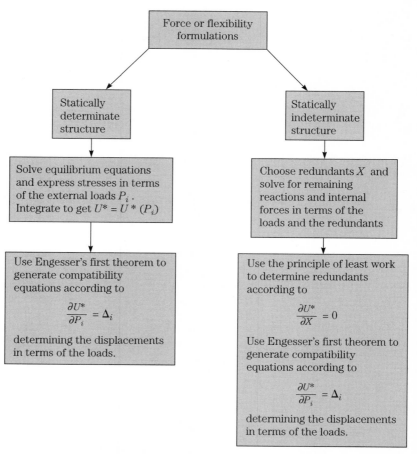

FIGURE 2-44 Flexibility formulations.

The stiffness and flexibility approaches are reciprocal in the sense that for stiffness formulations, compatibility is assured by virtue of taking the displacements as the independent variables. The governing variational principle then states that

> Of all possible displacements satisfying continuity and boundary conditions on B_u the correct ones are those that give the potential energy a stationary value and satisfy equilibrium in D and stress boundary conditions on B_t

where B_u indicates the portion of the boundary on which the displacements are prescribed and B_t the portion of the boundary on which the stresses are prescribed. Flexibility formulations, on the other hand, begin with equilibrium and the satisfaction of stress boundary conditions. The governing variational principle then states that

> Of all stress states satisfying equilbrium and stress boundary conditions on B_t the true stresses give the complementary energy a stationary value and satisfy compatibility in D and displacement boundary conditions on B_u.

In fact, the words stiffness and flexibility themselves indicate reciprocal concepts.

REFERENCES

1. Castigliano, A. *Theorie de l'equilibre des systemes elastiques et ses applications*. Turin, 1879.
2. Ritz, W. "Ueber eine neue Methode zure Losung gewisser Variationsprobleme der mathematischen physik." *J. Reine Angew. Math.*, vol. 135, pp. 1–61, 1908.
3. Moan, T. "On the local distribution of errors by finite element approximations," in *Theory and Practice in Finite Element Structural Analysis*, eds Y. Yamada and R. H. Galagher. University of Tokyo Press, Tokyo, 1973.
4. Bickford, W. B. *A First Course in the Finite Element Method*. 2nd ed. Irwin, Burr Ridge, Ill., 1994.
5. Engesser, F. "Ueber statisch unbestimmte träger bei beliebigen. . . ." *Z. Archit. Ing.*, vol. 35, 1889.
6. Langhaar, H. L. *Energy Methods in Applied Mechanics*. Wiley, New York, 1962.

EXERCISES

Section 2-2

1. Use the principle of stationary potential energy to derive the equilibrium equations. Take $k_1 = k_2 = 3\ kN/m$ and $P = 300\ N$. Solve the equations and determine the forces in the springs.

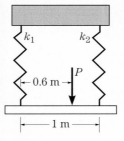

2. Use the principle of stationary potential energy to derive the equilibrium equations. Take $k_1 = k_2 = k_3 = 5\ kN/m$ and $P = 600\ N$. Solve the equations and determine the forces in the springs.

EXERCISES

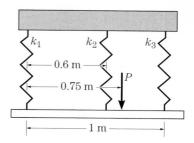

3. Rework problem 2 if the springs are replaced by thin linearly elastic wires (tension members) and a distance of 0.75 m, indicating that the position of the load is changed to a variable distance x. Are there any limits on x? What happens if these limits are exceeded? Assume that the magnitude of P is small enough so that the wires continue to behave elastically.

Section 2-3

4. Use the principle of stationary potential energy to derive the equations of equilibrium. With $A_1 = A_2 = A_3 = A, E_1 = E_2 = E_3 = E, L_1 = L_2 = L_3 = L, h/L = 0.5$ and $P = Q$, solve for the displacements and the internal forces in the three axial members.

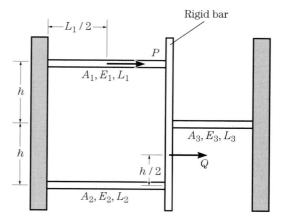

5. Use the principle of stationary potential energy to derive the equations of equilibrium. With $C_1 = C_2 = C_3 = C_4$, determine the rotations and the internal torques in the members. Take $r_1 = 1.5r_2$ and $r_3 = 1.2r_4$.

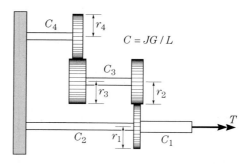

6. The rigid bar is supported by equal-area steel and aluminum bars and loaded as shown. Determine the elongations of and the forces in the bars. Take $E_{st}/E_{al} = 3.0$.

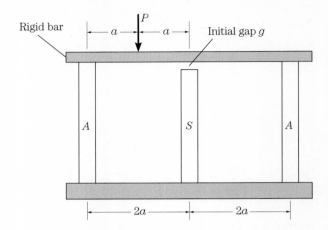

7. The bar shown is composed of an elastic material for which the stress-strain relation has the form $\sigma = k\epsilon^N$, where k and N are constants. Use stationary potential energy or Castigliano's first theorem to detemine the relation between the force P and the elongation Δ for the bar. Does your result reduce properly for a linearly elastic material?

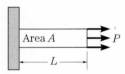

8. Take the displacement in the beam segment shown to be $w(x) = C_1 + C_2 x + C_3 x^2 + C_3 x^3$, that is, the homogeneous solution to the governing differential equation for the beam.

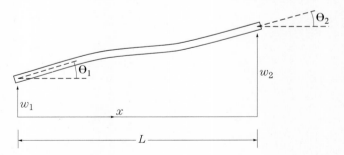

Choose the constants so as to satisfy $w(0) = w_1$, $w'(0) = \theta_1$, $w(L) = w_2$ and $w'(L) = \theta_2$ and hence show that $w = w_1 f_1(x) + \theta_1 f_2(x) + w_2 f_3(x) + \theta_2 f_4(x)$ where $f_1(x) = 1 - 3(x/L)^2 + 2(x/L)^3$, $f_2(x) = x(1 - 2(x/L) + (x/L)^2)$, $f_3(x) = 3(x/L)^2 - 2(x/L)^3$, and $f_4(x) = x((x/L)^2 - (x/L))$. Then show that the potential energy can be expressed as

$$U = \int_0^L \frac{EI w''^2}{2} dx = \frac{2EI}{L}\left[\theta_1^2 + \theta_1 \theta_2 + \theta_2^2 - 3\left(\frac{w_2 - w_1}{L}\right)(\theta_1 + \theta_2) + 3\left(\frac{w_2 - w_1}{L}\right)^2\right]$$

EXERCISES

9. Using the results of exercise 8, show that the moment and shear in the beam can be expressed as

$$M/EI = 6\left(\frac{w_2 - w_1}{L}\right)\left(1 - 2\frac{x}{L}\right) + \frac{6x}{L^2}(\theta_1 + \theta_2) - \frac{2}{L}(\theta_2 + 2\theta_1)$$

and

$$V/EI = -\frac{12}{L^3}(w_1 - w_2) - \frac{6}{L^2}(\theta_1 + \theta_2)$$

Section 2-4

10. Apply the ideas of section 2-4-2 to the constant area problem with the uniformly distributed loading q_0 shown below. Take two equal-length elements. Compare your results to the exact displacement $u(x) = q_0(2Lx - x^2)/2AE$ and $P(x) = q_0(L - x)$, where $P(x)$ is the internal force transmitted. Discuss the relationships between the differences in the internal forces transmitted between adjacent elements and the external loading at the corresponding nodes.

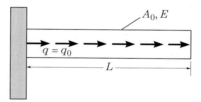

11. Repeat problem 10 using four equal-length elements.

12. Apply the ideas of section 2-4-2 to the problem of a uniform bar supported at the top and bottom and hanging under its own weight, as shown. Take two equal-length elements. Compare your solution to the exact solution $u = \gamma x(L - x)/2E$ and $P = A\gamma(L - 2x)/2$ where x is measured from the top of the bar. Discuss the relationships between the differences in the internal forces transmitted between adjacent elements and the external loading at the corresponding nodes. (Hint: $q(x) = A\gamma$.)

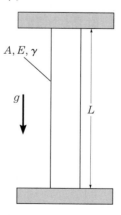

13. Repeat problem 12 using four equal-length elements.

14. Apply the ideas of section 2-4-2 to the problem of the circular cross-section bar with variable radius supported at the top and hanging under its own weight, as shown. Take two equal-length elements. Compare your solution to the exact solution. Discuss the relationship be-

tween the differences in the internal forces transmitted between adjacent elements and the external loading at the corresponding nodes. (Hint: $q(x) = A\gamma$.)

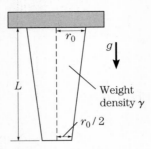

15. Repeat problem 14 using four equal-length elements.
16. Repeat problem 14 using eight equal-length elements.

Section 2-5

17. The bar shown is composed of an elastic material for which the stress-strain relation has the form $\sigma = k\epsilon^N$, where k and N are constants. Use complementary energy and Castigliano's second theorem to determine the elongation of the bar. Does your result reduce properly for a linearly elastic material?

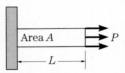

18. Use complementary energy and Castigliano's second theorem in connection with the idea of a dummy load to determine the displacement as a function of position ξ along the bar shown.

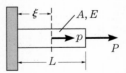

19. Use complementary energy and Castigliano's second theorem in connection with the idea of a dummy load to determine the displacement as a function of position ξ along the bar shown.

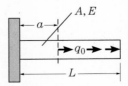

20. Use complementary energy and Castigliano's second theorem in connection with the idea of a dummy load to determine the angle of twist as a function of position ξ along the circular cross-section bar shown.

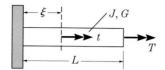

21. Use complementary energy and Castigliano's second theorem in connection with the idea of a dummy load to determine the angle of twist as a function of position ξ along the circular cross-section bar shown.

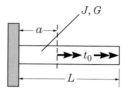

22. Determine the transverse displacement and rotation at the loaded point.

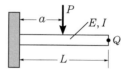

23. For the beam of problem 22, use dummy loads to determine the transverse displacement and rotation at point Q.

24. For the beam below, use dummy loads to determine the transverse displacement and rotation at point Q.

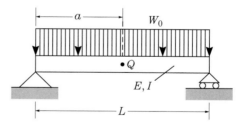

25. Including axial deformations, determine the displacement at the loaded point. If the cross section is square with sides h and $h/a = 0.1$, what percentage is due to the axial deformations?

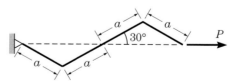

26. Determine the displacements in the x and y directions and the rotation about the z axis at the loaded point for the structure shown. Assume both members have circular cross sections and the same elastic properties. Neglect axial and shear deformations.

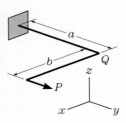

27. Determine the displacements in the x and y directions and the rotation about the z axis at point Q for the structure of problem 26. Assume both members have circular cross sections and the same elastic properties. Neglect axial and shear deformations.

28. Determine the displacement in the z direction and the rotations about the x and y axes at the loaded point for the structure shown. Assume both members have circular cross sections and equal elastic properties with $\nu = 0.3$. Neglect axial and shear deformations.

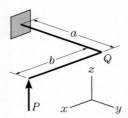

29. Determine the displacement in the z direction and the rotations about the x and y axes at point Q for the structure of problem 28. Assume both members have circular cross sections and equal elastic properties with $\nu = 0.3$. Neglect axial and shear deformations.

30. Determine the displacements in the x, y and z directions and the rotations about the x, y and z axes at the point of application of the load P for the structure shown. Assume all members have circular cross sections and equal elastic properties with $\nu = 0.3$. Neglect axial and shear deformations.

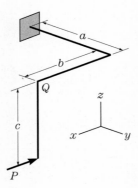

31. Determine the displacements in the x, y and z directions and the rotations about the x, y and z axes at point Q for the structure of problem 30. Assume all members have circular cross sections and equal elastic properties with $\nu = 0.3$. Neglect axial and shear deformations.

EXERCISES

32. Determine the horizontal displacement at the point of application of the load P for the structure shown. Neglect axial and shear deformations.

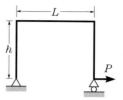

33. Use the principle of least work to determine the unknown redundants. Neglect axial and shear deformations and take $L = h$ and $H = 2h$.

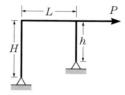

34. Use the principal of least work to determine the unknown redundants. Neglect axial and shear deformations and take $L = h$ and $H = 2h$.

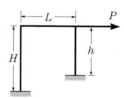

35. Use the principle of least work to determine the unknown redundants for the structure shown. Neglect axial and shear deformations.

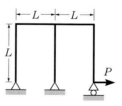

36. Use the principle of least work to determine the unknown redundants for the structure shown. Neglect axial and shear deformations.

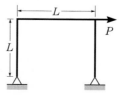

37. Use the principle of least work to determine the unknown redundants for the structure shown. Neglect axial and shear deformations.

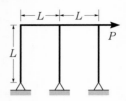

38. The structure is constrained against translation by a ball joint at A. Determine the reactions.

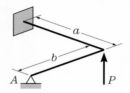

39. The structure is constrained against all motion at the supports. Determine the reactions.

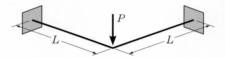

40. Assuming the structure of problem 39 pinned rather than clamped at the left end, determine the displacement at the loaded point.

41. For a two-dimensional problem the complementary energy density, the stresses and the strains are related according to
$$\varepsilon_x = \frac{\partial U_0^*}{\partial \sigma_x}, \quad \varepsilon_y = \frac{\partial U_0^*}{\partial \sigma_y} \quad \text{and} \quad \gamma_{xy} = \frac{\partial U_0^*}{\partial \tau_{xy}}$$
Show that if the stress strain relations are taken to be
$$\varepsilon_x = S_{11}\sigma_x + S_{12}\sigma_y + S_{13}\tau_{xy}$$
$$\varepsilon_y = S_{12}\sigma_x + S_{22}\sigma_y + S_{23}\tau_{xy}$$
$$\gamma_{xy} = S_{13}\sigma_x + S_{23}\sigma_y + S_{33}\tau_{xy}$$
there results
$$U_0^* = \frac{1}{2}[S_{11}\sigma_x^2 + 2S_{12}\sigma_x\sigma_y + 2S_{13}\sigma_x\tau_{xy}$$
$$+ S_{22}\sigma_y^2 + 2S_{23}\sigma_y\tau_{xy} + S_{33}\tau_{xy}^2]$$

42. Show that for the isotropic two-dimensional case where
$$\varepsilon_x = \frac{\sigma_x - \nu\sigma_y}{E}$$
$$\varepsilon_y = \frac{\sigma_y - \nu\sigma_x}{E}$$
$$\gamma_{xy} = \frac{\tau_{xy}}{G}$$

the internal complementary energy is

$$U_0^* = \frac{1}{2E}[\sigma_x^2 - 2\nu\sigma_x\sigma_y + \sigma_y^2] + \frac{\tau_{xy}^2}{2G}$$

In each of the following exercises, apply the ideas of the calculus of variations presented in Appendix A.

43. Show by requiring the potential energy of example 2-4 to be stationary that the equilibrium equation $(AEu')' + q = 0$ and the natural boundary condition $AEu'(L) = P_0$ result.

44. For the transverse bending problem shown, show that the total potential energy can be expressed as

$$V = \int_0^L \left(\frac{EIw''^2}{2} - w(x)q(x) \right) dx - M_0 w'(L) - Qw(L)$$

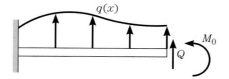

Use the calculus of variations to determine the Euler-Lagrange equations, and show that they represent the equilibrium equations and natural boundary conditions.

45. For the torsional problem shown, show that the total potential energy can be expressed as

$$V = \int_0^L \frac{JG\phi'^2}{2} dx - T_0\phi(L)$$

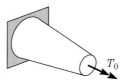

Use the calculus of variations to determine the Euler-Lagrange equations, and show that they represent the equilibrium equation and natural boundary condition.

3 Torsion

3-1 Introduction
3-2 Torsion of Circular Cross-Section Bars
3-3 Torsion of Noncircular Cross-Section Bars
    3-3-1   The Membrane Analogy
    3-3-2   Thin Rectangular Cross Section
    3-3-3   Thin-Walled Open Sections
    3-3-4   Finite-Element Models
3-4 Torsion of a Single Thin-Walled Tube
3-5 Torsion of Multicelled Tubes
3-6 Warping Restraint in Thin-Walled Open Sections
3-7 Alternative Formulation for Warping Restraint in Thin-Walled Open Sections
3-8 Summary
     References
     Exercises

3-1 Introduction

There are numerous applications in which a straight prismatic bar is subjected to external forces that produce a moment, frequently called a torque in a torsion situation, about the axis of the bar. This moment generally results in shear stresses in the interior of the bar. The corresponding deformations appear primarily as a twisting of the bar about its axis.

In this chapter the elementary theory of torsion of circular and annular cross-section bars will be reviewed. The problem of the torsion of noncircular bars will also be discussed. Approximate solutions for noncircular cross sections will be discussed in detail. Approximate theories for the torsion of thin-walled open sections, for the torsion of multitube sections and for the torsion of thin-walled sections in the presence of warping will also be considered.

3-2 Torsion of Circular Cross-Section Bars

Consider a straight, constant-diameter circular bar, as shown in Figure 3-1. The loading consists of torques T, applied in opposite directions at the ends of the bar, as shown. The specific manner in which the torque T must be applied to the faces at the ends of the bar will be discussed later in this section. The lateral or cylindrical surface is assumed to be stress-free.

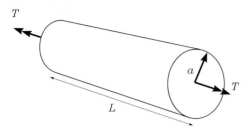
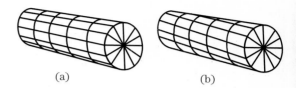

FIGURE 3-1 Torsion of a circular cross-section bar.

FIGURE 3-2 Observable deformations in torsion.

The observable deformations of the bar are shown in Figure 3-2. The lines originally parallel to the axis of the bar appear to have become helices with the circumferential lines remaining circumferential, that is, no distortion of a cross section, at least on the circumferential surface of the bar. On the basis of these observations, assume the following:

1. That any plane section perpendicular to the axis of the bar remains a plane section.
2. That each diameter of such a section rotates through the same angle $\phi(x)$.

Consulting Figure 3-3, the displacements v and w of point P can be expressed in terms of the angle ϕ through which the cross section rotates as

$$v(x, y, z) = y^* - y = r\cos(\beta + \phi) - r\cos\beta$$
$$= y(\cos\phi - 1) - z\sin\phi$$

and

$$w(x, y, z) = z^* - z = r\sin(\beta + \phi) - r\sin\beta$$
$$= z(\cos\phi - 1) + y\sin\phi$$

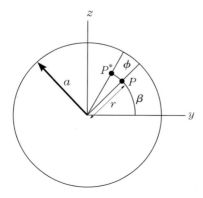

FIGURE 3-3 Torsional displacements in a circular bar.

The displacement in the direction of the axis of the bar is taken to be a constant u_0 that can be set equal to zero with no loss in generality. Using the linear strain-displacement relations then yields

$$\varepsilon_x = \frac{\partial u}{\partial x} = 0, \qquad \varepsilon_y = \frac{\partial v}{\partial y} = \cos\phi - 1, \qquad \varepsilon_z = \frac{\partial w}{\partial z} = \cos\phi - 1$$

$$\gamma_{xy} = \frac{\partial u}{\partial y} + \frac{\partial v}{\partial x} = -y\sin\phi \cdot \frac{d\phi}{dx} - z\cos\phi \cdot \frac{d\phi}{dx}$$

$$\gamma_{yz} = \frac{\partial w}{\partial y} + \frac{\partial v}{\partial z} = \sin\phi - \sin\phi = 0$$

$$\gamma_{xz} = \frac{\partial u}{\partial z} + \frac{\partial w}{\partial x} = -z\sin\phi \cdot \frac{d\phi}{dx} + y\cos\phi \cdot \frac{d\phi}{dx}$$

Now assume that $\phi' = d\phi/dx$ is small, that $\cos\phi \approx 1$ and $\sin\phi \approx \phi$ and that products of ϕ and $d\phi/dx$ can be discarded, leading to

$$\varepsilon_x = \varepsilon_y = \varepsilon_z = \gamma_{yz} = 0,$$

$$\gamma_{xy} = -z\frac{d\phi}{dx} \quad \text{and} \quad \gamma_{xz} = y\frac{d\phi}{dx}$$

Then assume that the material is linear and isotropic so that

$$\tau_{xy} = G\gamma_{xy} = -Gz\phi' \quad \text{and} \quad \tau_{xz} = G\gamma_{xz} = Gy\phi' \qquad (3\text{-}1)$$

Consulting Figure 3-4a, it is seen that the torque can be represented in terms of the stresses as

$$T = \int_{\text{Area}} (y\tau_{xz} - z\tau_{xy})\, dA$$
$$= G\phi' \int_{\text{Area}} (x^2 + y^2)\, dA = G\phi' \int_{\text{Area}} r^2\, dA = JG\phi'$$

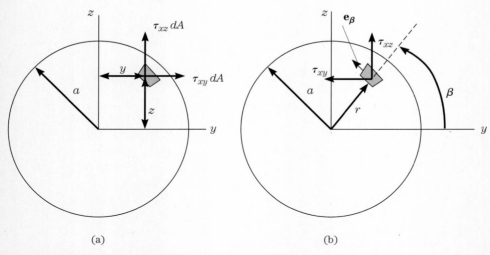

FIGURE 3-4 Shear stresses in the interior and on the boundary.

3-2 TORSION OF CIRCULAR CROSS-SECTION BARS

or
$$T = JG\phi' \tag{3-2}$$

where $J = \pi a^4/2$ is the polar moment of the area of the circular cross section. Equation (3-2) is the basic force-displacement relation for the elementary torsion problem. Eliminating $G\phi'$ between Equations (3-1) and (3-2), the stresses can be expressed in terms of the torque T as

$$\tau_{xy} = -\frac{Tz}{J} \quad \text{and} \quad \tau_{xz} = \frac{Ty}{J}$$

and are the only nonvanishing stresses based on the assumptions that have been made regarding the problem.

These stresses are shown at an arbitrary location in the cross section in Figure 3-4b. The stress vector can be expressed as

$$\mathbf{t} = -\mathbf{j}\frac{Tz}{J} + \mathbf{k}\frac{Ty}{J} = (-\mathbf{j}z + \mathbf{k}y)\frac{T}{J}$$

which, with $z = r \sin \beta$ and $y = r \cos \beta$, becomes

$$\mathbf{t} = (-\mathbf{j} \sin \beta + \mathbf{k} \cos \beta)\frac{Tr}{J}$$

The vector in the brackets is just the unit vector $\mathbf{e}_\beta$ in the direction shown. Thus the direction of the shear stress is everywhere tangent to the circle at that location, and with $\mathbf{t} = \tau \mathbf{e}_\beta$ it follows that the shear stress is given by

$$\tau = \frac{Tr}{J} \tag{3-3}$$

for the case of the torsion of a circular bar.

In summary, given the torque transmitted, the stresses are determined by the equation $\tau = Tr/J$ with the observable angle of twist ϕ obtained by integrating the force displacement relation $T = JG\phi'$.

It is also necessary to mention the manner in which the torque T is applied at the ends of the bar, that is, the manner in which the external loads are applied. The theory shows that the stresses must be applied at the ends according to $\tau = Tr/J$. Unless this is true there will be a region in the neighborhood of the ends of the bar where the solution is not given by the elementary theory. Analysis of this problem is beyond the scope of this text, and the student is referred to the literature for further information.

In the exercises the student is asked to show that the stresses satisfy the three-dimensional equations of equilibrium and the boundary conditions and hence that the solution obtained constitutes an exact solution to the linear equations of elasticity as long as the torque T is applied at the ends in accordance with the elementary theory.

The development in this section is typical of the approach that will be taken throughout the text, namely, to attempt to simplify an analysis on the basis of assumptions regarding kinematic or kinetic variables of the problem. Unlike this example, however, it will usually turn out that a theory developed in this manner will give only approximate solutions within the framework of the corresponding three-dimensional problem.

3-3 Torsion of Noncircular Cross-Section Bars

When a noncircular cross-section straight bar is twisted, the deformations are more complex than for the circular bar. Consider for instance the deformations in a bar with a square cross-section subjected to a torque T, as shown in Figure 3-5. For this square cross section it is quite clear from Figure 3-5b, which depicts the deformed bar, that plane sections do not remain plane, that is, that there is warping of the original section in the direction of the x axis. In Figure 3-5b the x component of displacement is positive in regions I and III and negative in regions II and IV owing to the warping. Further, as shown in Figure 3-5c the shear stress and hence the shear strain is zero at the corners, whereas as indicated in Figures 3-5d and e, there are nonzero shear stresses and strains at the midsides.

With these observations and comments in mind, assume the displacements to be of the form

$$u = \theta\psi(y, z), \quad v = -\theta \times z \quad \text{and} \quad w = \theta \times y \tag{3-4}$$

where ψ is the warping function representing the axial displacement and $\theta = \phi'$, assumed to be a constant, has the same meaning as for the circular cross section, that is, it is the rate of twist along the bar. Taking the angle of twist to be zero at $x = 0$, the angle ϕ through which a cross section at location x rotates is given by $x \cdot \theta = x \cdot \phi'$.

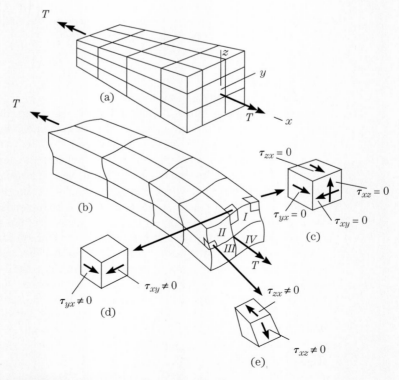

FIGURE 3-5 Torsion of a square cross-section bar.

3-3 TORSION OF NONCIRCULAR CROSS-SECTION BARS

Using the strain-displacement relations, the strains are computed according to

$$\varepsilon_x = \frac{\partial u}{\partial x} = 0, \qquad \varepsilon_y = \frac{\partial v}{\partial y} = 0, \qquad \varepsilon_z = \frac{\partial w}{\partial z} = 0$$

$$\gamma_{xy} = \frac{\partial u}{\partial y} + \frac{\partial v}{\partial x} = \theta\left(\frac{\partial \psi}{\partial y} - z\right)$$

$$\gamma_{yz} = \frac{\partial w}{\partial y} + \frac{\partial v}{\partial z} = \theta x - \theta x = 0$$

$$\gamma_{xz} = \frac{\partial u}{\partial z} + \frac{\partial w}{\partial x} = \theta\left(\frac{\partial \psi}{\partial z} + y\right)$$

For a linear isotropic material the two nonvanishing stresses are then

$$\tau_{xy} = G\gamma_{xy} = G\theta\left(\frac{\partial \psi}{\partial y} - z\right)$$

and

$$\tau_{xz} = G\gamma_{xz} = G\theta\left(\frac{\partial \psi}{\partial z} + y\right)$$

The equations of equilbrium reduce to

$$\frac{\partial \tau_{xy}}{\partial y} + \frac{\partial \tau_{xz}}{\partial z} = 0$$

$$\frac{\partial \tau_{xy}}{\partial x} = 0$$

$$\frac{\partial \tau_{xz}}{\partial x} = 0$$

The second and third of the equilibrium equations are satisfied identically with the first leading to

$$G\theta\left(\frac{\partial^2 \psi}{\partial y^2} + \frac{\partial^2 \psi}{\partial z^2}\right) = 0$$

or

$$\frac{\partial^2 \psi}{\partial y^2} + \frac{\partial^2 \psi}{\partial z^2} = 0$$

that is, the warping function ψ satisfies Laplace's equation. It is left in the homework exercises for the student to show that the boundary condition satisfied by the warping function ψ can be expressed in the form

$$\frac{\partial \psi}{\partial n} = \frac{1}{2}\frac{d}{ds}(x^2 + y^2)$$

Thus, the boundary value problem that must be satisfied by the warping function is

$$\nabla^2 \psi = 0 \quad \text{in } D$$

$$\frac{\partial \psi}{\partial n} = \frac{1}{2}\frac{d}{ds}(x^2 + y^2) \quad \text{on } B$$

where D corresponds to the region defined by the cross section of the bar and B its bounding curve. It is relatively difficult to determine solutions to Laplace's equation that also satisfy this boundary condition. An alternate formulation that results in a similar differential equation and in a boundary condition that is simpler in form will now be presented.

Consider again the first equation of equilibrium,

$$\frac{\partial \tau_{xy}}{\partial y} + \frac{\partial \tau_{xz}}{\partial z} = 0$$

Note that this equation (as well as the other two equations of equilibrium) is satisfied identically if we take the stresses τ_{xz} and τ_{xy} in terms of the Prandtl stress function $\Phi(y, z)$ as

$$\tau_{xz} = -\frac{\partial \Phi}{\partial y} \quad \text{and} \quad \tau_{xy} = \frac{\partial \Phi}{\partial z} \qquad (3\text{-}5)$$

Comparing these to the corresponding equations in terms of the warping function ψ yields

$$G\theta\left(\frac{\partial \psi}{\partial z} + y\right) = -\frac{\partial \Phi}{\partial y}$$

and

$$G\theta\left(\frac{\partial \psi}{\partial y} - z\right) = \frac{\partial \Phi}{\partial z}$$

Eliminating ψ results in

$$\frac{\partial^2 \Phi}{\partial y^2} + \frac{\partial^2 \Phi}{\partial z^2} = \nabla^2 \Phi = -2G\theta$$

This equation, referred to as *Poisson's equation*, must be satisfied by the stress function Φ in the region D corresponding to the shape of the cross section.

The condition to be satisfied by the stress function Φ on the boundary is deduced by considering Figure 3-6. In order that the shear stress τ_S shown in Figure 3-6a be zero on the lateral surface of the bar, the component of the shear stress vector in the

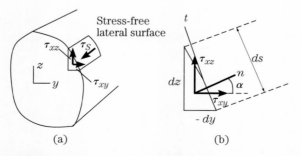

FIGURE 3-6 Boundary condition on the stresses.

n direction, that is, corresponding to τ_S, must vanish. Consulting Figure 3-6b it is seen that the shear stress in the n direction can be expressed as

$$\tau_{xn} = \tau_{xy} \cos \alpha + \tau_{xz} \sin \alpha = \tau_S = 0$$

or, with $\cos \alpha = dz/ds$ and $\sin \alpha = -dy/ds$, there results

$$\frac{\partial \Phi}{\partial z} \frac{dz}{ds} + \frac{\partial \Phi}{\partial y} \frac{dy}{ds} = \frac{d\Phi}{ds} = 0$$

which shows that Φ must be constant along the boundary of the region. For a simply connected region (no holes) this constant can be taken as zero with no loss in generality.

Last, the torque is given in terms of the stress function by

$$T = \int_{Area} (y\tau_{xz} - z\tau_{xy})\, dA$$

$$= \int_{Area} \left(-y\frac{\partial \Phi}{\partial y} - z\frac{\partial \Phi}{\partial z}\right) dA$$

Recall that the two-dimensional form of the divergence theorem states that

$$\int_{Area} \frac{\partial \Phi}{\partial y}\, dA = \oint \Phi \cos(n, y)\, ds$$

and

$$\int_{Area} \frac{\partial \Phi}{\partial z}\, dA = \oint \Phi \cos(n, z)\, ds$$

Writing

$$-y\frac{\partial \Phi}{\partial y} = -\frac{\partial}{\partial y}(y\Phi) + \Phi$$

and using the divergence theorem the first term can be converted to

$$\int_{Area} \left(-y\frac{\partial \Phi}{\partial y}\right) dA = -\oint \Phi\, y \cos(n, y)\, ds + \int_{Area} \Phi\, dA$$

Similarly,

$$\int_{Area} \left(-z\frac{\partial \Phi}{\partial y}\right) dA = -\oint \Phi\, z \cos(n, z)\, ds + \int_{Area} \Phi\, dA$$

Then, since $\Phi = 0$ on the boundary B, the line integrals vanish and it follows that the torque T can be expressed as

$$T = 2\int_{Area} \Phi\, dA$$

By virtue of the fact that Φ satisfies the linear differential equation $\nabla^2 \Phi = -2G\theta$, Φ depends linearly on $G\theta$ so that $T = \int\int \Phi\, dA$ produces an equation of the form $T = JG\theta$, where J is called the torsional constant and is the analog of the polar moment of the area for the circular cross section. This linear relationship can be used to eliminate $G\theta$

1. *Solve* the boundary value problem shown

$$\nabla^2 \phi = -2G\theta \text{ in } D$$
$$\phi = 0 \text{ on } \Gamma$$

Domain D, Boundary Γ

2. *Evaluate* $T = 2 \int_{Area} \phi dA$ $(= JG\theta)$
3. *Compute* the stresses from

$$\tau xz = -\frac{\partial \phi}{\partial y} \quad \text{and} \quad \tau xy = \frac{\partial \phi}{\partial z}$$

FIGURE 3-7 The steps in solving the torsion problem.

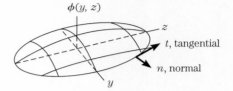

FIGURE 3-8 Geometry of the Φ surface.

and hence express the stresses in terms of the torque T. Thus the boundary value problem for the torsion problem is shown in Figure 3-7.

A good deal of insight into the relationship between the stress function and the stresses can be gained by considering the geometry of the Φ surface as shown in Figure 3-8. In a manner similar to that used to express the boundary condition in terms of $\tau_{xn} = d\Phi/ds = 0$, it can also be shown that in terms of the coordinate t shown in Figure 3-8, the tangential component of the shear stress is given by the directional derivative $\tau_{xt} = -d\Phi/dn$. Thus, the two sets of equations

$$\tau_{xz} = -\frac{\partial \Phi}{\partial y} \quad \text{and} \quad \tau_{xy} = \frac{\partial \Phi}{\partial z}$$

and

$$\tau_{xt} = -\frac{\partial \Phi}{\partial n} \quad \text{and} \quad \tau_{xn} = \frac{\partial \Phi}{\partial t} \tag{3-6}$$

are completely analogous, simply being expressed in two different orthogonal sets of coordinate, that is, in terms of y and z for the first pair of equations and in terms of n and t for the second. It is clear from these equations that the shear stress in a particular direction is proportional to the directional derivative of the stress function in a perpendicular direction and that the maximum shear stress occurs where this derivative has its largest value. For a simply connected bar, this maximum will occur on the boundary at the point where the slope of the Φ surface in the direction of the outward normal is maximum. This will be discussed further in the section where the membrane analogy is presented.

EXAMPLE 3-1 A cross section for which there is a simple closed-form solution to the torsion problem is the ellipse shown in Figure 3-9. The equation of the boundary of the ellipse is

$$f(y, z) = 1 - \frac{y^2}{a^2} - \frac{z^2}{b^2} = 0$$

Noting that $\nabla^2 f$ is a constant, we take the stress function $\Phi(y, z)$ as

$$\Phi(y, z) = C\left(1 - \frac{y^2}{a^2} - \frac{z^2}{b^2}\right)$$

3-3 TORSION OF NONCIRCULAR CROSS-SECTION BARS

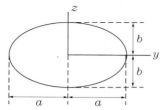

FIGURE 3-9 Elliptical cross section.

with

$$\nabla^2 \Phi = -2C\left(\frac{1}{a^2} + \frac{1}{b^2}\right) = -2G\theta$$

Thus, if we choose $C = G\theta a^2 b^2/(a^2 + b^2)$, the function Φ satisfies the differential equation and boundary condition. The stresses become

$$\tau_{xz} = -\frac{\partial \Phi}{\partial y} = \frac{2G\theta b^2}{a^2 + b^2} y$$

and

$$\tau_{xy} = \frac{\partial \Phi}{\partial z} = -\frac{2G\theta a^2}{a^2 + b^2} z$$

For $a > b$ the maximum occurs at $y = 0 \; z = \pm b$ with

$$|\tau_{\max}| = \frac{2G\theta a^2 b}{a^2 + b^2}$$

The torque is

$$T = \int_{\text{Area}} (y\tau_{xz} - z\tau_{xy}) \, dA$$

$$= \frac{2G\theta}{a^2 + b^2} \int_{\text{Area}} (y^2 b^2 + z^2 a^2) \, dA$$

Evaluating the integrals yields

$$T = \frac{\pi a^3 b^3}{a^2 + b^2} G\theta$$

from which the torsional constant is $J = \pi a^3 b^3/(a^2 + b^2)$. The stresses can then be expressed as

$$\tau_{xz} = \frac{2Ty}{\pi a^3 b} \quad \text{and} \quad \tau_{xy} = -\frac{2Tz}{\pi a b^3}$$

with

$$|\tau_{\max}| = \frac{2T}{\pi a b^2}$$

for $a > b$, that is, the maximum shear stress occurs at the ends of the semiminor axes. The rate of twist is given in terms of the torque as

$$\theta = \frac{T}{JG} = \frac{T(a^2 + b^2)}{G\pi a^3 b^3}$$

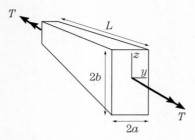

FIGURE 3-10 Rectangular cross-section bar.

EXAMPLE 3-2 A bar with a rectangular cross section is shown in Figure 3-10. Determining the deformations and stresses necessitates solving the boundary value problem

$$\nabla^2 \Phi + 2G\theta = 0 \qquad -\frac{a}{2} \le y \le \frac{a}{2} \qquad -\frac{b}{2} < z \le \frac{b}{2}$$

with

$$\Phi\left(\frac{a}{2}, z\right) = \Phi\left(-\frac{a}{2}, z\right) = 0 \qquad -\frac{b}{2} \le z \le \frac{b}{2}$$

and

$$\Phi\left(y, -\frac{b}{2}\right) = \Phi\left(y, \frac{b}{2}\right) = 0 \qquad -\frac{a}{2} \le y \le \frac{a}{2}$$

An assumption that automatically takes care of the boundary conditions on the y sides is

$$\phi(y, z) = \sum_{n \text{ odd}}^{\infty} \cos\left(\frac{n\pi y}{2a}\right) \psi_n(z)$$

where the $\psi_n(z)$ must be chosen so as to satisfy the differential equation and the remaining boundary conditions. Expanding $2G\theta$ in a like series results in

$$2G\theta = \sum_{n \text{ odd}}^{\infty} \frac{8G\theta}{n\pi} (-1)^{(n-1)/2} \cos\left(\frac{n\pi y}{2a}\right)$$

Substituting into the differential equation yields, with $\alpha_n = n\pi/2a$,

$$\psi_n''(z) - \alpha_n^2 \psi_n(z) = a_n$$

where $a_n = -(8G\theta/n\pi)(-1)^{(n-1)/2}$. The solution can be written as

$$\psi_n(z) = A_n \cosh \alpha_n z + B_n \sinh \alpha_n z - \frac{a_n}{\alpha_n^2}$$

Satisfying the boundary conditions at $z = \pm b/2$ yields $B_n = 0$ and

$$A_n = \frac{a_n}{\alpha_n^2 \cosh \alpha_n b}$$

3-3 TORSION OF NONCIRCULAR CROSS-SECTION BARS

The solution for the stress function is then

$$\phi(y, z) = \frac{32G\theta a^2}{\pi^3} \sum_{\substack{\text{nodd}}}^{\infty} \frac{(-1)^{(n-1)/2}}{n^3} \left[1 - \frac{\cosh \alpha_n z}{\cosh \alpha_n b}\right] \cos\left(\frac{n\pi y}{2a}\right)$$

The shear stresses are given by

$$\tau_{xz} = -\frac{\partial \phi}{\partial y} = \frac{16G\theta a}{\pi^2} \sum_{\substack{\text{nodd}}}^{\infty} \frac{(-1)^{(n-1)/2}}{n^2} \left[1 - \frac{\cosh \alpha_n z}{\cosh \alpha_n b}\right] \sin\left(\frac{n\pi y}{2a}\right)$$

$$\tau_{yz} = \frac{\partial \phi}{\partial z} = -\frac{16G\theta a}{\pi^2} \sum_{\substack{\text{nodd}}}^{\infty} \frac{(-1)^{(n-1)/2}}{n^2} \frac{\sinh \alpha_n z}{\cosh \alpha_n b} \cos\left(\frac{n\pi y}{2a}\right)$$

For $b > a$ the maximum shear stress occurs at $z = 0$, $y = \pm a$ with

$$\tau_{\max} = \frac{16G\theta a}{\pi^2} \sum_{\substack{\text{nodd}}}^{\infty} \frac{(-1)^{(n-1)}}{n^2} \left[1 - \frac{1}{\cosh \alpha_n b}\right]$$

For $b \gg a$, $\cosh \alpha_n b \gg 1$ and

$$\tau_{\max} \approx \frac{16G\theta a}{\pi^2} \sum_{\substack{\text{nodd}}}^{\infty} \frac{1}{n^2} = \frac{16G\theta a}{\pi^2} \frac{\pi^2}{8} = 2G\theta a$$

which coincides with the value given in section 3-3-2 for the thin rectangular cross section.

The torque is given by $T = 2 \iint \phi \, dA$, which can be evaluated to yield

$$T = \frac{G\theta (2a)^3 (2b)}{3} \left[1 - \frac{192}{\pi^5} \frac{a}{b} \sum_{\substack{\text{nodd}}}^{\infty} \frac{1}{n^5} \tanh \alpha_n b\right]$$

This series converges very rapidly and can be summed for any value of the aspect ratio a/b. For the case of a very narrow rectangle,

$$T \approx \frac{G\theta (2a)^3 (2b)}{3}$$

again coinciding with the result in section 3-2-2 for the thin rectangular cross section. In any event the maximum shear stress can be represented as

$$\tau_{\max} = \frac{T}{k_1 (2b)(2a)^2}$$

and the rate of twist as

$$\theta = \frac{T}{k_2 G (2b)(2a)^3}$$

where k_2 is determined by summing the series above for the torque and k_1 by summing the series for $\tau_{\max}$ and eliminating the $G\theta$ term from the torque equation. Values of k_1 and k_2 for a range of the aspect ratio are shown in Table 3-1. The significance of the limiting values of 0.333 for both k_1 and k_2 will be discussed further in the next section. In any event, the maximum shear stress and rate of twist for any rectangular bar can be

TABLE 3.1 Stress and Twist Coefficients

	$b/a = 1.0$	1.5	2.0	2.5	3.0	4.0	6.0	10.0	∞
k_1	0.208	0.231	0.246	0.256	0.267	0.282	0.299	0.312	0.333
k_2	0.141	0.196	0.229	0.249	0.263	0.281	0.299	0.312	0.333

determined solely on the basis of the aspect ratio b/a by interpolating the values in Table 3-1. ◆

3-3-1 The Membrane Analogy Consider the problem of the transverse deflections of a thin membrane, as shown in Figure 3-11. An easily understood example of a thin membrane is a soap film stretched over a hole in a plate, also shown in Figure 3-11. The actual experiment would consist of an airtight box into which a hole has been cut. The shape of the hole is the same as the cross section of the bar that is to be torqued. A soap film is then created over the hole, after which some additional air is introduced through an opening, resulting in a slightly larger pressure in the interior of the box. This pressure results in a transverse displacement of the soap film as shown. With S as the uniform tension in the soap film and $w(y, z)$ as the small transverse displacement, the governing equation of equilibrium is

$$S\left(\frac{\partial^2 w}{\partial y^2} + \frac{\partial^2 x}{\partial z^2}\right) + p = 0$$

or $\nabla^2 w = -p/S$ in the hole region and $w = 0$ on the boundary.

The form of the differential equation and boundary condition is precisely the same as for the stress function, namely, $\nabla^2 \phi = -2G\theta$ in the interior and $\phi = 0$ on the boundary so that the two problems are analagous, that is,

$$\frac{w}{\frac{p}{S}} = \frac{\phi}{2G\theta}$$

or

$$\phi = Cw$$

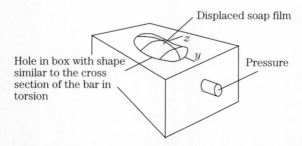

FIGURE 3-11 Soap film for the membrane analogy.

According to this analogy, any property of the stress function is proportional to the corresponding property of the membrane displacement. In particular, on the basis of Equations (3-5) the maximum stress in the bar is proportional to the slope of the ϕ surface in the perpendicular direction so that the location of the maximum stress in the bar coincides with the location of the maximum slope in the membrane. Although experiments have actually been conducted using the membrane in connection with the equation $\phi = Cw$, our purpose will be in visualizing the shape of the membrane to assist in deducing corresponding properties of the stress function and hence of the stresses and displacements.

3-3-2 Thin Rectangular Cross Section

Many structures consist of thin rectangular sheets or combinations of thin rectangular sheets carrying a combination of axial, shear, torsional and bending loads. Before dealing with this more general problem, the membrane analogy will be used to assist in constructing an approximate solution for the torsion of a single thin rectangle.

Consider then a thin rectangle subjected to a single torque T as shown in Figure 3-12. The shape of the corresponding membrane is shown in Figure 3-13. For $b \gg t$ it is seen that the shape of the membrane is essentially the same over a substantial portion of the length along the z axis. These observations suggest an approximate solution of the form $w = w(y)$, or, equivalently, a stress function of the form $\Phi = \Phi(y)$. This approximate solution will be accurate except in the neighborhood of the ends $z = \pm b/2$. With Φ a function only of y, the governing differential equation and boundary conditions then become

$$\frac{\partial^2 \Phi}{\partial y^2} = -2G\theta$$

$$\Phi\left(-\frac{t}{2}\right) = \Phi\left(\frac{t}{2}\right) = 0$$

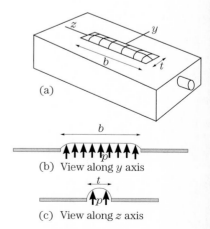

(a)

(b) View along y axis

(c) View along z axis

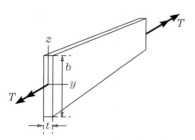

FIGURE 3-12 Thin rectangle subjected to a torque.

FIGURE 3-13 Shape of thin rectangular membrane.

Solving yields

$$\Phi = G\theta\left(\frac{t^2}{4} - y^2\right)$$

The stresses are given by

$$\tau_{xz} = -\frac{\partial \Phi}{\partial y} = -\frac{d\Phi}{dy} = 2G\theta y$$

and

$$\tau_{xy} = \frac{\partial \Phi}{\partial z} = 0$$

which are shown in Figure 3-14. The maximum value $(\tau_{xz})_{\max} = G\theta t$ occurs at $y = \pm t/2$ all along the edges parallel to the z axis.

The actual stress distribution is shown in Figure 3-15, where it is seen that for $b \gg t$ the maximum value of Tt/J acts over most of the length of the long side of the cross section. The value of the shear stress at the ends $z = \pm b/2$ is substantially smaller than the maximum along the long edges.

To relate the stresses to the torque we evaluate

$$T = 2\iint \phi \, dA = 2\int_{-b/2}^{b/2}\left[\int_{-t/2}^{t/2} G\theta\left(\frac{t^2}{4} - y^2\right) dy\right] dz$$

$$= \frac{bt^3}{3}G\theta = J_R G\theta$$

The quantity $J_R = bt^3/3$ is the torsional constant for the thin rectangle. Eliminating $G\theta$, the stress can be expressed in terms of the torque T as

$$\tau = \frac{Tt}{J_R} = \frac{3T}{bt^2}$$

Note that the two results $\theta = 3T/Gbt^3$ and $\tau = 3T/bt^2$ coincide with those given by Table 3-1 for the geometry $b/t = \infty$. Thus the current approximate theory is exact in the limiting case of a *very* thin rectangle. The validity of using the thin rectangle results for large values of $b/t < \infty$ can be judged by consulting Table 3-1. Owing to the t^3 in the denominator of θ and the t^2 in the denominator of τ, it follows that the rectangular section is weak in torsion.

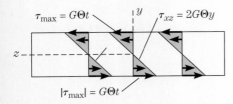

FIGURE 3-14 Shear stresses in a thin rectangular cross-section bar.

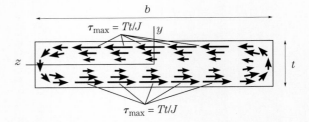

FIGURE 3-15 Actual stress distribution in a thin rectangle.

EXAMPLE 3-3 Consider a thin steel strip with a cross section with dimensions 2 in by 0.1 in that is subjected to a torque T. For an allowable shear stress of 10 ksi it is easily determined that the allowable torque that the thin rectangle can transmit is

$$T = \frac{10^4(2)(.01) \text{ in-lbf}}{3} = 66.67 \text{ in-lbf}$$

The student should show that the corresponding allowable torque in a solid circular section with the same area would be $T = 252$ in-lbf, approximately four times the value for the thin rectangle. The corresponding rates of twist for the thin rectangle and solid circle are

$$\theta = \frac{66.67(3)}{2(.001)11 \times 10^6} = 0.00909 \text{ rad/in} \quad \text{(for the rectangle)}$$

$$\theta = 0.00369 \text{ rad/in} \quad \text{(for the solid circle)}$$

showing the relative values. If the thickness of the rectangle is halved to $t = 0.05$ in, the corresponding torques and rates of twist in the rectangle and in a circle of equal area become

$$T = 16.67 \text{ in-lbf} \qquad \theta = 0.0727 \text{ rad/in} \qquad \text{(for the rectangle)}$$
$$T = 89.2 \text{ in-lbf} \qquad \theta = 0.00510 \text{ rad/in} \qquad \text{(for the solid circle)}$$

again showing the relative weakness of the thin rectangle.

3-3-3 Thin-Walled Open Sections There are many structural applications where a member that is composed of several interconnected thin rectangles transmits a torque. Several such commonly used structural shapes are shown in Figure 3-16.

The membrane analogy is used in connection with Figure 3-17, showing the shapes assumed by a membrane for an I-section and for a single arbitrarily shaped thin section. As was the case for a single rectangle, the corresponding membrane assumes a parabolic shape across the thickness at points along the length of any leg of the I-section or at any

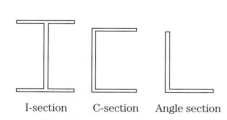

I-section   C-section   Angle section

FIGURE 3-16 Common structural shapes.

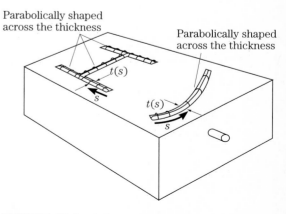

FIGURE 3-17 Parabolic membrane shapes.

point along the arc length s for the arbitrarily shaped thin section. Also as for the single rectangle, the shape deviates from parabolic near the ends of the flanges and at the intersections of the flanges and web in the I-section and at the two ends of the arbitrarily shaped thin section. As long as the thicknesses of the individual pieces are small compared to the lengths of the segments, these deviations do no negate the use of the membrane theory developed for the single thin rectangle. In any event, assume that the shape of the membrane is locally parabolic with the stress function Φ given by

$$\Phi(s, y) = G\theta\left(\frac{t(s)^2}{4} - y^2\right)$$

where y is a coordinate measured from the midline of the segment and $t(s)$ is the local thickness, as shown in Figure 3-17. The torque T is given by

$$T = 2\iint \phi\, dA = 2\int_0^L \left[\int_{-t(s)/2}^{t(s)/2} \Phi(s, y)\, dy\right] ds = \frac{G\theta}{3}\int_0^L t(s)^3\, ds = JG\theta$$

where

$$J = \frac{1}{3}\int_0^L t(s)^3\, ds$$

For a single constant-thickness segment $J = t^3 L/3$ as developed previously for the single rectangle. When the cross section consists of a number of segments of constant thickness t_i and length b_i the expression for the torque T becomes

$$T = \left(\sum \frac{b_i t_i^3}{3}\right) G\theta = J_o G\theta$$

where $J_o = \sum b_i t_i^3/3$ is the torsional constant for the thin-walled open section. The values used for the b_i should be measured along the center lines of the various segments that make up the section. The distribution of shear stress across the thickness is still given by

$$\tau = \frac{2Ty}{J_o}$$

with a maximum value at $y = \pm t(s)/2$ of

$$|\tau_{max}| = \frac{Tt(s)}{J_o}$$

Not accounting for stress concentrations, the maximum shear stress in an open section thus occurs at the location of the maximum thickness with

$$|\tau_{max}| = \frac{Tt_{max}}{J_0}$$

Note that these results for J_0 and τ_{max} obviously reduce properly in the case of a single thin rectangle of dimensions b by t.

Several of the common structural shapes consist of segments joining at right angles, as indicated in Figure 3-18. At the outside corner A it is easily shown that there can be

3-4 TORSION OF A SINGLE THIN-WALLED TUBE

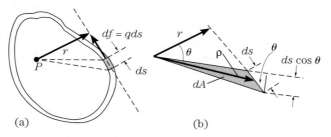

FIGURE 3-29 Equilibrium for relating the torque to the shear stress.

where since the shear flow q is a constant it can be brought outside the integral. Consulting Figure 3-29b, this can be expressed as

$$T = q \oint (\rho \cos \theta) \, ds = q \oint \rho(\cos \theta \, ds) = q \oint 2 \, dA = 2A_0 q$$

where A_0 is the area associated with the midline of the tube. Using the fact that $q = \tau t$, this relation can be expressed as

$$\tau = \frac{T}{2A_0 t} \tag{3-7}$$

verifying that the maximum shear stress occurs at the point around the periphery of the tube where the thickness is a minimum. The constant character of the shear flow, which is always tangent to the tube at the location in question, is indicated in Figure 3-30.

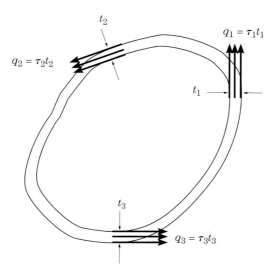

FIGURE 3-30 Shear stresses and shear flow in a single tube.

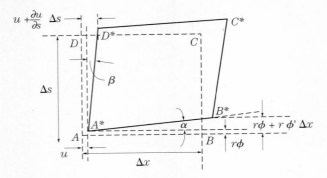

FIGURE 3-31 Undeformed and deformed elements in a thin-walled tube.

Deformations. The above development pertains to equations involving only the kinetics, that is, equilibrium, stress, shear flow, and so on. However, it is generally of interest to determine what deformations occur as a result of the loads. To this end consider the same differential element as that shown in Figure 3-28b, viewed, however, along a direction normal to the element, as shown in Figure 3-31. The dotted rectangle $ABCD$ indicates the undeformed configuration, and the solid quadrilateral $A^*B^*C^*D^*$ shows the deformed configuration. The displacements u in the axial direction and $r\phi$ in the tangential direction are shown. The displacements u are the warping displacements that occur in general for sections that are not circular in cross section.

The shear strain associated with the warping and the rotation ϕ is given by $\gamma = \alpha + \beta$ where

$$\beta = \lim_{\Delta s \to 0} \frac{u + \frac{\partial u}{\partial s}\Delta s - u}{\Delta s} = \frac{\partial u}{\partial s}$$

and

$$\alpha = \lim_{\Delta x \to 0} \frac{r\phi + r\phi'\Delta x - r\phi}{\Delta x} = r\phi'$$

so that

$$\gamma = \frac{\partial u}{\partial s} + r\phi' \tag{3-8}$$

Material Behavior. Assume a linearly elastic material with

$$\tau = G\gamma \tag{3-9}$$

Combination. The collection of equations from the three basic ideas is

$$q = \tau t, \quad T = 2A_0 q, \quad \gamma = \frac{\partial u}{\partial s} + r\phi' \quad \text{and} \quad \tau = G\gamma$$

3-4 TORSION OF A SINGLE THIN-WALLED TUBE

These equations can easily be combined to yield

$$\frac{T}{2A_0 Gt} = \frac{\partial u}{\partial s} + r\phi'$$

Integrating around the periphery of the tube results in

$$\oint \frac{T}{2A_0 Gt}\, ds = \oint du + \oint r\phi'\, ds$$

or with $\phi' = \text{const}$, $\oint du = 0$ and $\oint r\, ds = 2A_0$,

$$\phi' = \frac{T}{4A_0^2 G} \oint \frac{ds}{t}$$

which can be written as $\phi' = T/JG$ with

$$J = \frac{4A_0^2}{\oint \dfrac{ds}{t}} \qquad (3\text{-}10)$$

as the torsional constant. Integration of the torque-rate of twist equation then yields the usual relation $\phi = TL/JG$ for the relative angle of twist between the ends. It is left for the student to show in the exercises that the torque-rate of twist relation $\phi' = T/JG$, with the torsional constant $J = 4A_0^2/\oint (ds/t)$, can be developed using the complementary energy of the tube and Castigliano's second theorem.

EXAMPLE 3-6 A rectangular tube with a constant thickness t and outside dimensions as shown in Figure 3-32 is subjected to a torque of 3.0×10^5 in-lbf. Determine the required thickness if the allowable shear stress is 8 ksi and the allowable angle of twist is 1.5 degrees. Take $G = 4 \times 10^6$ psi and assume that the stress concentration factor at the inner reentrant corners is 1.4.

Solution: Using the basic expression $\tau = T/2A_0 t$ with the midline area as $A_0 = (15 - t)(12 - t)$ there results

$$\frac{8000 \text{ psi}}{1.4} = \frac{3.0 \times 10^5 \text{ in-lbf}}{2(15 - t)(12 - t)t}$$

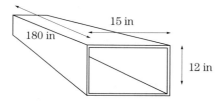

FIGURE 3-32 Torsion of a thin-walled rectangular tube.

from which we get $t(15 - t)(12 - t) = 26.25$ in^3 where t is measured in inches. This is a simple cubic equation with the solution $t \approx 0.149$ in. For the twist constraint we need to evaluate the torsional constant

$$J = \frac{4A_0^2}{\oint \frac{ds}{t}} = \frac{2[(15 - t)(12 - t)]^2 t}{(15 - t) + (12 - t)} \text{ in}^4$$

where t is measured in inches. Using the torque-rate of twist equation with $\phi' = [1.5\pi/(180)^2]$ in^{-1}, this leads to

$$t[(12 - t)(15 - t)]^2 - (27 - 2t)515.66 \text{ in}^4 = 0$$

with solution $t \approx 0.490$ in, which governs. ◆

It is instructive at this point to consider the torsion of two torsional members, one being a single closed tube and the other an identical tube with a longitudinal slit, as shown in Figures 3-33a and 3-33b, respectively. For the tube with the slit, there is a large axial displacement of one side of the slit relative to the other. This large axial displacement is related to the warping displacements mentioned above in the treatment of deformations. Warping displacements are also generally present in the closed tube but are not large enough to be observed with the naked eye. In both cases an original longitudinal line becomes helical but with the magnitude of the associated rate of rotation ϕ' being much much larger for the slit tube.

The corresponding shear stresses in both sections are shown in Figure 3-34. The quantitative differences between the stresses in the closed and open sections can be explained as follows. In either case the resultant associated with the stress distribution

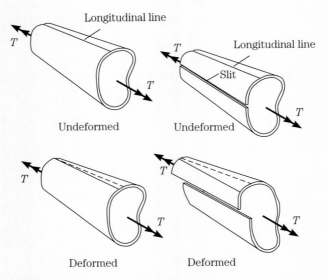

FIGURE 3-33 Deformations in tubes without and with a longitudinal slit.

3-5 TORSION OF MULTICELLED TUBES

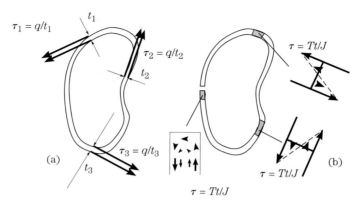

FIGURE 3-34 Stresses in tubes without and with a longitudinal slit.

must be statically equivalent to the applied torque T. For the closed tube the moment arm of a force $df = q\,ds$ is proportional to the diameter D of the tube, that is,

$$T = \sum q\,ds\,D \approx qLD = \tau_c tLD$$

from which $\tau_c \approx T/LDt$. For the open tube the resultant force arising from an element ds is zero and the total torque T must be generated by the sum of the moments associated with each of the ds's, that is,

$$T = \sum 2\tau_0 \left(\frac{t}{2}\right)\left(\frac{2}{3}\right)\left(\frac{t}{2}\right)\Delta s \approx \frac{L\tau_0 t^2}{3}$$

with $\tau_o \approx 3T/Lt^2$. The ratio is $\tau_o/\tau_c \approx 3D/t \gg 1$, essentially due to the difference in the moment arm of the maximum shear stress in the open and closed tubes.

3-5 Torsion of Multicelled Tubes

There are frequent situations in the shipbuilding and aerospace industries in which a geometry of a portion of the structure that transmits a torque consists of a number of connected tubes, as shown in Figure 3-35. Consider as typical of this general situation a structure with two interconnected tubes transmitting a torque as shown in Figure 3-36a.

Geometry. Assume that the cross section is constant with respect to distance along the axis of the tube and again make the assumption that all the thicknesses are small in comparison with the "diameter" of the tube. This enables us to conclude, based on the shape of the membrane and the ϕ surface indicated in Figures 3-36b and 3-36c, respectively, that the shear flow is constant in the web and in each of the two outer curved portions of the tube. Denote these shear flows as q_1, q_2 and q_3, as shown in Figure 3-37a.

Equilibrium. Consulting Figure 3-37b, showing a free body diagram of a portion of the tube containing one of the junctions where the shear flows intersect, it can be seen that

$$\sum F_x = q_1 \Delta x + q_3 \Delta x - q_2 \Delta x = 0$$

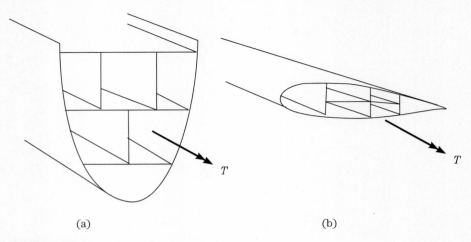

(a) (b)

FIGURE 3-35 Ship and wing cross sections.

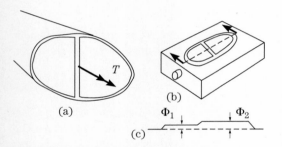

FIGURE 3-36 Geometry of two tubes and the corresponding Φ surface.

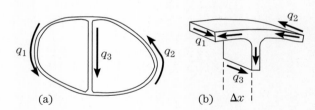

FIGURE 3-37 Shear flows in a multitube structure.

or

$$q_3 = q_2 - q_1$$

Essentially this says that at any junction "flow in equals flow out," that is, $q_2 = q_1 + q_3$. In addition, the sum of the torques provided by the collection of shear flows must be statically equivalent to the torque T. With O the point about which the torques are computed, as shown in Figure 3-38, the total torque provided is

$$T = \int_{ABC} rq_1 \, ds + \int_{CA} rq_3 \, ds + \int_{CDA} rq_2 \, ds$$

Recalling that each of the shear flows is assumed to be constant and that each of the integrals evaluates to twice the area swept out by the radius r, the torque becomes

$$T = q_1 2A_{ABCOA} + q_3 2A_{ACOA} + q_2 2A_{OCDAO}$$

3-5 TORSION OF MULTICELLED TUBES

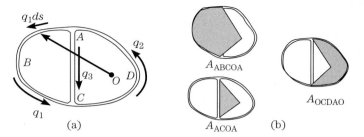

FIGURE 3-38 Shear flows and midline areas.

with each of the midline areas depicted in Figure 3-38b. Then with $q_3 = q_2 - q_1$ there results

$$T = q_1 2A_{ABCOA} + (q_2 - q_1)2A_{ACOA} + q_2 2A_{OCDOA}$$
$$= 2q_1(A_{ABCOA} - A_{ACOA}) + 2q_2(A_{OCDOA} + A_{ACOA})$$
$$= 2q_1 A_1 + 2q_2 A_2$$

where A_1 and A_2 are clearly the midline areas of the tubes 1 and 2, respectively. In general for a tube with N interconnected tubes,

$$T = 2 \sum_{i=1}^{N} A_i q_i$$

At this point in the development it is clear from the consideration of equilibrium that there is one equation, namely, $T = 2A_1 q_1 + 2A_2 q_2$, for the two unknown shear flows q_1 and q_2, that is, a statically indeterminate problem. The additional equations come from considering the deformations and the stress-strain relations.

Deformations and Stress Strain. The assumptions that were made in the above section regarding the deformations and material behavior continue to be valid for the multitube construction. The equation that resulted for the single tube, namely, $2A_0 G\phi' = \oint q\, ds/t$, will be used for each tube according to

$$\phi'_i = \frac{1}{2A_i G_i} \oint \frac{q\, ds}{t} \qquad i = 1, 2$$

leading to

$$2A_1 G_1 \phi'_1 = \int_{ABC} \frac{q_1\, ds}{t} - \int_{CA} \frac{q_3\, ds}{t}$$

and

$$2A_2 G_2 \phi'_2 = \int_{CDA} \frac{q_2\, ds}{t} + \int_{AC} \frac{q_3\, ds}{t}$$

Since the tubes are connected they must have the same rate of twist, that is, $\phi_1' = \phi_2' = \phi'$. It then follows with $q_3 = q_2 - q_1$ that

$$2A_1 G_1 \phi' = \left(\oint_{C_1} \frac{ds}{t}\right) q_1 - \left(\int_W \frac{ds}{t}\right) q_2$$

and

$$2A_2 G_2 \phi' = -\left(\int_W \frac{ds}{t}\right) q_1 + \left(\oint_{C_2} \frac{ds}{t}\right) q_2$$

where

$$\oint_{C_1} = \int_{ABC} + \int_{CA} = \int_{ABC} + \int_W$$

and

$$\oint_{C_2} = \int_{CDA} + \int_{AC} = \int_{CDA} + \int_W$$

and where the w subscript refers to the web. These simultaneous equations are solved for q_1 and q_2 in terms of the unknown ϕ', that is, $q_1 = c_1 G \phi'$ and $q_2 = c_2 G \phi'$, where c_1 and c_2 are constants involving the areas and arc lengths of the tubes. The expressions for q_1 and q_2 are then substituted into the torque equation to yield

$$T = (2A_1 c_1 + 2A_2 c_2) G \phi' = JG\phi' \tag{3-11}$$

where J is the torsional constant for the section. In terms of the torque T, the shear flows are then $q_1 = c_1 T/J$, $q_2 = c_2 T/J$ and $q_3 = (c_2 - c_1) T/J$ positive in the directions shown in Figure 3-38a.

EXAMPLE 3-7 For the two-tube section shown in Figure 3-39, the outer skins are 12 mm thick with the inner web 6 mm thick. If the allowable shear stress is 60 MPa and the allowable angle of twist for a 4 m length tube is 3 degrees, determine the allowable torque. Take $G = 28$ GPa and assume the stress concentration factor for the interior fillets to be 1.4.

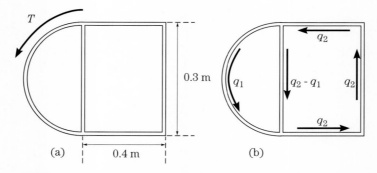

FIGURE 3-39 Two-tube section.

3-5 TORSION OF MULTICELLED TUBES

Solution: The assumed directions for the unknown shear flows q_1 and q_2 as well as the corresponding direction of the shear flow in the interior segment are shown in Figure 3-39b. With $A_1 = \pi(0.144 \text{ m})^2/2$ and $A_2 = (0.288 \text{ m})(0.391 \text{ m})$, the two deformation equations are

$$(0.06514 \text{ m}^2)G\phi' = \frac{0.45239}{0.012}q_1 - \frac{0.288}{0.006}(q_2 - q_1)$$

$$(0.22522 \text{ m}^2)G\phi' = \frac{1.070}{0.012}q_2 + \frac{0.288}{0.006}(q_2 - q_1)$$

or

$$(0.06514 \text{ m}^2)G\phi' = 85.699q_1 - 48q_2$$
$$(0.22522 \text{ m}^2)G\phi' = -48q_1 + 137.17q_2$$

Solving yields $q_1 = 2.089 \times 10^{-3} \text{ m}^2 \, G\phi'$, $q_2 = 2.373 \times 10^{-3} \text{ m}^2 \, G\phi'$. The torque is

$$T = \sum 2A_i q_i = 6.705 \times 10^{-4} \text{ m}^4 \, G\phi'$$

Based on the allowable angle of twist, the allowable torque is

$$T = 6.705 \times 10^{-4} \text{ m}^4 \left(\frac{28 \times 10^9 N}{\text{m}^2}\right)\left(\frac{3\pi}{180 \cdot 4 \text{ m}}\right) = 245.8 kN \cdot m$$

Eliminating $G\phi'$ there results $q_1 = 3.116 \, T/\text{m}^2$ and $q_2 = 3.539 \, T/\text{m}^2$, so that with $\tau = q/t$, the maximum allowable torque based on the allowable stress is given by

$$60 \times 10^6 \left(\frac{N}{\text{m}^2}\right) \approx 1.4 \frac{3.539T}{0.012 \text{ m}^3}$$

or $T = 145.3 kN \cdot m$, which governs. ◆

EXAMPLE 3-8 For the cross section shown in Figure 3-40 determine the shear stress in each segment and the rate of twist ϕ'. Take $G = 4 \times 10^6$ psi and $T = 10^6$ in-lbf. Take $t_1 = 0.05$ in, $t_2 = 0.15$ in and $t_3 = 0.2$ in. Rework the problem if $t_1 = t_2 = t_3 = 0.0625$ in. Also rework the problem if the two internal webs of thickness t_2 are removed and $t_1 = t_3 = 0.0625$ in.

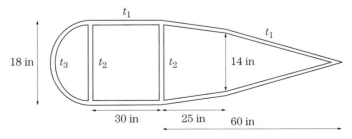

FIGURE 3-40 Multitube section.

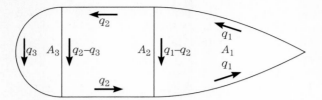

FIGURE 3-41 Assumed positive directions for shear flows.

Solution: For the first case we tabulate $A_1 = 645$ in^2, $A_2 = 540$ in^2 and $A_3 = 127$ in^2. Denoting the rate of twist by β and taking the positive shear flows in each of the tubes to be counterclockwise, as shown in Figure 3-41, the simultaneous equations for determining the shear flows are

$$2G\beta 645 = q_1 \frac{122}{0.05} + (q_1 - q_2)\frac{18}{0.15}$$

$$2G\beta 540 = q_2 \frac{60}{0.05} - (q_1 - q_2)\frac{18}{0.15} + (q_2 - q_3)\frac{18}{0.15}$$

$$2G\beta 127 = q_3 \frac{9\pi}{0.20} - (q_2 - q_3)\frac{18}{0.15}$$

or

$$2560 q_1 - 120 q_2 = 1290 G\beta$$
$$-120 q_1 + 1440 q_2 - 120 q_3 = 1080 G\beta$$
$$ -120 q_2 + 261 q_3 = 254 G\beta$$

with solution $q_1 = 0.547 G\beta$, $q_2 = 0.912 G\beta$ and $q_3 = 1.392 G\beta$. The torque is then given by

$$T = 2[645(0.547) + 540(0.912) + 127(1.392)]G\beta$$
$$= 2045 G\beta$$

so that $J = 2045$ in^4 and the rate of twist $\phi' = \beta = 10^6/(4 \times 10^6 (2045))$ rad/in = 1.22×10^{-4} rad/in. The values of the shear flows are given by

$$q_1 = 0.547 \frac{10^6}{2045} = 267 \text{ lbf/in}$$

$$q_2 = 0.912 \frac{10^6}{2045} = 446 \text{ lbf/in}$$

$$q_3 = 1.392 \frac{10^6}{2045} = 681 \text{ lbf/in}$$

The corresponding shear stresses are shown in Figure 3-42a.

If all the thicknesses are taken to be 0.0625 in, the equations for determining the shear flows can be written as

$$140 q_1 - 18 q_2 = 80.63 G\beta$$

3-5 TORSION OF MULTICELLED TUBES

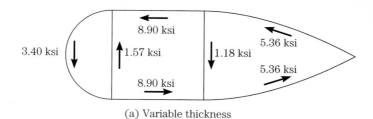

(a) Variable thickness

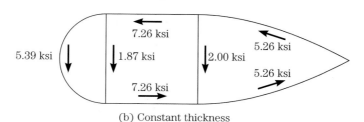

(b) Constant thickness

FIGURE 3-42 Comparison of shear stresses for variable and constant-thickness multitube sections.

and

$$-18q_1 + 96q_2 - 18q_3 = 67.5G\beta$$

and

$$-18q_2 + 46.3q_3 = 15.88G\beta$$

with solution $q_1 = 0.701G\beta$, $q_2 = 0.969G\beta$ and $q_3 = 0.720G\beta$. The torque is then given by

$$T = 2[645(0.701) + 540(0.969) + 127(0.720)]G\beta$$
$$= 2134G\beta$$

so that $J = 2134$ in^4 and the rate of twist $\phi' = \beta = 10^6/(4 \times 10^6(2134))$ rad/in $= 1.17 \times 10^{-4}$ rad/in. The values of the shear flows are given by

$$q_1 = 0.701 \frac{10^6}{2134} = 328 \text{ lbf/in}$$

$$q_2 = 0.969 \frac{10^6}{2134} = 454 \text{ lbf/in}$$

$$q_3 = 0.720 \frac{10^6}{2134} = 337 \text{ lbf/in}$$

The corresponding shear stresses are shown in Figure 3-42b.

If all the internal webs are removed, a single tube results with

$$A = (645 + 540 + 127) \text{ in}^2 = 1312 \text{ in}^2$$

$$J = \frac{4(1312)^2}{16(122 + 60 + 28.3)} \text{ in}^4 = 2046 \text{ in}^4$$

The uniform shear stress is given by

$$\tau = \frac{10^6}{2(1312)(0.0625)} \text{ psi} = 6.10 \text{ ksi}$$

with

$$\beta = \phi' = \frac{10^6}{4 \times 10^6 (2046)} = 1.22 \times 10^{-4} \text{ rad/in.} \qquad \blacklozenge$$

3-6 Warping Restraint in Thin-Walled Open Sections

In general the torsion of members other than those with solid and hollow circular sections involves deformations in the axial direction, that is, warping of the originally plane section. In particular there can be substantial warping in open thin-walled sections. A top view showing the warping in a wide flange or I-section, for example, is shown in Figure 3-43. The extent of the deformations is of course exaggerated. In the absence of any warping restraint, the torque is taken to be the usual Saint Venant torque $T_{sv} = J_0 G \phi'$ for a thin-walled open section. If the type of support at an end is such that this warping is prevented, there are additional stresses that result. When warping is restrained the total torque at any location along the beam can be represented as $T = T_{sv} + T_w$, where T_w is the part of the internal torque that arises due to the restraint against warping. This essentially assumes that the problem is linear with the deformations and stresses from the additional forces required to enforce the warping constraint added to those already present due to the usual theory. In this section the effect of warping for the case of an I-section with two axes of symmetry will be investigated in detail.

Equilibrium. Referring to Figure 3-44 showing the upper flange, it is clear that if warping is to be prevented at the end of the beam, a moment must be applied to the upper flange to rotate the end section back to its original position. An equal moment in

FIGURE 3-43 Warping deformations in an I-section.

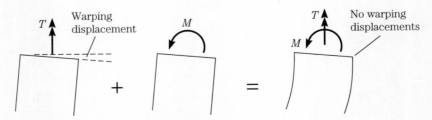

FIGURE 3-44 Moment to prevent warping.

3-6 WARPING RESTRAINT IN THIN-WALLED OPEN SECTIONS

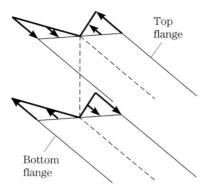

FIGURE 3-45 Bending stresses associated with the warping restraint.

the opposite direction would be applied to the lower flange. As shown in Figure 3-45, these moments will produce bending stresses in the flanges. These bending stresses and hence the moments must decay in some fashion to the known zero values at the unrestrained end of the beam. This decay in the moment is accompanied by shear forces. At any point along the beam the shear forces associated with the bending moments in the flanges will provide part of the internal torque necessary for equilibrium.

Consider then an FBD of a segment of the upper flange, as shown in Figure 3-46. Equilibrium in the x direction yields

$$t_f \frac{\partial \sigma_x}{\partial x} + \frac{\partial q}{\partial s} = 0 \qquad (3\text{-}12)$$

where t_f is the constant thickness of the upper flange. In addition, the relationship between the bending stress and the moment in the upper flange is

$$M = -\int_{A_{fl}} y \sigma_x \, dA \qquad (3\text{-}13)$$

As mentioned previously, the total torque T consists of the usual part due to the Saint Venant torsion, namely, $T_{sv} = J_0 G \phi'$, plus the part T_w, yet to be developed, due to the bending caused by the restraint against warping.

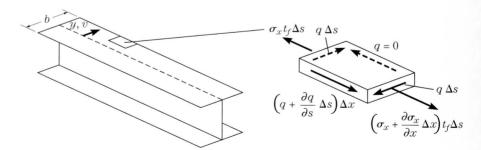

FIGURE 3-46 FBD of an element in bending.

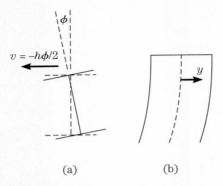

FIGURE 3-47 Torsional and bending deformations.

Deformations. The rotation of a cross section through the angle of twist ϕ results in a horizontal displacement of the upper flange, as shown in Figure 3-47a. As suggested in Figure 3-47b, for the case where the warping is restrained, there is a bending strain due to the associated bending curvature of the upper flange in the amount $\varepsilon_x = -yv''$. By virtue of the fact that in the upper flange, $v = -h\phi/2$, the bending strains can be expressed as

$$\varepsilon_x = -yv'' = \frac{yh\phi''}{2} \tag{3-14}$$

The sign of the bending strain in the lower flange is of course reversed.

Material Behavior. Assume that the beam is composed of a linearly elastic material according to

$$\sigma_x = E\varepsilon_x = \frac{Eyh\phi''}{2} \tag{3-15}$$

Combination. Combining Equations (3-12) and (3-15) leads to

$$\frac{\partial q}{\partial s} = -\frac{Et_f hy}{2} \phi'''$$

or, upon integration,

$$q(x, s) = -\frac{Et_f h}{2} \phi''' \int_0^s y \, ds$$

where the arbitrary function of integration has been discarded due to the fact that q must vanish at $s = 0$, the edge of the flange. With $y = b/2 - s$ there results

$$q(x, s) = -\frac{Et_f h \phi'''}{4} s(b - s) \tag{3-16}$$

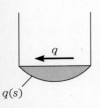

FIGURE 3-48 Shear flow associated with warping restraint.

having the parabolic distribution shown in Figure 3-48 across the thickness of the flange. Although q is a parabolic function of s at any location along the axis of the bar, that is, q

3-6 WARPING RESTRAINT IN THIN-WALLED OPEN SECTIONS

$\approx s(b - s)$, the manner in which the shear flow q depends on x will depend upon the dependent variable ϕ. From Equation (3-16) the maximum value of the shear flow at any position x is given by $q_{max} = E t_f h b^2 \phi'''/16$. The total force associated with this shear flow is

$$V(x) = -\int_0^b q(x, s)\, ds = \frac{E t_f h b^3 \phi'''}{24} \tag{3-17}$$

shown on an FBD of a portion of the beam shown in Figure 3-49, for both upper and lower flanges. The torque provided by the warping restraint is clearly

$$T_w = Vh = \frac{E t_f h^2 b^3 \phi'''}{24} = J_w E \phi''' \tag{3-18}$$

where $J_w = t_f h^2 b^3/24$ is the *warping constant* for the section.

Consulting Figure 3-49, the torque equation is written as

$$T + T_w = T_{sv}$$

which, with $T_{sv} = J_0 G \phi'$ and $T_w = J_w E \phi'''$, can be written as

$$T = J_0 G \phi' - J_w E \phi''' \tag{3-19}$$

This equation clearly states that the total torque T consists of the portion $T_{sv} = J_0 G \phi'$ due to the Saint Venant theory plus the portion $T_w = -J_w E \phi'''$ due to the warping restraint. The shear stresses associated with the Saint Venant portion of the torque transmitted are given by

$$\tau_{sv} = \frac{T_{sv} t}{J_0} \tag{3-20}$$

acting in both the flanges and the web. The shear stresses associated with the warping restraint are given by $\tau = q/t_f$ and can be represented as

$$\tau_w = \frac{T_w h s(b - s)}{4 J_w} \tag{3-21}$$

acting only in the flanges.

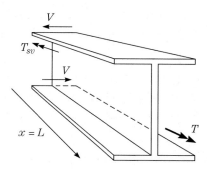

FIGURE 3-49 Free-body diagram for the torque equation.

The boundary conditions for the case where the warping is restrained at $x = 0$ are explained as follows. At the end where the warping is restrained, the moment at the support is such so as to produce $v = 0$ and $v' = 0$, that is, a clamped condition. With $v' = 0$ it follows from $v'(0) = h\phi'(0)/2$ that $\phi'(0) = 0$. At the end $x = L$ the bending stresses in the flanges must vanish, leading to

$$\sigma_x(L) = -\frac{Eyh\phi''(L)}{2} = 0 \Rightarrow \phi''(L) = 0$$

With $\theta = \phi'$ and $\alpha^2 = J_0 G/J_w E$, rewrite Equation (3-19) as

$$\theta'' - \alpha^2 \theta = -\frac{T}{J_w E} \tag{3-22}$$

with the boundary conditions as $\theta(0) = 0$ and $\theta'(L) = 0$. The solution can be expressed as

$$\theta(x) = C_1 \cosh \alpha x + C_2 \sinh \alpha x + \frac{T}{J_0 G}$$

Satisfying the two boundary conditions yields $C_1 = -T/J_0 G$ and $C_2 = -C_1 \sinh \alpha L/\cosh \alpha L$ with

$$\theta(x) = \frac{T}{J_0 G}\left[1 - \frac{\cosh \alpha(L-x)}{\cosh \alpha L}\right] \tag{3-23}$$

The portions of the total torque arising from the Saint Venant and warping restraint

$$T_{sv} = J_0 G \phi' = T\left[1 - \frac{\cosh \alpha(L-x)}{\cosh \alpha L}\right] \tag{3-24}$$

and

$$T_w = T - T_{sv} = \frac{\cosh \alpha(L-x)}{\cosh \alpha L} \tag{3-25}$$

respectively, are shown for a typical value of α in Figure 3-50a. Figure 3-50b, where again the region above the curve indicates T_{sv} and below the curve T_w, shows the relative values of the two torques for several values of α. This indicates that with increasing values of α the effect of the warping restraint decays more quickly with the distance away from the support. Note in all instances that at the support the entire torque is supplied by the warping restraint, and that except for very large values of α, a portion of the total torque is supplied by the warping at all points along the beam. This indicates that unless the external torque T is applied at $x = L$ in accordance with the split indicated by Figure 3-53b, the assumptions implicit in the theory are not met. As long as αL is reasonably large, the errors involved are small. In general the theory does a better job of predicting the effect of the warping restraint near the support than at locations near the unrestrained end. These issues will be further explored in the examples.

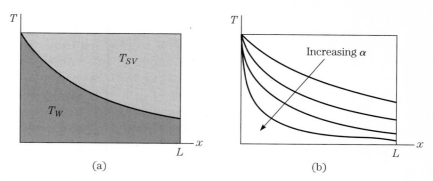

FIGURE 3-50 Saint Venant and warping torques.

From Equation (3-21), the maximum warping shear stress occurring in the flanges is

$$\tau_{max} = \frac{q_{max}}{t_f} = \frac{T_w h b^2}{16 J_w} = \frac{3 T_w}{2 b h t_f} \qquad (3\text{-}26)$$

The maximum value along the bar is at the support with $\tau_{max} = 3T/2bht_f$. At any point along the length of the bar the total shear stress consists of the sum of the Saint Venant shear stress $\tau = T_{sv} t/J_0$ in the flanges and web plus the warping shear stress $\tau = q/t$ in the flanges, that is,

$$\tau = \frac{T(1 - f(x))t}{J_0} + \frac{3Tf(x)}{2bht_f} \quad \text{(in the flanges)}$$

$$\tau = \frac{T(1 - f(x))t}{J_0} \quad \text{(in the web)}$$

where $f(x) = \cosh \alpha(L - x)/\cosh \alpha L$.

The bending stress $\sigma_x = Ehy\phi''/2$ can be expressed in the upper flange as

$$\sigma_x = \frac{Ehy\alpha}{2} \frac{\sinh \alpha(L - x)}{\cosh \alpha L} \frac{T}{J_0 G} \qquad (3\text{-}27)$$

with maximum values $|\sigma_{max}| = TEbh\alpha \tanh \alpha L/4 J_0 G$ at $x = 0$ and $y = \pm b/2$, that is at the edges of the flanges.

Recalling that $\phi' = \theta$, Equation (3-23) can be integrated to yield

$$\phi = \frac{T}{J_0 G}\left[x + \frac{\sinh \alpha(L - x) - \sinh \alpha L}{\alpha \cosh \alpha L} \right]$$

and is shown plotted in Figure (3-51) for several different values of αL. The maximum occurs at $x = L$ with

$$\phi(L) = \frac{TL}{J_0 G}\left[1 - \frac{\tanh \alpha L}{\alpha L} \right]$$

and is clearly reduced by the amount $\tanh \alpha L/\alpha L$ from the case for no restraint. Figure (3-51) clearly shows that as αL becomes larger, the warping effect is increasingly more localized near the restrained end.

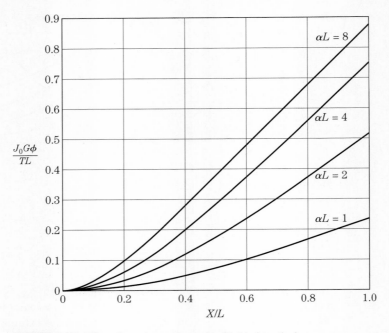

FIGURE 3-51 Effect of warping on the angle of twist along the bar.

EXAMPLE 3-9 The section shown in Figure 3-52 is subjected to a torque of 120 in-lbf. Analyze the stresses and deformations. Take $E = 10^7$ psi and $G = E/2(1 + \nu)$ with $\nu = 0.3$. With $J_0 = 7(0.1)^3$ in^4/3 = 2.333×10^{-3} in^4, the Saint Venant stress is given by

$$\tau_{sv} = \frac{120(0.1)}{2.333 \times 10^{-3}} \text{ psi} = 5.14 \text{ ksi}$$

With $J_w = (0.1)(9)(8)$ in^6/24 = 0.3 in^6 the parameter α is given by $\alpha = \sqrt{GJ_0/EJ_w} = [\sqrt{2.333 \times 10^{-3}/(2.6)(0.3)}]$/in = 0.0547/in with $\alpha L = 1.969$, tanh $\alpha L = 0.962$ and

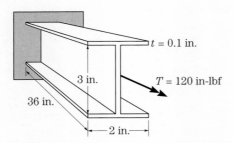

FIGURE 3-52 Torsion of a thin-walled beam with warping restraint.

cosh αL = 3.652. As given by Equation (3-27) the maximum bending stress at the support is

$$\sigma_x(x=0) = \frac{(120)(2.6)(3)(2)(0.0547)(0.962)}{4(2.333 \times 10^{-3})} \text{ psi} = 10.55 \text{ ksi}$$

At the midpoint along the length of the beam this maximum has decayed to

$$\sigma_x\left(x=\frac{L}{2}\right) = \frac{(120)(2.6)(3)(2)(0.0547)(0.962)}{4(2.333 \times 10^{-3})} \frac{1.152}{3.652} \text{ psi} = 3.33 \text{ ksi}$$

Using Equation (3-26) the maximum shear stress at the support occurs at the midpoints of the flanges with the values

$$\tau_{max} = \frac{(3)(120)}{2(2)(3)(0.1)} \text{ psi} = 0.3 \text{ ksi}$$

Since $\phi'(0) = 0$ there is no Saint Venant shear stress at the support and hence no shear stress in the web. At the midpoint $x = L/2$ where $f(L/2) = 0.418$ the shear stress at the middle of either flange is given by $\tau = \tau_{sv} + \tau_w$.

$$\tau = \tau_{sv} + \tau_w$$
$$= \frac{120(1 - 0.418)(0.1)}{2.333 \times 10^{-3}} + \frac{3(120)(0.418)}{2(3)(2)(0.1)}$$
$$= 2.15 \text{ ksi} + 0.13 \text{ ksi} = 2.28 \text{ ksi} \quad \text{(at the midpoint of the flanges)}$$

$$\tau = \tau_{sv} = \frac{120(1-0.418)(0.1)}{2.333 \times 10^{-3}} = 2.15 \text{ ksi} \quad \text{(in the web)}$$

At the end $x = L$ where $f(L) = 0.274$ the theory states that the shear stresses are given by

$$\tau = \tau_{sv} + \tau_w$$
$$= \frac{120(1-0.274)(0.1)}{2.333 \times 10^{-3}} + \frac{3(120)(0.274)}{2(3)(2)(0.1)}$$
$$= 3.73 \text{ ksi} + 0.08 \text{ ksi} = 3.81 \text{ ksi} \quad \text{(at the midpoint of the flanges)}$$

$$\tau = \tau_{sv} = \frac{120(1-0.274)(0.1)}{2.333 \times 10^{-3}} = 3.73 \text{ ksi} \quad \text{(in the web)}$$

showing that the shear stresses due to warping have not completely decayed to zero at the end $x = L$. The angle of twist at $x = L$ is given by

$$\phi(L) = \frac{2.6(120)(36)}{10^7(2.333 \times 10^{-3})}\left(1 - \frac{1}{3.652}\right) = 0.481(0.726) = 0.349 \text{ rad}$$

showing that the warping restraint at $x = 0$ has reduced the rotation at $x = L$ by about 27%.

3-7 Alternate Formulation for Warping Restraint in Thin-Walled Open Sections

The purpose of this section is to introduce an alternate derivation of the warping torque that is somewhat easier to apply in the case of sections that lack any symmetry. The same basic assumptions are made as for the previous development, namely, that the total torque equals the sum of the usual Saint Venant torque $T_{sv} = J_R G \phi'$ plus the torque arising from the restraint against warping T_w. The development of the warping torque will involve consideration and combination of the three basic ideas of equilibrium, deformation and material behavior.

Equilibrium. With t the thickness as a function of the coordinate s, the equilibrium equation is as derived above in section 3-6, namely,

$$t_f \frac{\partial \sigma_x}{\partial x} + \frac{\partial q}{\partial s} = 0 \tag{3-12}$$

Deformations. The reader is referred to section 3-4 for the development of the strain-displacement relation in an open section, namely, Equation (3-8)

$$\gamma = \frac{\partial u}{\partial s} + r\phi' \tag{3-8}$$

For an open section the extreme flexibility precludes the possibility of any substantial shear strain, allowing us to state that

$$\gamma \approx 0 = \frac{\partial u}{\partial s} + r\phi'$$

Integration yields

$$\begin{aligned} u &= u_0 - \phi' \int r \, ds \\ &= u_0 - \phi' \omega(s) \end{aligned} \tag{3-28}$$

where $\omega = \int r \, ds$ is defined as the sectorial area and is discussed in detail in Appendix C. For an open section, as shown in Figure 3-53, the sectorial area $\omega(s)$ is determined by integrating $d\omega = r \, ds$. The result depends on the point I chosen as the beginning point of the integration and the point P, called the *pole*, used to determine the perpendicular distance r to the directed line segment ds. For a given choice of I and P, the result of the integration can be expressed as

$$\omega(s) = \omega_0 + \int_I^s r \, ds \tag{3-29}$$

In any event, the constant of integration u_0 in Equation (3-28) represents a rigid body motion and can be set equal to zero without loss in generality so that

$$u(x, s) = -\phi'(x)\omega(s) \tag{3-30}$$

3-7 ALTERNATE FORMULATION FOR WARPING RESTRAINT IN THIN-WALLED OPEN SECTIONS

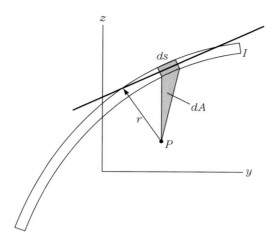

FIGURE 3-53 Definition of sectorial area.

The other strain-displacement relation is for the extensional strain associated with the warping, namely,

$$\varepsilon_x = \frac{\partial u}{\partial x} \tag{3-31}$$

Material Behavior. Assume a linearly elastic one-dimensional solid for which

$$\sigma_x = E\varepsilon_x \tag{3-32}$$

Combining Equations (3-30), (3-31) and (3-32) leads to

$$\sigma_x = -E\phi''\omega \tag{3-33}$$

Note that since only the torque T is being transmitted along the axis of the open section, any internal force resultant associated with the stress σ_x must vanish. In particular σ_x must be such that the equilibrium equations

$$P = \int_A \sigma_x \, dA = 0, \quad M_z = -\int_A y\sigma_x \, dA = 0 \quad \text{and} \quad M_y = \int_A z\sigma_x \, dA = 0$$

are all satisfied. Using Equation (3-31), it can be seen that these equations imply that

$$\int_A \omega \, dA = 0, \quad \int_A y\omega \, dA = 0 \quad \text{and} \quad \int_A z\omega \, dA = 0$$

The points P and I and the constant ω_0 must be chosen so as to satisfy these three equations. When these choices are made, the sectorial properties are referred to as *principal sectorial properties*. As indicated in Appendix C, once the initial point I has been chosen, these conditions lead to three equations for determining the location of the pole P and the value of the constant ω_0.

Substituting Equation (3-33) into (3-12) yields

$$\frac{\partial q}{\partial s} = Et\phi'''\omega \qquad (3\text{-}34)$$

which can be integrated to determine the shear flow. The torque arising from this shear flow is then given by

$$T_\omega = \int qr\, ds = \int q\, d\omega$$

Integration by parts yields

$$T_\omega = (q\omega)\bigg|_{s_0}^{s_f} - \int \omega \frac{\partial q}{\partial s}\, ds$$

Due to the fact that the shear flow vanishes at the beginning and ending free edges for an open section, the integrated term vanishes. Substituting Equation (3-34) then yields

$$T_\omega = -E\phi''' \int \omega^2 t\, ds = -E\phi''' \int_{\text{Area}} \omega^2\, dA$$

Defining $\int \omega^2\, dA = J_\omega$, there results

$$T_\omega = -EJ_\omega \phi''' \qquad (3\text{-}35)$$

As will be seen presently, J_ω will coincide with the torsional constant developed in the section above.

Thus the governing equation for the torque is

$$T = T_{sv} + T_\omega = J_R G\phi' - J_\omega E\phi'''$$

coinciding with the equation developed in the section above. The shear flows associated with the warping restraint are obtained by integrating Equation (3-34), namely,

$$q = \int_0^s Et\phi'''\omega\, ds = E\phi''' \int_0^s \omega\, dA \qquad (3\text{-}36)$$

which will turn out to coincide with result of the development in the section above. The reader is referred to Appendix C for the equations pertaining to the torsion of an *I*-section using the ideas associated with the sectorial area.

EXAMPLE 3-10 Consider again the problem of the torsion and associated warping restraint for the doubly symmetric section of Example 3-7, which is shown in Figure 3-54. Take $E = 10^7$ psi and $G = E/2(1 + \nu)$ with $\nu = 0.3$.

Solution: It is left to the reader to show that the pole should be taken at the centroid of the section and that the constant ω_0 is zero. As indicated in Figure 3-55a, the integrations are begun at the pole P. The corresponding results for the sectorial area ω are indicated in Figure 3-55b. According to Equation (3-33) the normal stresses are given by $\sigma_x = -E\phi''\omega$ so that at point A, for instance, where $\omega = hb/4$,

$$\sigma_x = -\frac{E\phi'' hb}{4}$$

which coincides with Equation (3-15) when $y = -b/2$, showing that the two formulations produce the same results for the normal stresses. Equation (3-36) can be used

3-7 ALTERNATE FORMULATION FOR WARPING RESTRAINT IN THIN-WALLED OPEN SECTIONS

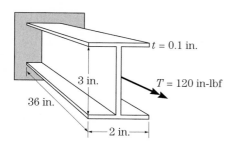

FIGURE 3-54 Torsion of a thin-walled beam with warping restraint.

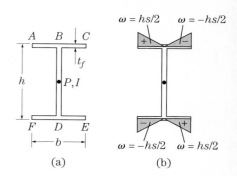

FIGURE 3-55 Warping function.

to determine the expression for the shear stresses in the upper flange. Measuring s to the left from point C in the upper flange (where $q = 0$), the sectorial area is $\omega = -h(b/2 - s)/2$, resulting in

$$q = E\phi''' \int_0^s \omega \, dA = E\phi''' \int_0^s \left[-\frac{h}{2}\left(\frac{b}{2} - s\right) \right] t_f \, ds = -\frac{E t_f h \phi'''}{4} s(b - s)$$

again consistent with the result given in Equation (3-16). Last, the expression for the sectorial second moment given by $J_w = \int \omega^2 \, dA$ becomes

$$J_w = \int \omega^2 \, dA = 4 \int_0^{b/2} \left(\frac{hs}{2}\right)^2 t_f \, ds = \frac{b^3 h^2 t_f}{24}$$

as was developed using the other formulation. ◆

The utility of the alternate formulation for the effect of the torsional restraint is in terms of its use in situations where the cross section has no plane of symmetry or only one plane of symmetry.

EXAMPLE 3-11 Consider the problem of determining the warping function and related quantities for a channel section. Let the base be b, the height h, the web thickness t_w and the flange thickness t_f. For the purpose of establishing the warping function ω, locate I and P as shown in Figure 3-56. The data for determining the warping function are shown in

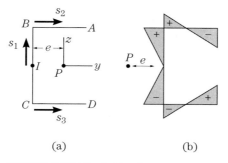

FIGURE 3-56 Warping function.

TABLE 3-4 Warping Function Data

Segment	ω	y	z	
CB	$\omega_0 - es_1$	$-e$	s_1	$-h/2 \leq s_1 \leq h/2$
BA	$\omega_0 - h(e + s_2)/2$	$-e + s_2$	$h/2$	$0 \leq s_2 \leq b$
CD	$\omega_0 + h(e + s_3)/2$	$-e + s_3$	$-h/2$	$0 \leq s_3 \leq b$

Table 3-4. To determine the location of the pole P and the value of the constant ω_0, evaluate the integrals:

$$\int \omega \, dA = 0 = \int_{-h/2}^{h/2} (\omega_0 - es_1) t_w \, ds_1 + \int_0^b \frac{\omega_0 - h(e + s_2)}{2} t_f \, ds_2$$
$$+ \int_0^b + \frac{\omega_0 + h(e + s_3)}{2} t_f \, ds_3$$

from which $\omega_0 = 0$. Then, with $\omega_0 = 0$, the student should show that $\int \omega y \, dA = 0$. Finally

$$\int \omega z \, dA = \int_{-h/2}^{h/2} (-es_1) s_1 t_w \, ds_1 + \int_0^b \frac{-h(e + s_2)}{2} t_f \left(\frac{h}{2}\right) ds_2$$
$$+ \int_0^b + \frac{h(e + s_3)}{2} t_f \left[-\left(\frac{h}{2}\right)\right] ds_3 = 0$$

Evaluating this integral and simplifying yields $e = -3b^2 t_f/(ht_w + 6bt_f)$, showing that the pole actually lies to the left of the flange with the location of P and the resulting warping function appearing in Figure 3-56b. Finally, then, with $e = 3b^2 t_f/(ht_w + 6bt_f)$ the equations that define the warping function are shown in Table 3-5. The torsional constant $J_\omega = \int \omega^2 \, dA$ then becomes

$$J_\omega = \int_{-h/2}^{h/2} t_w (es)^2 \, ds + 2t_f \int_0^b \left[\frac{h(e - s)}{2}\right]^2 ds$$
$$= \frac{e^2 t_w h^3}{12} + \left(\frac{t_f h^2}{6}\right)[(e + b)^3 - e^3]$$

After substituting for e and simplifying, there results

$$J_\omega = \frac{t_f h^2 b^3}{12} \frac{2ht_w + 3bt_f}{ht_w + 6bt_f}$$

Note that the location of the pole P is precisely the location for the so-called shear center, which will be discussed in Chapter 4.

TABLE 3-5 Warping Function Data

Segment	ω	
CB	es_1	$-h/2 \leq s_1 \leq h/2$
BA	$h(e - s_2)/2$	$0 \leq s_2 \leq b$
CD	$-h(e - s_3)/2$	$0 \leq s_3 \leq b$

In any event, the torque transmitted by the transverse shear stresses associated with the warping restraint is

$$T_\omega = -EJ_\omega \phi'''$$

with the governing equation for the torque given as

$$T = T_{sv} + T_\omega = J_0 G\phi' - J_\omega E\phi'''$$

where $J_0 = \Sigma\, b_i t_i^3/3$ and $J_\omega = t_f h^2 b^3 (2ht_w + 3bt_f)/[12(ht_w + 6bt_f)]$.

The shear flow in the upper flange is determined by evaluating Equation (3-36), namely,

$$q = \int Et_f \phi''' \omega\, ds$$

Beginning at the outer edge of the upper flange where $q = 0$ and using the fact that $s_2 + s = b$,

$$q(s) = Et_f \phi''' \int_0^s \frac{h(e - b + s)}{2}\, ds = \frac{Et_f \phi''' h s}{4}[s - 2(b - e)]$$

which is shown in Figure 3-52. In the web, the shear flow is

$$q(s) = \frac{Et_f \phi''' h b}{4}[b - 2(b - e)] + Et_f \phi''' \int_0^s e\left(\frac{h}{2} - s\right) ds$$

$$= \frac{Et_f \phi''' h b}{4}[b - 2(b - e)] + Et_f \phi''' \frac{e}{2} s(h - s)$$

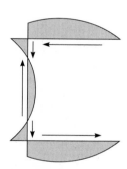

FIGURE 3-57 Warping shear flows in a channel section.

which is shown in Figure 3-57. The student should show that there is no resultant force associated with the shear flow in the web and that the resultant forces associated with the shear flow in the flanges give rise to the part of the torque produced by the warping restraint.

3-8 Summary

The problem of the pure torsion of a linearly isotropic, homogeneous, straight solid prismatic bar admits of the mathematical formulation

$$\nabla^2 \Phi = -2G\theta \quad \text{in } D$$
$$\Phi = 0 \quad \text{on } B$$

where Φ is the Prandtl stress function. Solutions for the stresses and displacements are exact within the theory, which assumes that warping of the cross section is free to occur.

Except in relatively few instances, however, it is generally not possible to obtain exact solutions and thus approximate approaches must be used to estimate stresses and displacements. Results from the finite-element method applied to the problem of the torsion of a rectangular section were presented and discussed in this regard. For thin-walled sections the membrane analogy was used to assist in visualization of the appropriate Prandtl stress function to be used for a single thin rectangular section, giving rise to approximate stresses and displacements. This solution was then extended to collections of interconnected thin-walled segments commonly used in applications. These same ideas were also used to assist in developing approximate analyses for single and multiple tubes subjected to pure torsional loadings.

To complete the treatment of torsion, the problem associated with the restraint of warping of thin-walled open sections was addressed. There it was seen that substantial normal stresses can arise in the course of enforcing the warping restraint. The normal stresses associated with the warping restraint are localized at the boundary where the restraint is enforced.

REFERENCES

1. Timoshenko, S. P., and Goodier, J. N. *Theory of Elasticity*, 2nd ed. McGraw-Hill, New York, 1951.
2. Boresi, A. P., and Lynn, P. P. *Elasticity in Engineering Mechanics*. Prentice-Hall, Englewood Cliffs, N.J., 1974.
3. Timoshenko, S. P. *Strength of Materials: Part I, Elementary Theory and Problems & Part II, Advanced Theory and Problems*, 3rd ed. Krieger, New York, 1955.
4. Oden, J. T., and Ripperger, E. A. *Mechanics of Elastic Structures*. McGraw-Hill, New York, 1981.
5. Wang, C. T. *Applied Elasticity*. McGraw-Hill, New York, 1953.
6. Roark, R. J., and Young, W. C. *Formulas for Stress and Strain*, 5th ed. McGraw-Hill, New York, 1975.
7. Bickford, W. B. *A First Course in the Finite Element Method*, 2nd ed. Irwin, Burr Ridge, Ill., 1994.

EXERCISES

Section 3-2

1. A linearly elastic shaft with a circular cross section of radius a is to transmit a torque with a maximum torsional shear stress τ_0. What would be outside radius of a hollow shaft having its outside radius twice the inside radius and with the same maximum stress τ_0?

2. A linearly elastic shaft with a circular cross section of radius a is to transmit a torque with a maximum torsional shear stress τ_0. If the maximum torsional stress is to be τ_0, what would be the inner and outer radii of a corresponding hollow shaft with average radius R and a thickness given by $t/R = 0.2$? What would be the ratio of the rates of twist?

3. Solve the torsion problem for a solid circular shaft, that is,
$$\nabla^2 \phi(r) + 2G\theta = 0 \qquad 0 \leq r \leq a$$
$$\phi(a) = 0$$
and hence show that all the equations for stresses and deformations coincide with the results from the elementary theory.

Section 3-3

4. A linearly elastic shaft with a circular cross section of radius a is to transmit a torque with a maximum torsional shear stress τ_0. What would be the maximum torsional shear stress in a solid square cross-section shaft having an equal area? What would be the ratio of the rates of twist?

5. A linearly elastic shaft with a circular cross section of radius a is to transmit a torque with a maximum torsional shear stress τ_0. What would be the maximum torsional shear stress in a solid square cross-section shaft with $B/A = 2$ and having an equal area? What would be the ratio of the rates of twist?

6. A linearly elastic shaft with a circular cross section of radius a is to transmit a torque with a maximum torsional shear stress τ_0. What would be the maximum torsional shear stress in a solid elliptical cross-section shaft with $B/A = 2$ and having an equal area? What would be the ratio of the rates of twist?

7. A geometry for which there is an exact solution to the torsion problem is the equilateral triangle shown below. Take the origin of the coordinates to be at the centroid of the triangle, and

EXERCISES

show by forming the stress function according to $\phi = C\ell_1\ell_2\ell_3$, where $\ell_1 = 0$, $\ell_2 = 0$ and $\ell_3 = 0$ are the equations for the sides of the triangle, that it is possible to choose C so as to satisfy $\nabla^2\phi + 2G\theta = 0$. Calculate the location and magnitude of the maximum shear stress and the torsional constant J.

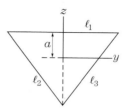

8. The thin ($a \ll b$) rectangular section is loaded as shown below. On the basis of the theory for thin-walled sections, what is the maximum torsional shear stress that results? Where does it occur?

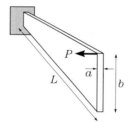

9. Assuming that the shape of the ϕ function is locally parabolic across the thickness, develop approximate expressions for the shear stress as a function of s, the maximum shear stress and the torsional constant J for the teardrop-shaped member shown below.

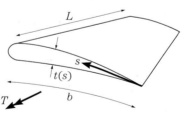

10. Apply the results of problem 9 to the case where $t(s) = t_0(s/b)$, that is, where the thickness varies linearly with position.

11. Develop approximate expressions for the maximum shear stress and torsional constant for the section shown below. All dimensions are in millimeters. If the allowable torsional stress is 70 MPa, what is the allowable torque T? What is the corresponding rate of twist?

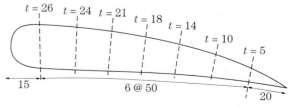

12. Repeat problem 11 by attempting to represent the thickness as a linear function using a least-squares fit. How do the results compare with those from problem 11?

13. Determine J for the section shown. Take $a = 20$ mm, $b = 30$ mm, $t_1 = t_2 = 4$ mm and $R = 30$ mm. If $\tau_{all} = 50$ MPa, what is the allowable torque? Neglect stress concentrations.

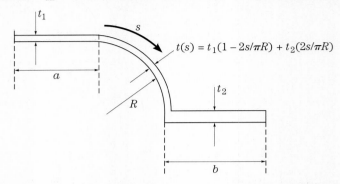

14. Repeat problem 13 for $t_1 = 4$ mm, $t_2 = 8$ mm and the rest of the data unchanged.

Section 3-4

15. Two members, one a hollow circular tube and the other a tube with a slit, are subjected to a torque T, as shown below. In terms of the average radius R and the thickness t, determine the ratios of the maximum shear stress and the rate of twist.

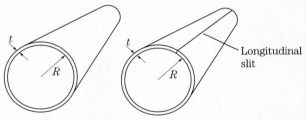

16. Two tubes, one a circle and the other a square, are subjected to a torque T. If the thickness t and the total area of the region occupied by the material are the same, compare the maximum shear stresses and the rates of twist. Neglect stress concentrations and assume $t \ll b$.

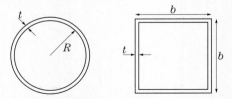

17. Reconsider problem 16 if the thickness t and the interior cavity areas of the two tubes are to be the same.

18. Compare the maximum stresses in two tubes with equal thicknesses and material areas, one of which is circular in cross section and the other elliptic with $b/a = 2$, as shown below. Assume $t \ll b$.

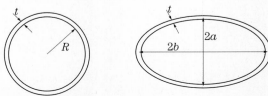

EXERCISES

19. Reconsider problem 18 if the thickness t and the interior cavity areas of the two tubes are to be the same.

20. Structures that transmit torques can be configured so that in addition to a tube there are one or more attached "fins" that are integral parts of the structure, as shown in the figure below.

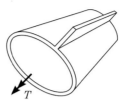

By appealing to the membrane analogy and the fact that the rate of twist of each of the parts of the structure must be the same, show that the torque T can be represented as

$$T = T_{\text{tube}} + T_{\text{fins}} = G\theta \left(\frac{4A_0^2}{\oint \frac{ds}{t}} + \sum J_{\text{fin}} \right) = J_T G\theta$$

where for a constant thickness fin, $J_{\text{fin}} = bt^3/3$. Also show that

$$\tau_{\text{tube}} = \frac{J_{\text{tube}}}{J_T} \frac{T}{2A_0 t}$$

and

$$\tau_{\text{fin}} = \frac{J_{\text{fin}}}{J_T} \frac{Tt_{\max}}{J_{\text{fin}}} = \frac{Tt_{\max}}{J_T}$$

21. A circular tube has four radial fins, as shown below. Determine the maximum allowable torque if $\tau_{\text{all}} = 10$ ksi. What is the corresponding rate of twist? Take $R = 3$ in, $t_t = 0.04$ in, $t_1 = t_2 = 0.04$ in, $a = b = 2$ in and $G = 4 \times 10^6$ psi. Neglect stress concentrations and assume $t \ll R$.

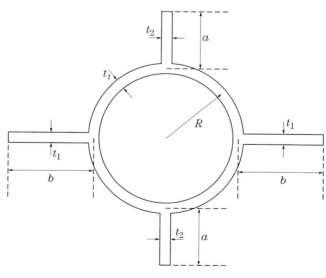

22. Repeat problem 21 with $\tau_{\text{all}} = 10$ ksi, $R = 3$ in, $t_t = 0.03$ in, $t_1 = 0.04$ in, $t_2 = 0.08$ in, $a = 2$ in, $b = 4$ in and $G = 4 \times 10^6$ psi.

23. A circular tube has four radial fins, as shown below. Determine the maximum allowable torque if $\tau_{all} = 10$ ksi. What is the corresponding rate of twist? Take $R = 3$ in, $t_t = 0.04$ in, $t_1 = t_2 = 0.04$ in, $a = b = 2$ in and $G = 4 \times 10^6$ psi. Neglect stress concentrations and assume $t \ll R$.

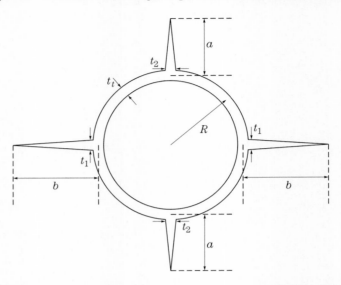

24. Repeat problem 23 with $\tau_{all} = 10$ ksi, $R = 3$ in, $t_t = 0.04$ in, $t_1 = 0.04$ in, $t_2 = 0.08$ in, $a = 2$ in, $b = 4$ in and $G = 4 \times 10^6$ psi.

Section 3-5

25. Determine the shear flows, the maximum torsional stress, the rate of twist and the torsional constant for the section shown below. Take $t_1 = t_2 = t$ and $t_3 = 1.5t$. Repeat the problem assuming the vertical web is removed. Compare the results. Neglect stress concentrations and assume $t \ll R$.

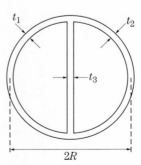

26. Repeat problem 25 for $t_1 = t$, $t_2 = 1.5t$ and $t_3 = 2t$.

27. Determine the shear flows, the maximum torsional stress, the rate of twist and the torsional constant for the section shown. Take $t_1 = t_2 = t$, $t_3 = 1.5t$, $b_1 = h = b$ and $b_2 = 2b$. Repeat the problem assuming the vertical web is removed. Compare the results. Neglect stress concentrations and assume $t \ll b$.

EXERCISES

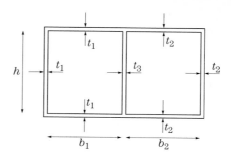

28. Repeat problem 27 if $t_1 = t_2 = t_3 = t$ and $b_1 = b_2 = h$.

29. Repeat problem 27 if $t_1 = t$, $t_2 = 2t$, $t_3 = t$ and $b_1 = b_2 = h$.

30. Determine the shear flows, the maximum torsional stress, the rate of twist and the torsional constant for the section shown below. Take $t_1 = t_2 = t_3 = t_4 = t$ and $b = 2h$. Rework the problem assuming the web is removed. Neglect stress concentrations and assume $t_1, t_2, t_3, t_4 \ll h$.

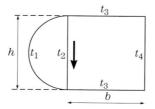

31. Repeat problem 30 for $t_1 = t_2 = 2t$ and $h = b$.

32. Determine the shear flows, the maximum torsional stress, the rate of twist and the torsional constant for the section shown below. Take $t_1 = t$, $t_2 = 2t$, $t_3 = t_4 = t$ and $B = H$. Repeat the problem assuming the webs are removed. Compare the results. Neglect stress concentrations and assume $t \ll B$.

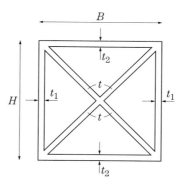

33. Repeat problem 32 for $t_1 = t_2 = 2t$, $t_3 = t_4 = t$ and $B = H$.

34. Determine the shear flows and the torsional constant for the section shown below. Take $t_1 = t_2 = t_3 = t$, $t_4 = t_5 = 2t$ and $b_1 = b_2 = h = b$. If $b = 10$ in, $\tau_{\text{all}} = 8$ ksi and $t = 0.1$ in, what torque can be transmitted? Neglect stress concentrations and assume $t \ll b$. Repeat assuming all the internal webs are removed.

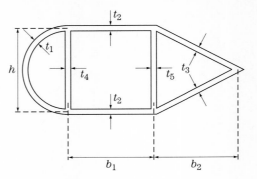

35. Repeat problem 34 if $t_1 = t_2 = 2t$, $t_3 = t_4 = t_5 = t$ and $b_1 = b_2 = 2h$.
36. Determine the shear flows and the torsional constant for the section shown below. Take $t_1 = t_6 = 2t$, $t_2 = t_3 = t_4 = t_5 = t$, $b_1 = b$, $b_2 = 2b$, and $h_1 = h_2 = b$. If $b = 200$ mm, $\tau_{all} = 60$ MPa and $t = 8$ mm, what torque can be transmitted? Neglect stress concentrations and assume $t \ll b$.

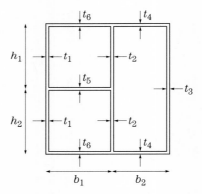

37. Repeat problem 36 assuming the horizontal interior web is removed. Compare the results. Rework assuming the vertical web is also removed.

Sections 3-6 and 3-7

38. Determine ω_0 and the location of the pole for generating the warping function for the section shown below. Assume $t_f \ll b$ and $t_w \ll b$.

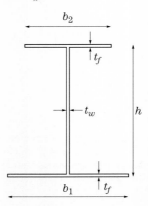

EXERCISES

39. Determine the warping constant $J_\omega = \int \omega^2 \, dA$ for the section shown in problem 38.

40. A beam whose cross section is of the type given in problem 39 is acted upon by a torque T and is restrained against warping at $x = 0$, as shown in the figure below. For $T = 250$ in-lbf, $t_w = t_f = 0.075$, $b_1 = h = 5.0$ in, $b_2 = 4.0$ in and the length $L = 80$ in, determine (a) the maximum normal stress σ_x at the restrained end, (b) the maximum warping shear stress at the restrained end, (c) the Saint Venant shear stress and (d) the angle of rotation at the loaded end. Take $E = 3 \times 10^7$ psi and $\nu = 0.3$.

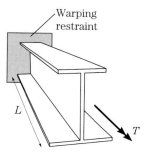

41. Show that the location of the pole P is located a distance $e = 3b^2 t_f / (ht_w + 6bt_f)$ to the left of the web for the C-section shown below. Assume $t_f \ll b$ and $t_w \ll b$.

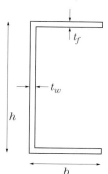

42. Show that the warping constant $J_\omega = \int \omega^2 \, dA$ for the C-section shown in problem 41 is $J_\omega = (t_f h^2 b^3)(2ht_f + 3bt_f)/(12ht_f + 6bt_f)$.

43. A beam whose cross section is of the type given in problem 41 is acted upon by a torque T and is restrained against warping at $x = 0$. For $T = 200$ in-lbf, $t_w = t_f = 0.0625$, $b = 7.0$ in, $h = 10$ in and the length $L = 120$ in, determine (a) the maximum normal stress σ_x at the restrained end, (b) the maximum warping shear stress at the restrained end, (c) the Saint Venant shear stress and (d) the angle of rotation at the loaded end. Take $E = 3 \times 10^7$ psi and $\nu = 0.3$.

44. Determine ω_0 and the location of the pole for generating the warping function for the section shown below. Take $b_1 = b_2 = h$ and assume a uniform thickness $t \ll b$.

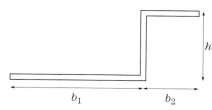

45. Determine the warping constant $J_\omega = \int \omega^2 \, dA$ for the section shown in problem 44.

46. A beam whose cross section is of the type given in problem 44 is acted upon by a torque T and is restrained against warping at $x = 0$. For $T = 240$ in-lbf, $t = 0.0625$, $b_1 = b_2 = 6.0$ in, $h = 6.0$ in and the length $L = 90$ in, determine (a) the maximum normal stress σ_x at the restrained end, (b) the maximum warping shear stress at the restrained end, (c) the Saint Venant shear stress and (d) the angle of rotation at the loaded end. Take $E = 3 \times 10^7$ psi and $\nu = 0$.

47. A WF section restrained against warping at both ends and twist at the left end transmits a torque T, as shown below. Determine (a) the maximum normal stress at $x = 0$ and $x = L$ and (b) the angle of twist at $x = L$. With $\alpha = \sqrt{JG/EJ_\omega}$, discuss any approximations that can be made if αL is large. Refer to problem 53 for the proper boundary conditions.

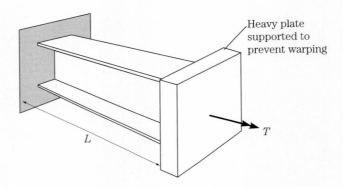

48. A WF section restrained against warping and twist at one end is subjected to a uniformly distributed torque t_0, as shown below. Determine (a) the maximum normal stress at $x = 0$ and (b) the angle of twist at $x = L$. With $\alpha = \sqrt{JG/EJ_\omega}$, discuss any approximations that can be made if αL is large. Refer to problem 53 for the proper boundary conditions.

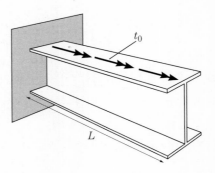

49. A WF section is restrained against warping and twist at both ends and is subjected to a torque T_0 at the midpoint, as shown below. Determine (a) the maximum normal stress at $x = 0$, $x = L/2$ and $x = L$ and (b) the angle of twist at $x = L/2$. With $\alpha = \sqrt{JG/EJ_\omega}$, discuss any approximations that can be made if αL is large. Refer to problem 53 for the proper boundary conditions.

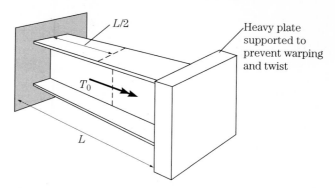

Exercises for Variational Formulations

50. Show that with ϕ taken to be zero on the boundary, the Euler-Lagrange equation for the functional

$$F = \int\int \left[\frac{1}{2}\left\{\left(\frac{\partial\phi}{\partial y}\right)^2 + \left(\frac{\partial\phi}{\partial z}\right)^2\right\} - 2G\theta\phi\right] dA$$

is the governing equation for the Prandtl stress function, namely, $\nabla^2\phi + 2G\theta = 0$.

51. For the rectangular region shown below, take $\phi = C_1(y^2 - a^2)(z^2 - a^2)$ and use the Ritz method in connection with the functional of problem 50 to determine C_1. Use the relation $T = 2\int\int \phi\, dA$ to express C_1 in terms of T and hence investigate the maximum shear stress in the section. For $a = b$, compare your results for T and the maximum shear stress with the exact solution.

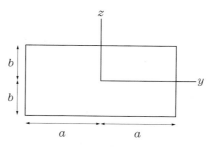

52. Repeat problem 51 by taking $\phi = C_2 \cos(\pi y/2a)\cos(\pi z/2b)$.

53. For the problem of the torsion of a thin-walled section with warping restraint, the potential energy can be expressed as

$$V = \int_0^L \left\{\frac{EJ_\omega}{2}\phi''^2 + \frac{GJ_0}{2}\phi'^2 - t(x)\phi\right\} dx$$

Using the principle of stationary potential energy, show that the governing differential and boundary conditions are

$$EJ_\omega \phi'''' - GJ_0\phi'' - t(x) = 0 \qquad 0 \le x \le L$$

$$EJ_\omega \phi''\, \delta\phi' \big|_0^L = 0$$

$$(JG\phi' - EJ_\omega \phi''')\, \delta\phi \big|_0^L = 0$$

Interpret the boundary conditions.

4 Transverse Loading of Unsymmetrical Beams

4-1 Introduction
4-2 Unsymmetrical Bending
    4-2-1 Stresses in Unsymmetrical Bending
    4-2-2 Displacements in Unsymmetrical Bending
4-3 Shear Stresses in Thin-Walled Open Sections
4-4 Shear Center for Thin-Walled Open Sections
4-5 Bending of a Single Thin-Walled Tube
4-6 Thin-Walled Multitube Bending
4-7 Combined Bending and Torsion
4-8 Summary
    References
    Exercises

4-1 Introduction

The elementary theory of the bending of beams is based on the following assumptions (refer to Figure 4-1):

1. The axis of the beam is straight with area $A(x)$ and second moment $I_{yy}(x)$ that are constant or that vary slowly with x with no initial twist of the cross section.
2. The cross section is such that the xz-plane is a plane of symmetry.
3. The loading consists of concentrated forces P or distributed forces q lying in the xz-plane and parallel to the z-axis, and of moments M parallel to the y axis.
4. The transverse displacements $w(x)$ in the z direction are such that the rotations or slopes $w'(x)$ are small.
5. The beam is composed of a homogeneous linearly elastic material.

Based on these assumptions, the equations that can be developed are

$$\sigma = \frac{M_y z}{I_{yy}} \tag{4-1}$$

for the bending stress and

$$EI_{yy}\frac{d^2w}{dx^2} = -M_y \tag{4-2}$$

4-1 INTRODUCTION

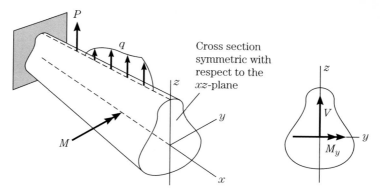

FIGURE 4-1 Symmetric bending.

for determining the transverse displacements $w(x)$. M_y represents the moment transmitted and I_{yy} the second moment of the area about the y axis at the x location in question. Further, the shear stresses usually associated with the elementary theory are computed according to

$$\tau_{\text{avg}}(z) = \frac{VQ(z)}{I_{yy}b(z)} \tag{4-3}$$

where V is the shear force transmitted at the section, $Q(z)$ is the first moment of the area above the position z at which the average shear stress τ_{avg} is being computed and $b(z)$ is the thickness at that location.

◆ EXAMPLE 4-1   A cantilever beam with a T-section shape is loaded as shown in Figure 4-2. Determine the bending stresses, the transverse shear stresses and the transverse displacement.

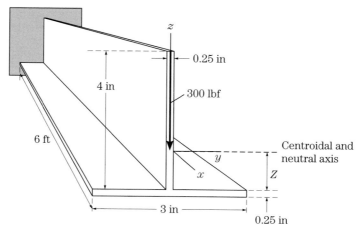

FIGURE 4-2 T-section cantilever beam.

Solution: The cross section data required are as follows:

$$Z = \frac{3(0.25)(0) + 4(0.25)(2.0)}{7(0.25)} = 1.143 \text{ in}$$

and

$$I_{yy} = [3(0.25)(1.143)^2 + \frac{0.25(4)^3}{12} + 4(0.25)(0.857)^2] \text{ in}^4$$

$$= 3.048 \text{ in}^4$$

The bending stress distribution, shown in Figure 4-4a, is largest at the top of the cross section. The maximum value occurs at the support with

$$\sigma_{\max} = \sigma_x(x=0) = \frac{300(72)(4 - 1.143)}{3.048} \text{ psi} = 20.2 \text{ ksi}$$

For the shear-stress calculations, consult Figure 4-3. The integration is carried out beginning at point A, proceeding to point B and then to points C and D.

In leg AB: In terms of the coordinate s_1 shown in Figure 4-3a, the area $A = s_1 \cdot t$ and $\bar{z} = 2.857 - s_1/2$ so that with $Q = A\bar{z}$

$$\tau_{\text{avg}} = \frac{300(s_1 t)(2.857 - 0.5 s_1)}{3.048 t} = 98.4 s_1 (2.857 - 0.5 s_1)$$

which is a parabolic distribution, as shown in Figure 4-4b. The maximum of 402 psi occurs at the neutral axis. At the intersection with the horizontal leg the value of the shear stress is 338 psi, as shown in Figure 4-4b.

In legs BC and BD: From symmetry the shear stresses in legs BC and BD will be the same. Writing the first moment as $Q = \Sigma A_i \bar{z}_i$ there results

$$Q = 4(t)(1.857) + 2 s_2 t(-1.143)$$

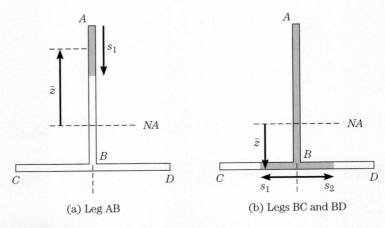

(a) Leg AB   (b) Legs BC and BD

FIGURE 4-3 Shear-stress calculations.

4-2 UNSYMMETRICAL BENDING

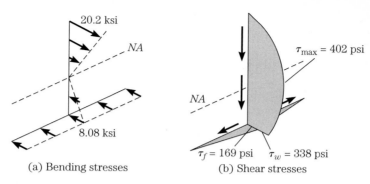

(a) Bending stresses
(b) Shear stresses

FIGURE 4-4 Bending and shear stresses.

with

$$\tau_{\text{avg}} = \frac{300[4(t)(1.857) + 2s_2 t(-1.143)]}{3.048(2t)} = 49.2(3.428 - 2.286 s_2)$$

which is also shown in Figure 4-4b. The maximum value of 169 psi occurs at the junction of the vertical and horizontal portions. Note that each of these values is half of the flow from the vertical leg into the junction. ◆

If any of the above assumptions regarding cross-sectional properties and loadings are not satisfied, the stresses and displacements must be determined from a more general theory based on relaxing certain of these assumptions. In this chapter a theory will be developed that is capable of predicting stresses and displacements in straight beams with cross sections that have no symmetry and that are subjected to external forces resulting in shears and moments in both the y and z directions. This theory is usually referred to as *unsymmetric bending*. Having developed the theories pertaining to the torsion of straight bars in the last chapter, the manner in which the loads must be applied in order to avoid any twisting of the beam—that is, so that only bending deformations result—will be discussed. This necessitates discussion of the shear center, defined as the point in the cross section through which the resultant force must pass in order to avoid any twisting. All these ideas will be applied in the appropriate fashion to open sections, to tubes and to sections consisting of several interconnected tubes.

4-2 Unsymmetrical Bending

Consider a straight, untwisted prismatic bar of arbitrary cross section, as shown in Figure 4-5. As shown, the x axis is taken to lie along the centroid of the cross section. The orientations of the y and z coordinates must be such so as to constitute a right-handed system but otherwise can be selected for ease in formulating and solving the problem in question.

With $\mathbf{j}$ and $\mathbf{k}$ the unit vectors in the y and z directions, respectively, the external loading can consist of concentrated forces of the form $\mathbf{P} = \mathbf{j}P_y + \mathbf{k}P_z$, distributed

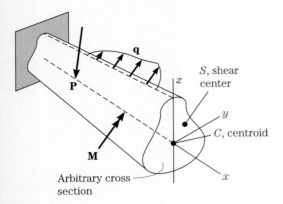

FIGURE 4-5 Coordinate system and loading.

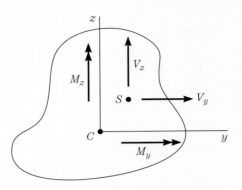

FIGURE 4-6 Internal force resultants for unsymmetrical bending.

loadings of the form $\mathbf{j}q_y(x) + \mathbf{k}q_z(x)$, and concentrated moments of the form $\mathbf{M} = \mathbf{j}M_y + \mathbf{k}M_z$. The lines of actions of the external forces must be such that at all points along the axis of the beam, *the resultant force passes through the shear center S*. The shear center S, whose location depends on the specific cross section, is the point through which the resultant shear force must pass in order that there be no twisting of the beam about the x axis. The ideas behind and the means for calculating the shear center will be discussed in greater detail later in this chapter. However, if either the y or z axis is a plane of symmetry of the cross section, the shear center will lie somewhere on that plane of symmetry. And for a cross section where both the y and z axes are planes of symmetry, the shear center coincides with the centroid.

The result of these assumptions is that at any point along the axis of the beam the internal force resultants are the transverse forces V_y and V_z whose lines of action pass through the shear center S, together with the moments M_y and M_z. The positive directions for V_y, V_z, M_y and M_z are indicated in Figure 4-6 where the face shown is the face whose normal is in the positive x direction. (The directions for V_y, V_z, M_y and M_z are reversed for faces whose normal is in the negative x direction.) For a statically determinate problem, the quantities V_y, V_z, M_y and M_z are determined in the usual manner by determining the reactions, drawing free-body diagrams that expose the internal force resultants at the desired locations and then requiring equilibrium. For statically indeterminate problems, any of the force or displacement formulations discussed in Chapter 2 can be used to determine all the unknown reactions, after which the internal force resultants can be determined.

The theory for unsymmetrical bending will be developed by considering each of the ideas of equilibrium, deformation and material behavior in turn. This will allow the internal force resultants—and hence the stresses—and the displacements to be determined.

Equilibrium. Consider an FBD of a segment Δx of the beam, as shown in Figure 4-7. Shown are the distributed loadings and the incremented force resultants acting on the face at location $x + \Delta x$. Forces V_y and V_z and moments M_y and M_z, not shown, act in

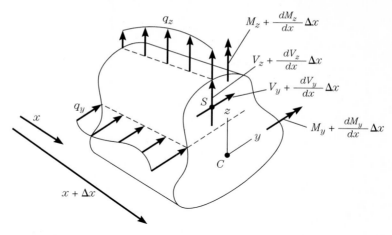

FIGURE 4-7 Equilibrium for unsymmetrical bending.

the negative y and z directions on the face at location x. Summing forces in the y and z directions, respectively, yields

$$\sum F_y = -V_y + \left(V_y + \frac{dV_y}{dx}\Delta x\right) + q_y \Delta x = 0$$

and

$$\sum F_z = -V_z + \left(V_z + \frac{dV_z}{dx}\Delta x\right) + q_z \Delta x = 0$$

from which

$$\frac{dV_y}{dx} + q_y = 0 \tag{4-4}$$

and

$$\frac{dV_z}{dx} + q_z = 0 \tag{4-5}$$

Summing moments about the y and z axes yields the two additional equations

$$\frac{dM_y}{dx} - V_z = 0 \tag{4-6}$$

and

$$\frac{dM_z}{dx} + V_y = 0 \tag{4-7}$$

In addition to these four differential equations of equilibrium, there are five relations between the stress distributions and the internal force resultants that must be introduced. The reader is asked to verify that these are

$$\int_A \sigma_x \, dA = 0, \qquad M_y = \int_A z\sigma_x \, dA, \qquad M_z = -\int_A y\sigma_x \, dA$$

$$V_y = \int_A \tau_{xy} \, dA \quad \text{and} \quad V_z = \int_A \tau_{xz} \, dA$$

(4-8a–e)

Strain-Displacement. Consistent with assuming that it is appropriate to formulate the kinetics of the problem in terms of the internal force resultants $(V_y, V_z, M_y \text{ and } M_z)$ that depend only on x, simplifying assumptions are also made that result in the displacement components depending essentially on x. This is a typical "mechanics of materials" approach in which simplifying assumptions are made to replace stresses by force resultants and in which the kinematic variables are assumed to have specific forms depending on the geometry of the region in order to reduce the governing differential equations from the partial differential equations of solid mechanics to a set of ordinary differential equations.

To this end assume that the axial component of displacement $U(x, y, z)$ can be approximated by

$$U(x, y, z) \approx u(x) - yu_1(x) + zu_2(x) \qquad (4\text{-}9a)$$

and that the transverse components are

$$V(x, y, z) \approx v(x) \qquad (4\text{-}9b)$$
$$W(x, y, z) \approx w(x) \qquad (4\text{-}9c)$$

where u, u_1 and u_2, shown in Figures 4-8a, 4-8b and 4-8c, respectively, represent a translation in the direction of the axis of the bar and small rotations of the cross section about the positive z and y axes. The linear dependence on the in-plane coordinates y and z amounts to assuming that a section that was originally perpendicular to the x axis remains plane. The plane section displaces an amount u and rotates through the small angles u_2 and u_1 about the y and z axes, that is, *plane sections remain plane*. On the

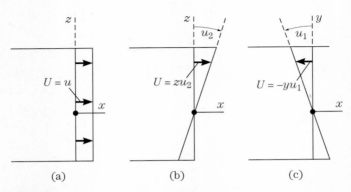

FIGURE 4-8 Displacements and rotations in bending.

basis of the above approximations, all points in the particular cross section located a distance x along the beam displace the same amounts $v(x)$ and $w(x)$ in the y and z directions, respectively.

The strains associated with these displacements are

$$\varepsilon_x = \frac{\partial U}{\partial x} = u'(x) - yu_1'(x) + zu_2'(x)$$

$$\gamma_{xy} = \frac{\partial U}{\partial y} + \frac{\partial V}{\partial x} = -u_1(x) + v'(x)$$

$$\gamma_{xz} = \frac{\partial U}{\partial z} + \frac{\partial W}{\partial x} = u_2(x) + w'(x)$$

$$\gamma_{yz} = \varepsilon_y = \varepsilon_z = 0$$

At this point the decision is made to neglect shear deformation by setting $\gamma_{xy} = \gamma_{xz} = 0$, leading to the relationships $u_1 = v'$ and $u_2 = -w'$, as shown in Figure 4-9.

Kinematically, these two constraints require that the plane section rotates through the same angle as the does the centerline of the beam, that is, that in addition to a plane section remaining plane, it must also remain perpendicular to the deformed axis of the beam. This is frequently referred to as the *Kirchoff assumption*. The strain then becomes

$$\varepsilon_x = u'(x) - yv''(x) - zw''(x) \tag{4-10}$$

$$\gamma_{xy} = \gamma_{xz} = 0$$

where $v''(x)$ and $w''(x)$ are the approximate expressions for the curvatures associated with the deformations in the y and z directions, respectively. Implicit in this development is the assumption that the slopes v' and w' are small.

For the development that neglects the shear deformation, proceed as follows. Eliminate V_y and V_z from Equations (4-4–4-7) to obtain

$$\frac{d^2 M_y}{dx^2} = -q_z \quad \text{and} \quad \frac{d^2 M_z}{dx^2} = q_y$$

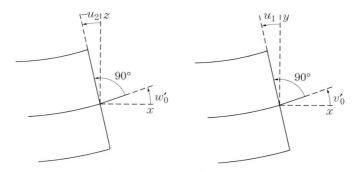

FIGURE 4-9 Plane sections remain plane and perpendicular to centroidal axis.

from which M_y and M_z can be determined. The four equations

$$\int_A \sigma_x \, dA = 0, \qquad M_y = \int_A z\sigma_x \, dA,$$

$$M_z = -\int_A y\sigma_x \, dA \quad \text{and} \quad \varepsilon_x = u'(x) - yv''(x) - zw''(x)$$

involve the five unknowns $\sigma_x, \varepsilon_x, u, v$ and w.

Material Behavior. The additional equation that enables us to combine the kinetic variable σ_x with the kinematic variables ε_x, u, v and w in the above four equations is taken to be $\sigma_x = E\varepsilon_x$, the usual assumption for a one-dimensional elastic solid. The constant E is Young's modulus or the modulus of elasticity.

Combination. Eliminate ε_x from the above equations to obtain

$$\sigma_x = E(u' - yv'' - zw'') \tag{4-11}$$

and substitute into the second, third and fourth equations to obtain

$$0 = E\int_A (u' - yv'' - zw'') \, dA$$

$$M_y = E\int_A z(u' - yv'' - zw'') \, dA$$

$$M_z = -E\int_A y(u' - yv'' - zw'') \, dA$$

Recalling that the origin of the coordinate system is at the centroid it follows with $I_{yz} = \int yz \, dA, I_{yy} = \int z^2 \, dA$ and $I_{zz} = \int y^2 \, dA$ that

$$0 = EAu'$$
$$M_y = E(-I_{yz}v'' - I_{yy}w'')$$
$$M_z = E(I_{zz}v'' + I_{yz}w'')$$

with solutions $u' = 0$,

$$v'' = \frac{M_z I_{yy} + M_y I_{yz}}{E\Delta} \tag{4-12a}$$

and

$$w'' = -\frac{M_y I_{zz} + M_z I_{yz}}{E\Delta} \tag{4-12b}$$

where $\Delta = I_{yy}I_{zz} - I_{yz}^2$. Given M_y and M_z and suitable boundary conditions on v and w, Equations (4-12) can be integrated to determine the displacements.

The bending stresses are now completely determined by substituting Equations (4-12) for v'' and w'' into Equation (4-11) to obtain

$$\sigma_x = -y\frac{M_z I_{yy} + M_y I_{yz}}{\Delta} + z\frac{M_y I_{zz} + M_z I_{yz}}{\Delta} \tag{4-13}$$

4-2 UNSYMMETRICAL BENDING

The reader should show that when an axial resultant P is transmitted, $\int \sigma_x \, dA = P$, and it would follow that $u' = P/AE$ with the normal stresses given by

$$\sigma_x = \frac{P}{A} - y\frac{M_z I_{yy} + M_y I_{yz}}{\Delta} + z\frac{M_y I_{zz} + M_z I_{yz}}{\Delta} \qquad (4\text{-}14)$$

Thus the formulation of the unsymmetrical bending problem has been completed. Equation (4-13) or (4-14) can be used to evaluate the stresses with integration of Equations (4-12) yielding the displacements. In the rest of this section these two issues will be investigated in detail.

4-2-1 Stresses in Unsymmetrical Bending
When evaluating the stresses it is convenient to classify the type of loading, that is, (1) loading in a single plane or (2) general loading. By loading in a single plane we mean that all loads are applied parallel to a single plane and all moments have directions perpendicular to that plane. An example would be the loading of the angle section shown in Figure 4-10a. Here all the loads lie in a plane making an angle of 30° with the z axis with the moment applied perpendicular to that plane. This is contrasted with the case shown in Figure 4-10b, where loads are not applied in a single plane.

The reason for this distinction arises in connection with the evaluation of the stresses. Of interest is the location of the neutral axis for the case represented by Equation (4-13), that is, where there is no axial load being transmitted. The location of the neutral axis is determined by setting the bending stress equal to zero. This results in

$$\frac{z}{y} = \frac{M_z I_{yy} + M_y I_{yz}}{M_y I_{zz} + M_z I_{yz}} = \tan\alpha \qquad (4\text{-}15)$$

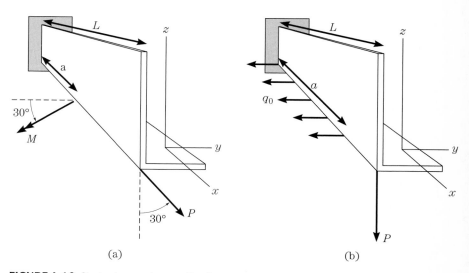

FIGURE 4-10 Single plane and general loadings.

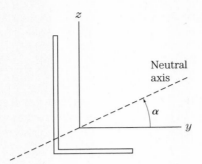

FIGURE 4-11 Location of the neutral axis.

which is the equation for a straight line passing through the centroid and having the orientation shown in Figure 4-11.

In the case of the single plane of loading shown in Figure 4-10a, the moments to be used in Equation (4-13) are

$$0 \leq x \leq a: \quad \begin{aligned} M_y &= P(L - x)\cos 30° - M \cos 30° \\ M_z &= P(L - x)\sin 30° - M \sin 30° \end{aligned}$$

so that the orientation of the neutral axis is given by

$$\tan \alpha = \frac{(P(L - x)\sin 30° - M \sin 30°)I_{yy} + (P(L - x)\cos 30° - M \cos 30°)I_{yz}}{(P(L - x)\cos 30° - M \cos 30°)I_{zz} + (P(L - x)\sin 30° - M \sin 30°)I_{yz}}$$

$$= \frac{I_{yy} \sin 30° + I_{yz} \cos 30°}{I_{zz} \cos 30° + I_{yz} \sin 30°}$$

showing that the angle of inclination of the neutral axis is constant along the portion $0 \leq x \leq a$. The reader should verify that a similar result obtains for $a \leq x \leq L$. The essence of the single plane of loading case is that both M_y and M_z depend in exactly the same manner on x and so cancel each other in the expression for $\tan \alpha$, yielding $\alpha =$ constant.

On the other hand, for the general loading case shown in Figure 4-10b, there results

$$0 \leq x \leq a: \quad \begin{aligned} M_y &= P(L - x) \\ M_z &= -\frac{q_0(a - x)^2}{2} \end{aligned}$$

so that

$$\tan \alpha = \frac{-\dfrac{q_0(a - x)^2}{2} I_{yy} + P(L - x) I_{yz}}{P(L - x) I_{zz} - \dfrac{q_0(a - x)^2}{2} I_{yz}}$$

showing clearly that α is a function of x. These two cases will be demonstrated in what follows.

Single Plane of Loading. As indicated in Figure 4-12, the bending stresses are positive on one side of the neutral axis and negative on the other at any point along the beam. The maximum bending stress occurs at the point farthest from the neutral axis. When in addition to the transverse loading there are axial loads present, the neutral axis is a straight line shifted from the origin of the yz coordinate system an amount proportional to the magnitudes of the axial loads. The maximum normal stress still occurs at the point farthest from the neutral axis.

For the symmetric problem the bending axis is parallel to the neutral axis. For the unsymmetrical bending problem, the bending axis makes an angle β, which is constant, with the y axis as shown in Figure 4-12. Taking $\tan \beta = M_z/M_y$, as indicated in Figure 4-12, Equation (4-15) can be rewritten as

$$\frac{z}{y} = \frac{\tan \beta \, I_{yy} + I_{yz}}{I_{zz} + \tan \beta \, I_{yz}} = \tan \alpha$$

showing that in general the neutral axis is not parallel to the bending axis for the unsymmetrical case. Even when $M_z = 0$ so that $\beta = 0$, $\tan \alpha = I_{yz}/I_{zz}$, which can be large depending on the specific cross section. Note that if $I_{yz} = 0$ (xz a plane of symmetry, for instance) and $M_z = 0$ it follows that $z = 0$ is the neutral axis—essentially, the symmetric bending case considered in the introduction to this chapter.

Principal Axis Representation. In the event that the y and z axes are chosen to be the principal axes, $I_{yz} = 0$ and the expression for the bending stress becomes

$$\sigma_x = -y\frac{M_z}{I_{zz}} + z\frac{M_y}{I_{yy}} \tag{4-16}$$

that is, the expression for the stress decouples into two terms, the first of which gives the stresses for bending about the principal z axis with the second one giving the stresses for bending about the principal y axis. Although this form for the bending

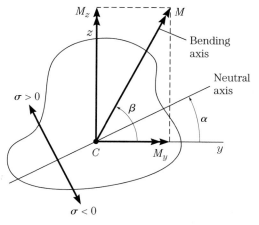

FIGURE 4-12 Location of neutral axis relative to bending axis.

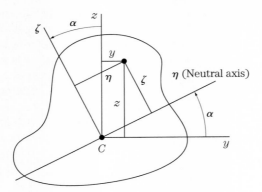

FIGURE 4-13 Neutral axis coordinates.

stresses is relatively simple, it must be remembered that after the location of the principal axes has been determined, the moments M_y and M_z as well as the principal coordinate locations y and z must be evaluated—so in fact it may be less work to deal with Equation (4-13) for evaluating the bending stresses.

Neutral Axis Coordinate Representation. Figure 4-13 suggests still another representation for the bending stresses in terms of the rectangular set of coordinates η and ζ. As shown, the η coordinate lies along the neutral axis with the ζ coordinate measured perpendicular to the neutral axis. Combining Equations (4-13) and (4-15), the bending stress can be represented as

$$\sigma_x = \frac{(M_y I_{zz} + M_z I_{yz})(z - y \tan \alpha)}{\Delta}$$
$$= \left(\frac{M_y I_{zz} + M_z I_{yz}}{\Delta}\right)\left(\frac{z \cos \alpha - y \sin \alpha}{\cos \alpha}\right) \quad (4\text{-}17)$$
$$= \frac{(M_y I_{zz} + M_z I_{yz})}{\Delta \cos \alpha} \zeta$$

expressing the bending stress in terms of the coordinate ζ perpendicular to the neutral axis. Irrespective of which of Equations (4-13), (4-16) or (4-17) is used to evaluate the bending stress, there is at any point along the axis of the beam a neutral axis making an angle α with the y axis such that the stresses are positive on one side of the neutral axis and negative on the other and such that the maximum bending stress occurs at the point farthest from the neutral axis.

EXAMPLE 4-2 The cantilever beam constructed from an angle section is loaded by a single load P, as shown in Figure 4-14. Analyze the beam for stresses using each of the three formulations discussed above.

Solution: The student should verify that the centroid is located as shown in Figure 4-14 and that the section properties are $I_{yy} = 1.0234$ in^4, $I_{zz} = 0.1863$ in^4 and

4-2 UNSYMMETRICAL BENDING

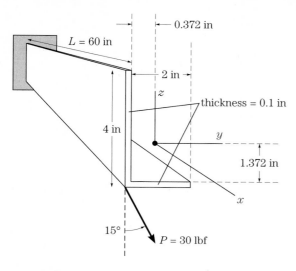

FIGURE 4-14 End-loaded cantilever beam.

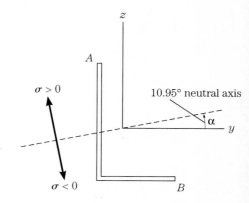

FIGURE 4-15 Location of neutral axis.

$I_{yz} = -0.2512$ in^4. With $x = 0$ at the support, it is a straightforward matter to compute that $M_z = P(L - x)\sin 15°$ and $M_y = P(L - x)\cos 15°$, showing clearly that M_z and M_y depend in precisely the same manner on x. The maximum values of both these moments occur at the support with $M_z(0) = PL \sin 15°$ and $M_y(0) = PL \cos 15°$. Substituting into Equation (4-13) for the bending stress yields

$$\sigma_x(0) = \frac{PL}{\Delta}[-y\{1.0234(0.2588) - 0.9659(0.2512)\}$$
$$+ z\{0.9659(0.1863) - 0.2588(0.2512)\}] \text{ in}^5$$
$$= (-313y + 1621z) \text{ psi}$$

where y and z are measured in inches. Setting this expression equal to zero gives the location of the neutral axis as $z/y = 0.1935 = \tan \alpha$ so that $\alpha = 10.95°$, as shown in Figure 4-15a. The stresses at points A and B are given by

$$\sigma(A) = [-(314)(-0.372) + 1622(2.628)] \text{ psi} = 4.38 \text{ ksi}$$
$$\sigma(B) = [-(314)(1.628) + 1622(-1.372)] \text{ psi} = -2.74 \text{ ksi}$$

The maximum bending stress at any other location along the beam would occur at point A in the cross section with the magnitude being proportional to $L - x$.

Principal axis formulation: For the principal axis representation, the location of the principal axes and the principal second moments of the area are required. These are determined from

$$\tan 2\theta = \frac{2I_{yz}}{I_{zz} - I_{yy}} = 0.6002$$

so that $\theta = 15.5°$. The corresponding principal second moments are then given by

$$I_{YY} = \frac{I_{yy} + I_{zz}}{2} + \sqrt{\left(\frac{I_{yy} - I_{zz}}{2}\right)^2 - I_{yz}^2} = 1.0930 \text{ in}^4$$

$$I_{ZZ} = +\frac{I_{yy} + I_{zz}}{2} + \sqrt{\left(\frac{I_{yy} - I_{zz}}{2}\right)^2 - I_{yz}^2} = 0.1168 \text{ in}^4$$

where θ and the YZ coordinates are as shown in Figure 4-16. The moments M_Y and M_Z are given by

$$M_Y = M_y \cos\theta + M_z \sin\theta = 1800 \text{ in-lbf}$$

and

$$M_Z = -M_y \sin\theta + M_z \cos\theta = -15 \text{ in-lbf}$$

The coordinates of point A are given by

$$Y_A = z_A \sin\theta + y_A \cos\theta = 0.3434 \text{ in}$$

and

$$Z_A = z_A \cos\theta - y_A \sin\theta = 2.6319 \text{ in}$$

so that

$$\sigma_x(A) = -0.3434 \frac{-15}{0.1168} + 2.6319 \frac{1800}{1.0930} = 4.38 \text{ ksi}$$

The coordinates of point B are given by

$$Y_B = z_B \sin\theta + y_B \cos\theta = 1.2025 \text{ in}$$

and

$$Z_B = z_B \cos\theta - y_B \sin\theta = -1.7570 \text{ in}$$

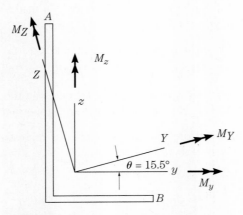

FIGURE 4-16 Location of principal axes.

4-2 UNSYMMETRICAL BENDING

so that

$$\sigma_x(B) = -1.2025\frac{-15}{0.1168} + (-1.7570)\frac{1800}{1.0930} = -2.74 \text{ ksi}$$

as computed using the xy coordinates in the previous example.

Neutral axis representation: The neutral axis representation for the bending stress is

$$\sigma_x = \frac{(M_y I_{zz} + M_z I_{yz})(z - y \tan \alpha)}{\Delta}$$

where α, computed above as 10.96°, is the angle between the y-axis and the neutral axis and where all the other terms have their usual meanings. The stresses at the root are easily computed as

$$\sigma_x(A) = \frac{(PL \cos 15° \,(0.1863) + PL \sin 15° \,(-0.2512))(2.628 - (-0.372)(0.1935))}{1.0234(0.18673) - 0.2512^2}$$

$$= 4.38 \text{ ksi}$$

and

$$\sigma_x(B) = \frac{(PL \cos 15° \,(0.1863) + PL \sin 15° \,(-0.2512))(-1.372 - (1.628)(0.1935))}{1.0234(0.18673) - 0.2512^2}$$

$$= -2.74 \text{ ksi}$$

coinciding with the values given by the two previous approaches. ◆

General Loading. For the case of a general loading the orientation of the neutral axis changes from point to point along the axis of the beam with the location within the cross section of the maximum stress also changing. This case is best illustrated by example.

◆ **EXAMPLE 4-3** A beam whose cross section consists of the angle section described in Example 4-2 is simply supported at both ends and loaded as shown in Figure 4-17. From Exam-

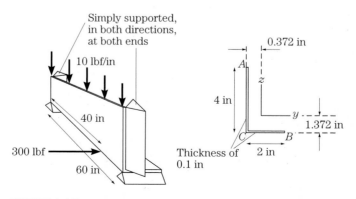

FIGURE 4-17 Multiplane loading of a simply supported beam.

ple 4-2 the required area properties are $I_{yy} = 1.0234$ in^4, $I_{zz} = 0.1863$ in^4 and $I_{yz} = -0.2512$ in^4. With x measured in inches from the left end of the beam, the bending moments are $M_y = -5x(60 - x)$ in-lbf and $M_z = -100x$ in-lbf, $0 \leq x \leq 40$ and $M_z = -200(60 - x)$ in-lbf, $40 \leq x \leq 60$. It is appropriate to consider the bending stresses in the portions to the left and to the right of $x = 40$ in. Substituting into Equation (4-13) gives

For $0 \leq x \leq 40$: $\sigma_x = -y(-212x - 9.847x^2) - z(212x - 7.303x^2)$

For $40 \leq x \leq 60$: $\sigma_x = y(96276 - 2196x + 9.847x^2) + z(23632 - 832x + 7.303x^2)$

The angle between the y axis and the neutral axis is obtained by equating each of the above expressions to zero. The results, shown in Figure 4-18, show a large variation in the location of the neutral axis, from approximately -30 degrees near the left end to nearly 80 degrees at the location of the concentrated load.

Shown in Figure 4-19 are the bending stresses at points A, B and C (from Figure 4-17) along the beam. Figure 4-19 shows that between $x = 0$ and a point $x \approx 31$ inches, the bending stress of largest magnitude occurs at point A. Beyond that point the bending stress of largest magnitude occurs at point B. ◆

4-2-2 Displacements in Unsymmetrical Bending

The equations that are used to determine the displacements are

$$v'' = \frac{M_z I_{yy} + M_y I_{yz}}{E \Delta} \tag{4-12a}$$

and

$$w'' = -\frac{M_y I_{zz} + M_z I_{yz}}{E \Delta} \tag{4-12b}$$

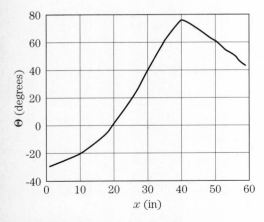

FIGURE 4-18 Angle between neutral axis and y axis as a function of position along the beam.

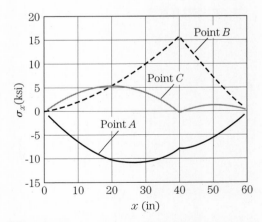

FIGURE 4-19 Bending stresses at various locations in the cross section as functions of position along the beam.

4-2 UNSYMMETRICAL BENDING

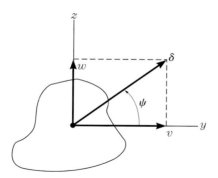

FIGURE 4-20 Displacement magnitude and direction.

These equations can be integrated with the four constants of integration evaluated using the appropriate boundary conditions on v and w at the ends of the beam. The results will be of the form $w = w(x)$ and $v = v(x)$, which are depicted in Figure 4-20 and which show that at any point along the beam the magnitude of the transverse displacement $\delta(x)$ is given by $\delta(x) = \sqrt{v^2 + w^2}$, having the orientation $\Psi(x)$ given by $\tan(\Psi(x)) = w(x)/v(x)$.

Single Plane of Loading. For the special case of loading in a single plane, Equation (4-15) can be used to rewrite these as

$$v'' = \frac{M_y I_{zz} + M_z I_{yz}}{E\Delta} \tan \alpha = F(x) \tan \alpha$$

and

$$w'' = -\frac{M_y I_{zz} + M_z I_{yz}}{E\Delta} = -F(x)$$

where $F(x) = (M_y I_{zz} + M_z I_{yz})/E\Delta$. Integration then yields

$$v = G(x) \tan \alpha + C_1 x + C_2$$

and

$$w = -G(x) + C_3 x + C_4$$

where $G(x)$ is the second indefinite integral of $F(x)$. If, further, the boundary conditions on v and w are identical, then it will follow that $C_1 = C_3$ and $C_2 = C_4$ with the result that the displacements can be represented as

$$v = H(x) \tan \alpha$$

and

$$w = -H(x)$$

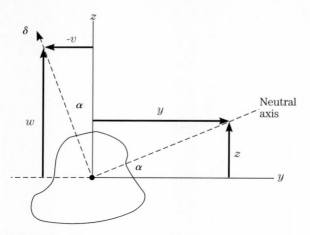

FIGURE 4-21 Relation between neutral axis and displacement components for single plane loading and like support conditions.

Thus, at any point along the beam

$$\tan(\Psi(x)) = \frac{w}{v} = -\frac{1}{\tan \alpha}$$

or

$$\tan \alpha = -\frac{v}{w}$$

Recalling that the location of the neutral axis is given by $\tan \alpha = z/y$, it is seen from Figure 4-21 that the direction of the displacement $\delta = \sqrt{(v^2 + w^2)}$ is perpendicular to direction of the neutral axis. For this case it is possible using Equations (4-13) and (4-15) to eliminate M_z, with the result that

$$w'' = -\frac{M_y}{E(I_{yy} - I_{yz} \tan \alpha)} \tag{4-18}$$

where only M_y depends on x. Note that this is of the same form as the corresponding equation for the symmetric case with the second moment I_{yy} replaced by $I_{yy} - I_{yz} \tan \alpha$ so that any known results for transverse displacements that are available from the theory for bending of symmetrical beams can be used directly in the present unsymmetric case with a single plane of loading as long as the appropriate $M_y/(I_{yy} - I_{yz} \tan \alpha)$ is used. After integrating Equation (4-18) and determining w, v can then be evaluated according to $v = -w \tan \alpha$. This idea will be discussed further in the examples.

EXAMPLE 4-4 For the cantilever beam described in Example 4-2 (see Figure 4-22), determine the displacements. Take $E = 10^7$ psi.

Solution: From Example 4-2 the section properties are $I_{yy} = 1.0234$ in^4, $I_{zz} = 0.1863$ in^4 and $I_{yz} = -0.2512$ in^4. Also from Example 4-2, $M_z = P(L - x)\sin 15°$ and

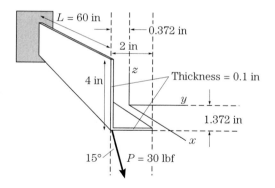

FIGURE 4-22 End-loaded cantilever beam.

$M_y = P(L - x)\cos 15°$. For this loading both M_y and M_z are multiples of $L - x$ so that the location of the neutral axis does not change along the length of the bar, that is, it is a single plane of loading. Substitution into Equations (4-12a and b), namely

$$v'' = \frac{M_z I_{yy} + M_y I_{yz}}{E\Delta}$$

and

$$w'' = -\frac{M_y I_{zz} + M_z I_{yz}}{E\Delta}$$

yields

$$v'' = 0.523 \times 10^{-6} \frac{L - x}{\text{in}^2}$$

and

$$w'' = -2.703 \times 10^{-6} \frac{L - x}{\text{in}^2}$$

where x is measured in inches. Integrating and satisfying the boundary conditions $w(0) = w'(0) = 0$ and $v(0) = v'(0) = 0$ gives

$$v(x) = 0.0873 \times 10^{-6} x^2 (3L - x) \text{ in}$$

and

$$w(x) = -0.451 \times 10^{-6} x^2 (3L - x) \text{ in}$$

where x is measured in inches. The maximum value occurs at $x = L = 60$ in with $w(L) = -0.195$ in and $v(L) = 0.0377$ in. Note that

$$\frac{-v(L)}{w(L)} = \frac{0.0377}{0.195} = 0.1933 \approx \tan(10.96°) = 0.1935$$

verifying that the displacement is perpendicular to the neutral axis.

Alternate representation: Employing Equation (4-18), namely,

$$w'' = -\frac{M_y}{E(I_{yy} - I_{yz}\tan\alpha)} \quad (4\text{-}18)$$

and the result from Example 4-2 that $\tan\alpha = 0.1935$, it follows immediately that

$$w'' = -\frac{30(L-x)\cos 15°}{10^7(1.0234 - (-0.2512)(0.1935))}$$

$$= -2.703 \times 10^{-6}\frac{L-x}{\text{in}^2}$$

which is identical to what was determined above. ◆

General Loading. If the support conditions for displacement in the y and z directions are different at the end(s) and/or if the loading is not in a single plane, the magnitude of the displacement δ will have an arbitrary direction relative to the neutral axis. Both of Equations (4-12) must then be integrated to determine the displacements.

EXAMPLE 4-5 For the cantilever beam described in Examples 4-2 and 4-4, loaded as shown here in Figure 4-23, determine the displacements. Take $E = 10^7$ psi.

Solution: From Example 4-2 the section properties are $I_{yy} = 1.0234$ in^4, $I_{zz} = 0.1863$ in^4 and $I_{yz} = -0.2512$ in^4. With $q_0 = 2$ lbf/in and $P = 55$ lbf the reader should show that the moments are given by $M_z = -q_0(L-x)^2/2$ and $M_y = P(L-x)$. Substitution into Equations (4-12a and b), namely,

$$v'' = \frac{M_z I_{yy} + M_y I_{yz}}{E\Delta}$$

and

$$w'' = -\frac{M_y I_{zz} + M_z I_{yz}}{E\Delta}$$

yields

$$E\Delta v'' = -q_0 I_{yy}\frac{(L-x)^2}{2} + PI_{yz}(L-x)$$

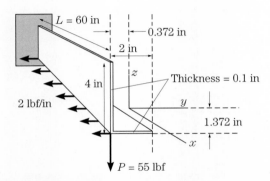

FIGURE 4-23 Cantilever beam.

and

$$E\Delta w'' = -P(L-x)I_{zz} + q_0 I_{yz}\frac{(L-x)^2}{2}$$

Integrating and satisfying the boundary conditions $w(0) = w'(0) = 0$ and $v(0) = v'(0) = 0$ gives

$$E\Delta v(x) = -\frac{q_0 I_{yy}[x^4 - 4Lx^3 + 6L^2x^2]}{24} + \frac{PI_{yz}[3Lx^2 - x^3]}{6}$$

and

$$E\Delta w(x) = \frac{q_0 I_{yz}[x^4 - 4Lx^3 + 6L^2x^2]}{24} - \frac{PI_{zz}[3Lx^2 - x^3]}{6}$$

Substituting the values and performing the calculations, the displacements v and w appear as shown in Figure 4-24. The maximum values occur at $x = L = 60$ in with $w(L) = -1.216$ in and $v(L) = -3.379$ in. Shown in Figures 4-25a and 4-25b are the

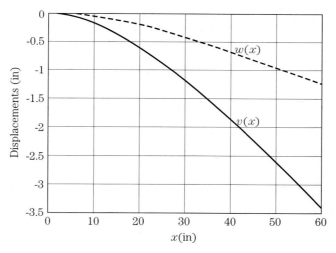

FIGURE 4-24 Displacement components $v(x)$ and $w(x)$.

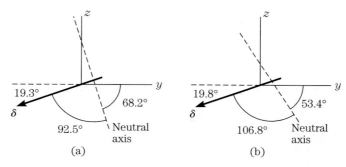

FIGURE 4-25 Neutral and displacement axes.

directions of the neutral axis and of the displacement δ at $x = L/2$ and $x = L$, respectively. It is seen that these two directions are *not* perpendicular as in the case of the single plane of loading. ◆

4-3 Shear Stresses in Thin-Walled Open Sections

As discussed in the previous section, an unsymmetrical cross-section beam subjected to a general loading transmits bending moments M_y and M_z as well as transverse shear forces V_y and V_z. The moments result in the normal or bending stresses considered in the above section, whereas the transverse shear forces result in shear stresses acting on transverse faces. This section will be used to develop and apply a theory that allows for the approximate calculation of such shear stresses. As presented, the theory will be applicable only to thin-walled open sections. The treatment of shear stresses in an arbitrary solid cross section is beyond the scope of this text. In all instances it will be assumed that the resultant of the external transverse load acts through the so-called shear center, to be discussed in section 4-4.

Consider then a thin-walled open section acted upon by an external transverse shear force, as shown in Figure 4-26a. The lateral surface of such a thin-walled section transmitting shear forces and bending moments is stress-free. In particular, there is no shear stress on the lateral surface and hence, as indicated in Figure 4-24b for stresses in the interior of the member, there cannot be any shear stress normal to the direction of the member at the edges. When the member is thin, this essentailly constrains any nonzero shear stress to have the direction tangent to the lateral surfaces.

Consider then an arbitrary thin-walled section transmitting positive shear force components V_y and V_z, as shown in Figure 4-27a. The shear stress is determined on the basis of an equilibrium argument for the free-body diagram shown in Figure 4-27b. Summing forces in the x direction yields

$$\sum F_x = \tau t + F' = 0 \tag{4-19}$$

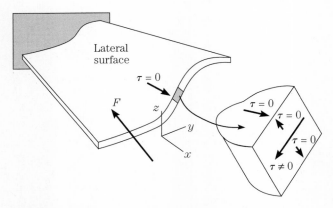

FIGURE 4-26 Direction of shear stresses in thin-walled members.

4-4 SHEAR STRESSES IN THIN-WALLED OPEN SECTIONS

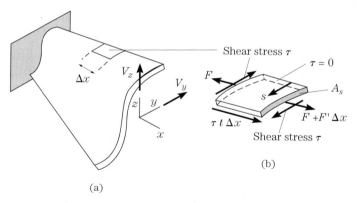

FIGURE 4-27 Shear forces transmitted in a thin-walled section.

where F, arising from the bending stresses, is the normal force on the area A_s. The force F can be represented as

$$F = \int_{A_s} \sigma_x \, dA \tag{4-20}$$

Note that the integration begins at a point where the shear stress is known to be zero. Recall from Equation (4-13) that

$$\sigma_x = -y\frac{M_z I_{yy} + M_y I_{yz}}{\Delta} + z\frac{M_y I_{zz} + M_z I_{yz}}{\Delta}$$

and from Equations (4-3) and (4-4) that $M'_y = V_z$ and $M'_z = -V_y$. Combining these equations leads, after a bit of algebra, to

$$\tau t = q = \frac{V_y I_{yz} - V_z I_{zz}}{\Delta} A_s \bar{z} + \frac{V_z I_{yz} - V_y I_{yy}}{\Delta} A_s \bar{y} \tag{4-21}$$

for the shear flow $q = \tau t$. The shear flow q is a function of the coordinate s measured from the stress-free edge with $A_s \bar{z}$ and $A_s \bar{y}$, respectively, representing the first moments with respect to the centroidal y and z axes of the shaded area in Figure 4-27b. As indicated in Figure 4-27b, the positive direction for the shear flow is in the direction of integration, that is, in the positive s direction. For a single member such as that shown in Figure 4-27 the calculations are carried out by beginning at a free edge and integrating along the centerline of the member until the other free edge is reached. A very valuable check on the calculations is that by the time the integrations have proceeded to the final edge the shear flow should again assume the value zero. For more complicated open sections it may be necessary to begin integrations from several free edges.

Recall that in the development of the equation for computing the bending stresses (in the previous section), we chose to neglect shear by taking the shear strains as zero. For a linearly elastic isotropic material ($\tau = G\gamma$), this then requires that the shear stresses and hence the shear flows be zero, inconsistent with the above development. In spite of this contradiction, Equation (4-21) gives a good approximation to shear stresses acting in thin-walled open sections.

EXAMPLE 4-6 For the beam and loading described in Example 4-1, pictured again here in Figure 4-28, determine the shear stresses.

Solution: Recall from Example 4-1 that $I_{yy} = 1.0234$ in^4, $I_{zz} = 0.1863$ in^4 and $I_{yz} = -0.2512$ in^4. The shear forces transmitted are $V_y = P \sin 15°$ and $V_z = -P \cos 15°$. As indicated in Figure 4-29, the integration is begun at A, the top of the vertical leg, proceeds along the midline to point B at the intersection of the two midlines and then along the horizontal leg to the end at point C. Substituting into Equation (4-21), the shear flow is

$$q(s) = \frac{P \sin 15°(-0.2512) - (-P \cos 15°)(0.1863)}{\Delta} A_s \bar{z}$$

$$+ \frac{-P \cos 15°(-0.2512) - P \sin 15°(1.0234)}{\Delta} A_s \bar{y}$$

$$= \frac{P}{\Delta}(0.1149 A_s \bar{z} - 0.0222 A_s \bar{y})$$

On AB: $A_s = 0.1s$, $\bar{y} = -0.322$ and $\bar{z} = 2.628 - 0.5s$ where $0 \leq s \leq 3.95$ in. The shear flow can be expressed as

$$q_{AB}(s) = \frac{P(0.03091s - 0.00575s^2)}{\Delta}$$

with $q_{AB}(3.95) = 0.03239 P/\Delta$.

On BC: $A_s = 0.1s$, $\bar{y} = -0.322 + 0.5s$ and $\bar{z} = -1.322$ where $0 \leq s \leq 1.95$ in. The shear flow can be expressed as

$$q_{BC}(s) = \frac{P(0.03239 - 0.01447s - 0.00111s^2)}{\Delta}$$

with $q_{BC}(1.95) = -0.00005 P/\Delta \approx 0$, as required. The shear flows along with the corresponding directions are plotted in Figure 4-30. The maximum shear stress clearly occurs in the vertical leg where q is maximum with

$$\tau_{max} = \frac{0.04154 P}{t \cdot \Delta} = \frac{0.04154(30)}{0.1(0.1275)} \text{ psi} = 97.7 \text{ psi}$$

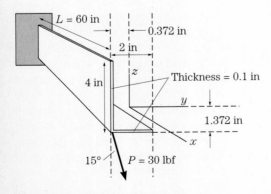

FIGURE 4-28 End-loaded cantilever beam.

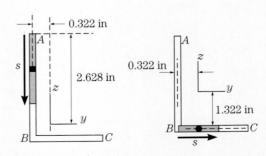

FIGURE 4-29 Integrations in an angle section.

4-4 SHEAR STRESSES IN THIN-WALLED OPEN SECTIONS

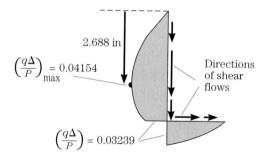

FIGURE 4-30 Shear flows in an angle section.

The reader should verify that the integrals of the shear flows over the vertical and horizontal legs give, respectively, $P \cos 15°$ and $P \sin 15°$.

EXAMPLE 4-7 Determine the shear flows in the unequal-leg Z-section transmitting a shear force P, as shown in Figure 4-31. Assume that the load P is applied through the shear center.

Solution: The reader should verify that $I_{yy} = 8.625b^3 t$, $I_{zz} = 2.625b^3 t$, $I_{yz} = -3.375b^3 t$ and $\Delta = 11.25b^6 t^2$. Then, with $V_y = 0.8P$ and $V_z = 0.6P$, substitution into Equation (4-21) yields

$$q = \frac{[0.6P(-3.375) - 0.8P(8.625)]A_s \bar{y} - [0.6P(2.625) + 0.8P(3.375)]A_s \bar{z}}{11.25 b^3 t}$$

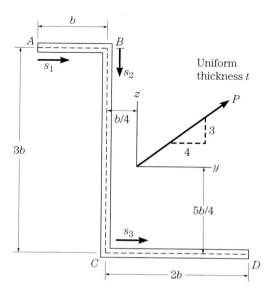

FIGURE 4-31 Unequal-leg Z-section transmitting a shear force P.

which can be simplified to

$$\frac{qb^3 t}{P} = -0.7933 A_s \bar{y} - 0.3800 A_s \bar{z}$$

On AB: With $0 \le s_1 \le b$ measured as shown in Figure 4-31, $A_s = t \cdot s_1$, $\bar{y} = -5b/4 + s_1/2$ and $\bar{z} = 7b/4$ with

$$\frac{qb}{P} = \frac{s_1}{b}\left(0.3266 - 0.3967\frac{s_1}{b}\right)$$

as shown in Figure 4-32. Note that at B, $qb/P = -0.0701$.

On BC: With $0 \le s_2 \le 3b$ measured as shown, $A_s = t \cdot s_2$, $\bar{y} = -b/4$ and $\bar{z} = 7b/4 - s_2/2$ with

$$\frac{qb}{P} = -0.0701 - \frac{s_2}{b}\left(0.4667 - 0.1900\frac{s_2}{b}\right)$$

with $qb/P = 0.2398$ at C. This is again shown in Figure 4-32, where the maximum, also shown, occurs approximately 1.2 in from B.

On CD: With $0 \le s_3 \le 2b$ measured as shown, $A_s = t \cdot s_3$, $\bar{y} = -b/4 + s_3/2$ and $\bar{z} = -5b/4$ with

$$\frac{qb}{P} = 0.2398 + \frac{s_3}{b}\left(0.6733 - 0.3967\frac{s_3}{b}\right)$$

as shown in Figure 4-32. The maximum of 0.5255 occurs at approximately 0.82 in from C. As a check, $qb/P = -0.0004 \approx 0$ at C.

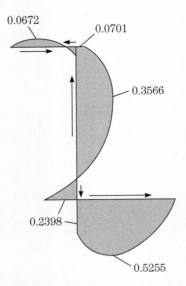

FIGURE 4-32 Shear flows (qb/P) in an unequal-leg Z-section.

4-4 Shear Center for Thin-Walled Open Sections

When a beam is subjected to transverse forces, such as the cantilever beam loaded by an end force F shown in Figure 4-33a, the deformations generally consist of both bending deformations and twisting. There is a special point known as the shear center where the application of the load F results in no twist, as indicated in Figure 4-33b. It is important to know the location of this point in the cross section in order to determine the internal force resultants that are transmitted for arbitrarily applied loads. When the loads pass through the shear center, the only stresses that arise are bending stresses and the transverse shear stresses associated with bending. When the loads *don't* pass through the shear center, there is a torque about the axis of the beam that produces additional torsional shear stresses.

For cross sections with a single plane of symmetry, such as those indicated in Figures 4-34a and 4-34b, the shear center will lie somewhere on that plane of symmetry. When there are two planes of symmetry, such as is indicated in Figure 4-34c, the shear center coincides with the centroid.

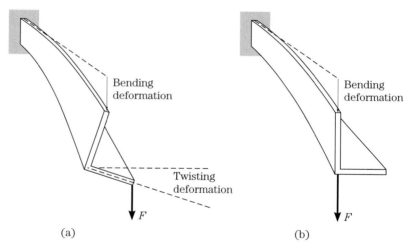

FIGURE 4-33 Deformations in transverse loading.

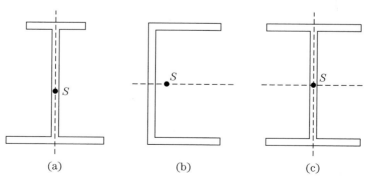

FIGURE 4-34 Cross sections with one or two planes of symmetry.

For cross sections that consist of thin, straight segments that all intersect at a point, such as those indicated in Figures 4-35a, 4-35b and 4-35c, the shear center is located at the point of intersection of the segments. In most other thin-walled open cross sections it is necessary to compute the location of the shear center. The computations, presented in the remainder of this section, are based purely on statics.

Consider then the case of an arbitrary thin-walled cross section cantilever beam loaded by shear forces V_y and V_z passing through the shear center, as shown in Figure 4-36a. The basic idea is that the external loads V_y and V_z must be applied so that the resultant moment about any axis parallel to the x axis vanishes, that is, so that there is no twisting moment transmitted along the beam. Summing moments about the x axis for the free-body diagram shown in Figure 4-36b yields

$$\sum M_x = V_z e_y - V_y e_z - \int r(q\ ds) = 0 \qquad (4\text{-}22)$$

where the shear flow at the section in question has the negative direction consistent with Newton's third law. Recalling from Equation (4-21) that

$$q = \frac{V_y I_{yz} - V_z I_{zz}}{\Delta} A_s \bar{z} + \frac{V_z I_{yz} - V_y I_{yy}}{\Delta} A_s \bar{y}$$

$$= Z A_s \bar{z} + Y A_s \bar{y}$$

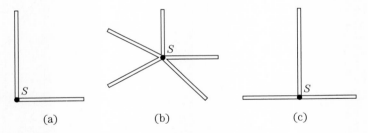

(a)　　　(b)　　　(c)

FIGURE 4-35 Cross sections with intersecting straight segments.

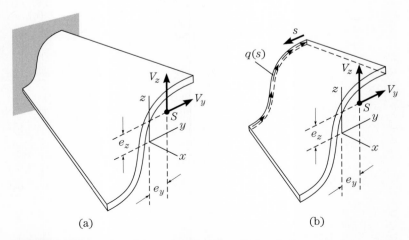

(a)　　　(b)

FIGURE 4-36 Location of shear center for thin-walled sections.

and that $A_s \bar{z} = \int_{A_s} z \, dA$ and that $A_s \bar{y} = \int_{A_s} y \, dA$, it follows that

$$V_z e_y - V_y e_z - \int r \left(Z \int_{A_s} z \, dA + Y \int_{A_s} y \, dA \right) ds = 0$$

Interchange the order of integration to write

$$\int r \int_{A_s} z \, dA \, ds = \int_{A_s} z \left(\int r \, ds \right) dA = \int_{A_s} z\omega \, dA = A_{\omega z}$$

and

$$\int r \int_{A_s} y \, dA \, ds = \int_{A_s} y \left(\int r \, ds \right) dA = \int_{A_s} y\omega \, dA = A_{\omega y}$$

from which

$$V_z e_y - V_y e_z = Z A_{\omega z} + Y A_{\omega y}$$

The quantities $A_{\omega y}$ and $A_{\omega z}$ are termed the *first moments of the sectorial area* and are discussed in Appendix C. Substituting for Z and Y yields

$$V_z \left(e_y + \frac{A_{\omega z} I_{zz} - A_{\omega y} I_{yz}}{\Delta} \right) - V_y \left(e_z + \frac{A_{\omega z} I_{yz} - A_{\omega y} I_{yy}}{\Delta} \right) = 0$$

Since V_z and V_y are arbitrary, it follows that

$$e_y = \frac{A_{\omega y} I_{yz} - A_{\omega z} I_{zz}}{\Delta} \tag{4-23a}$$

and

$$e_z = \frac{A_{\omega y} I_{yy} - A_{\omega z} I_{yz}}{\Delta} \tag{4-23b}$$

giving the location of the shear center with respect to the centroid of the cross section entirely in terms of the geometrical properties of the cross section. In practice it is much easier to simply evaluate $\int r(q \, ds)$ using Equation (4-22) on the basis of the specifics of the problem at hand. This is demonstrated in the examples that follow.

EXAMPLE 4-8 The cantilever beam is constructed from a C-section, as shown in Figure 4-37. Determine the shear flow distribution and the location of the shear center S.

Solution: It was stated above that for a section with a plane of symmetry the shear center lies on that plane. Loads in the y direction would be applied along the y axis at all points along the beam. When $V_y = 0$ and there is a plane of symmetry so that $I_{yz} = 0$, the shear flow is given by

$$q(s) = -\frac{V_z}{I_{yy}} A_s \bar{z}$$

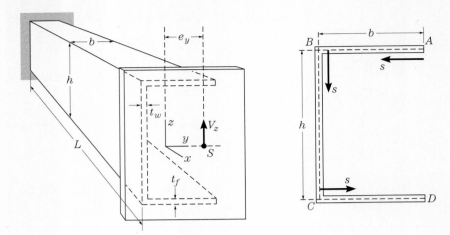

FIGURE 4-37 C-section loaded in shear.

Assuming that $t_f \ll h$, it follows that $I_{yy} = h^2(ht_w + 6bt_w)/12$. With the points A, B, C and D as shown in the figure and $I_{yy}q(s)/V_z = Q(s) = -A_s \bar{z}$, the shear flows are calculated according to the following equations.

On AB: $A_s = st_f, \bar{z} = h/2, Q(s) = -st_f h/2, 0 \leq s \leq b$ with $Q(b) = -bt_f h/2$.

On BC: $A_s = st_w, \bar{z} = (h-s)/2, Q(s) = -bt_f h/2 - st_w(h-s)/2, 0 \leq s \leq h$ with $Q(h) = -bt_f h/2$.

On CD: $A_s = st_f, \bar{z} = -h/2, Q(s) = -bt_f h/2 + st_f h/2, 0 \leq s \leq b$ with $Q(b) = 0$.

The shear flow distribution is as shown in Figure 4-38a. Note again that $q < 0$ means that the direction of the shear flow is opposite to the direction of integration. The cor-

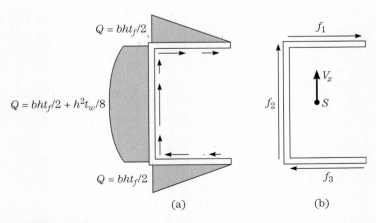

FIGURE 4-38 Shear flows and associated resultant forces.

4-4 SHEAR CENTER FOR THIN-WALLED OPEN SECTIONS

responding forces in the flanges and web with the directions shown in Figure 4-38b are computed according to

$$f_1 = \int_0^b q_{AB}\, ds = \left(\frac{V_z}{I_{yy}}\right)\left(\frac{t_f h b^2}{4}\right) = f_3$$

and with $I_{yy} = b t_f h^2/2 + t_w h^3/12$

$$f_2 = \int_0^h q_{BC}\, ds = \frac{V_z}{I_{yy}}\left(\frac{b t_f h^2}{2} + \frac{t_w h^3}{12}\right) = V_z$$

Consider then the free-body diagram shown in Figure 4-39, namely, of a section consisting of the loaded end of the cantilever with the internal forces acting as shown at a typical interior section. Note that since the normal to the face asssociated with the section on which f_1, f_2 and f_3 act in Figure 4-39 is in the negative x direction, the directions of the internal forces in Figure 4-39 must be reversed from those shown in Figure 4-38b. Note also that the distance e_y is being measured from the centerline of the web rather than from the centroid for this calculation. Summing moments about an axis parallel to the x axis, passing through the center of the web and midway between the flanges yields

$$\sum M_x = V_z e_y + \frac{f_1 h}{2} + \frac{f_3 h}{2} = 0$$

Substituting for $f_1 = f_3$ yields

$$V_z e_y + \frac{V_z}{I_{yy}} \frac{t_f h b^2}{4} h = 0$$

or

$$e_y = -\frac{t_f h^2 b^2}{4 I_{yy}} = -\frac{3 b^2 t_f}{h t_w + 6 b t_f}$$

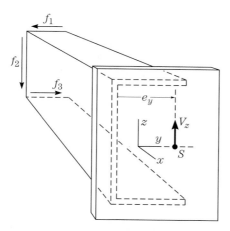

FIGURE 4-39 Free-body diagram for determining e_y.

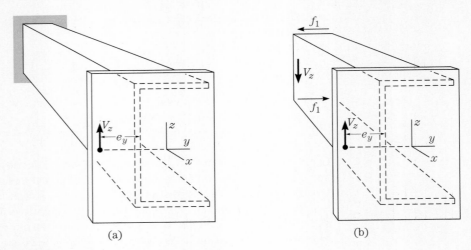

FIGURE 4-40 Shear center location for a C-section.

that is, to the left of the web, as shown in Figure 4-40. As shown, this necessitates the addition of a plate or some other equivalent piece of structure since the shear center doesn't lie at a point physically on the cross section. The mechanics of the situation are that the two couples shown in Figure 4-40b balance, namely, $-V_z e_y$ and $+f_1 h$, with no resultant torque about the x axis as long as the load V_z is applied the distance e_y to the left of the web.

Thus for the C-section any external force F must be applied through the shear center, as shown in Figure 4-41a, in order that only the shear stresses associated with

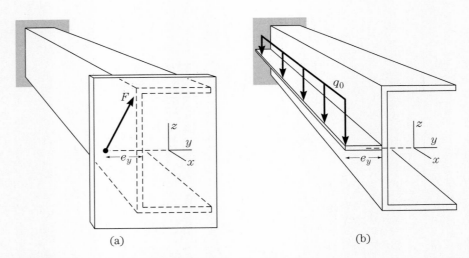

FIGURE 4-41 Loading of a cantilevered C-section for no twisting.

4-4 SHEAR CENTER FOR THIN-WALLED OPEN SECTIONS

the bending be present. The load F can be resolved into components V_y and V_z and the shear flows computed according to

$$q(s) = -\frac{V_z}{I_{yy}}A_s\bar{z} - \frac{V_y}{I_{zz}}A_s\bar{y}$$

For a C-section loaded by distributed loads such as are pictured in Figure 4-41b, it would be necessary to attach a *thin* plate of width e_y along the axis of symmetry as shown in order to avoid any twisting. For distributed loading in the z direction, the thin plate would have a negligible effect on either the bending stresses or the shear stresses. For distributed loadings parallel to the y axis, the additional thin plate would of course alter the value of I_{zz} and hence also alter the values of the bending stresses and shear stresses associated with bending about the z axis.

In Example 4-8 the plane of symmetry substantially simplified the process of determining the shear flows and the location of the shear center. In the example that follows analogous calculations will be carried out in a situation where no such simplification is possible.

EXAMPLE 4-9 The cantilever beam is constructed from a nonstandard built-up section as shown in Figure 4-42. Determine the shear flow distribution and the location of the shear center S.

Solution: The reader should verify the location shown for the centroid of the section and that $I_{yy} = 0.2 \text{ in}^4$, $I_{zz} = 0.3333 \text{ in}^4$ and $I_{yz} = -0.1833 \text{ in}^4$. In general the expression for the shear flow is

$$q(s) = \frac{V_y I_{yz} - V_z I_{zz}}{\Delta}A_s\bar{z} + \frac{V_z I_{yz} - V_y I_{yy}}{\Delta}A_s\bar{y}$$

which can be rewritten as

$$q(s) = V_y F(s) + V_z G(s)$$

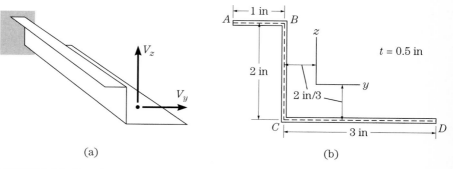

FIGURE 4-42 Shear flow and shear center for a nonstandard section.

where $F(s) = A_s(I_{yz}\bar{z} - I_{yy}\bar{y})/\Delta$ and $G(s) = A_s(I_{yz}\bar{y} - I_{zz}\bar{z})/\Delta$. With locations A, B, C and D as shown in Figure 4-42b, the shear flows are calculated according to the following equations.

On AB: $A_s = 0.05s_1$, $\bar{y} = -1.6667 + 0.5s_1$ and $\bar{z} = 1.3333$ with $0 \leq s_1 \leq 1.0$.

$$F(s) = 1.5126s_1(0.0889 - 0.1s_1)$$
$$G(s) = 1.5126s_1(-0.1389 - 0.0917s_1)$$

with $F(1) = -0.0168$ and $G(1) = -0.3487$ so that the shear flow at point B is $q(B) = -0.0168V_y - 0.3487V_z$. The portions of the shear flow arising from V_y and V_z in each portion of the cross section are shown in Figures 4-43a and 4-43b, respectively.

On BC: $A_s = 0.05s_2$, $\bar{y} = -0.6667$ and $\bar{z} = 1.3333 - 0.5s_2$ with $0 \leq s_2 \leq 2.0$.

$$F(s) = 1.5126s_2(-0.1111 + 0.0917s_2)$$
$$G(s) = 1.5126s_2(-0.3222 + 0.1667s_2)$$

with $F(2) = 0.2187$ and $G(2) = 0.0339$ so that the shear flow at point C is $q(C) = (0.2187 - 0.0168)V_y - (0.3488 - 0.0339)V_z = 0.2019V_y - 0.3149V_z$.

On CD: $A_s = 0.05s_3$, $\bar{y} = -0.6667 + 0.5s_3$ and $\bar{z} = -0.6667$ with $0 \leq s_3 \leq 3.0$.

$$F(s) = 1.5126s_3(0.2555 - 0.1s_3)$$
$$G(s) = 1.5126s_3(0.3444 - 0.0917s_3)$$

with $F(3) = -0.2019$ and $G(2) = 0.3144$ so that the shear flow at D is $q(D) = (0.2019 - 0.2019)V_y - (0.3149 - 0.3144)V_z = 0.0000V_y - 0.0005V_z \approx 0$.

The forces in each of the segments due to the shear flows are determined in the usual manner. For instance,

$$f_1 = \int_A^B (V_y F(s) + V_z G(s))\, ds = 0.0168V_y - 0.1513V_z$$

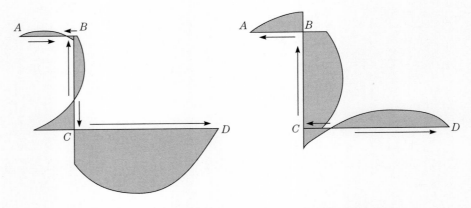

(a) Shear flow for V_y (b) Shear flow for V_z

FIGURE 4-43 Shear flow distributions.

4-4 SHEAR CENTER FOR THIN-WALLED OPEN SECTIONS

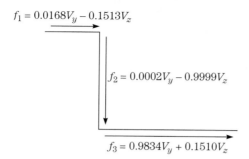

FIGURE 4-44 Resultant forces in flanges and web.

as shown in Figure 4-44 together with the resultant forces in each of the segments. Note that from Figure 4-44, discounting the part of f_2 attributable to V_y, which should actually be zero,

$$\sum F_y = 0.0168V_y + 0.9834V_y = 1.0002V_y \approx V_y$$

and

$$\sum F_z = -(-0.9999V_z) = 0.9999V_z \approx V_z$$

so that equilibrium of forces is essentially verified. The student should show that retention of more significant figures in representing the intermediate calculations results in smaller errors in the equilibrium checks.

To determine the location of the shear center, we proceed as in the previous example by constructing the free-body diagram shown in Figure 4-45. Note again that the directions for the resultant forces must be reversed from those shown in Figure 4-44 since the normal to the face is now in the negative x direction. Summing moments about an x axis through 0 yields

$$V_z e_y - V_y e_z + f_1 \cdot 2 = 0$$

or

$$V_z e_y - V_y e_z + 2(0.0168V_y - 0.1513V_z) = 0$$

from which $e_y = 0.3026$ in and $e_z = 0.0336$ in, as shown in Figure 4-46.

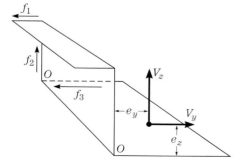

FIGURE 4-45 Free-body diagram for determining location of the shear center.

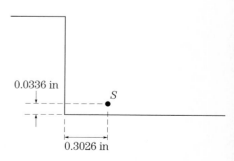

FIGURE 4-46 Location of shear center.

To reiterate, if there is to be no twisting, the line of action of any external loads must pass through S at all points along the axis of the beam. ◆

4-5 Bending of a Single Thin-Walled Tube

For an *open* thin-walled section, the value of the shear flow is known at all edges to be zero. This corresponds to a statically determinate problem in that the shear flows can be determined by integration at all points in the cross section on the basis of statics alone. Statics alone corresponds to having developed the expression for the shear flow, namely,

$$q(s) = \frac{V_y I_{yz} - V_z I_{zz}}{\Delta} A_s \bar{z} + \frac{V_z I_{yz} - V_y I_{yy}}{\Delta} A_s \bar{y} \tag{4-24}$$

on the basis of requiring equilibrium for the free-body diagram of Figure 4-27b. Knowing the shear flow, the shear center can be determined and hence the problem of the bending, without twisting, of an unsymmetrical section can be considered solved.

Consider now the problem of a *closed* tube undergoing bending, as indicated in Figure 4-47a. The coordinate system is centroidal. For simplicity we again assume a cantilever beam subjected to end loads V_y and V_z acting at the shear center, that is, at the point that results in no twisting of the tube. As in the case of the open section, the shear center will be determined on the basis of a statics argument once the shear flows are known. For the closed tube there is no free edge where the shear flow is known so that in general there is an unknown value of the shear flow at any point around the periphery of the tube. As indicated in Figure 4-47b, select a convenient point in the tube, make a longitudinal cut and designate the unknown value of the shear flow at that point as $\bar{q}$.

With $\bar{q}$ as an unknown, proceed with the integration around the tube to determine

$$q(s) = \bar{q} + \frac{V_y I_{yz} - V_z I_{zz}}{\Delta} A_s \bar{z} + \frac{V_z I_{yz} - V_y I_{yy}}{\Delta} A_s \bar{y} = \bar{q} + q_0(s)$$

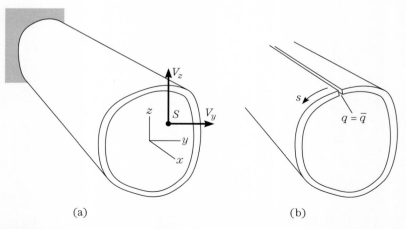

(a) (b)

FIGURE 4-47 Shear flows in a closed tube undergoing bending.

4-5 BENDING OF A SINGLE THIN-WALLED TUBE

where $\bar{q}$ is an unknown force variable. The problem is clearly statically indeterminate in that the equation of equilbrium involves the unknown $\bar{q}$. Consistent with the idea that unknown force variables in statically indeterminate situations are determined on the basis of displacement requirements, the unknown $\bar{q}$ is determined by requiring that there be no twisting deformation associated with the resultant shear flow $q(s)$. Recall from Chapter 3 that in terms of the shear flow, the rate of twist is given by

$$\phi' = \frac{1}{2A_0 G} \oint \frac{q\,ds}{t}$$

Forcing ϕ' to be zero results in no twisting and leads to

$$0 = \oint \frac{q\,ds}{t} = \oint \frac{(\bar{q} + q_0(s))\,ds}{t}$$

or

$$\bar{q} = -\frac{\oint \dfrac{q_0(s)\,ds}{t}}{\oint \dfrac{ds}{t}} \tag{4-25}$$

from which the shear flow $\bar{q}$ can be determined. The shear center can then be determined in the usual manner.

EXAMPLE 4-10 For the rectangular tube shown in Figure 4-48, determine the location of the shear center and the resultant shear stresses. Take $t_1 = 4$ mm, $t_2 = 6$ mm, $t_3 = 10$ mm, $h = 150$ mm, $b = 200$ mm and $V_z = 50$ kN.

Solution: With a horizontal plane of symmetry, $I_{yz} = 0$ and the shear flow in general is given by Equation (4-21) as

$$q(s) = -\frac{V_z}{I_{yy}} A_s \bar{z}$$

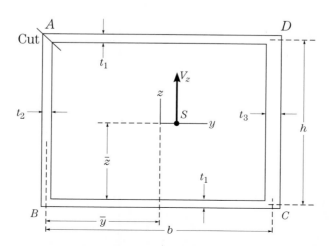

FIGURE 4-48 Bending of a single rectangular tube.

The reader should verify that $\bar{z} = 75$ mm, $\bar{y} = 115$ mm and $I_{yy} = 13.5 \times 10^6$ mm^4. For brevity, denote $Q(s) = I_{yy}q(s)/V_z = -A_s\bar{z}$ and integrate in a counterclockwise direction around the tube from the cut at A. Lengths, areas and first moments have units mm, mm^2 and mm^3, respectively.

On AB: $A_s = 6s$, $\bar{z} = 75 - 0.5s$, $0 \leq s \leq 150$ so that $Q(s) = -6s(75 - 0.5s)$ with $Q(150) = 0$.

On BC: $A_s = 4s$, $\bar{z} = -75$, $0 \leq s \leq 200$ so that $Q(s) = 300s$ with $Q(200) = 60000$.

On CD: $A_s = 10s$, $\bar{z} = -75 + 0.5s$, $0 \leq s \leq 150$ so that $Q(s) = 60000 - 10s(-75 + 0.5s)$ with $Q(150) = 60000$.

On DA: $A_s = 4s$, $\bar{z} = 75$, $0 \leq s \leq 200$ so that $Q(s) = 60000 - 300s$ with $Q(200) = 0$.

These values, all of which must be multiplied by V_z/I_{yy}, are shown in Figure 4-49. The value of $\bar{q}$, the unknown shear flow at A is determined by evaluating Equation (4-25). Recalling that the area of a symmetric parabola is $2bh/3$, the numerator of Equation (4-25) is

$$\oint q_0 \frac{ds}{t} = \left[-\frac{2}{3}\frac{16875(150)}{6} + 2\frac{\frac{60000(200)}{2}}{4} + \frac{\frac{60000 + 2(28125)}{3}}{10} 150 \right] \frac{V_z}{I_{yy}} = 3.9 \times 10^6 \frac{V_z}{I_{yy}} \text{ mm}^3$$

Then with $\oint ds/t = 140$,

$$\bar{q} = -3.9 \times \frac{10^6 \frac{V_z}{I_{yy}} \text{ mm}^3}{140} = -27857 \frac{V_z}{I_{yy}}$$

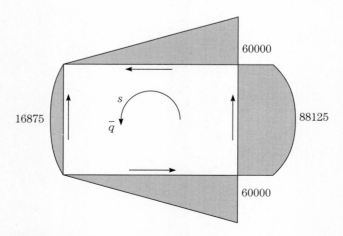

FIGURE 4-49 Shear flows $q_0(s)$ with cut at A.

4-5 BENDING OF A SINGLE THIN-WALLED TUBE

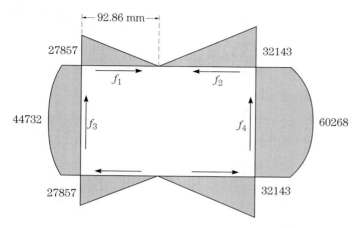

FIGURE 4-50 Resultant shear flows.

FIGURE 4-51 Free-body diagram for shear center calculations.

The resultant shear flows, which must be multiplied by V_z/I_{yy}, are shown in Figure 4-50. The forces associated with each of the distributions shown in Figure 4-50 are

$$[f_1 \ \ f_2 \ \ f_3 \ \ f_4] = [1.293 \ \ 1.722 \ \ 5.866 \ \ 7.634] \times \frac{10^6 V_z}{I_{yy}}$$

Note that as a check, $\Sigma F_z = 13.5 \times 10^6 V_z/I_{yy} = V_z$, since $I_{yy} = 13.5 \times 10^6$ mm^4.

To determine the location of the shear center, construct the free-body diagram shown in Figure 4-51. Summing moments about axis OO yields

$$\Sigma M_o = V_z e_y + (f_1 - f_2)150 - f_4(200) = 0$$

from which

$$e_y = [(1.722 - 1.293)(150) + 7.634(200)] \times \frac{10^6}{I_{yy}} = 117.9 \text{ mm}$$

that is, approximately 3 mm to the right of the centroid.

It is also possible to determine the location of the shear center as follows. Recall from Chapter 3 that the moment of a shear flow $\bar{q}$ in a closed tube is a pure torque given by $T = 2A\bar{q}$, where A is the area of the tube. Using the shear flow distribution from the analysis of the open tube, shown again in Figure 4-52, compute $f_1 = 6 \times 10^6 V_z/I_{yy}$,

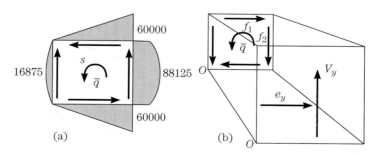

FIGURE 4-52 Shear flows $q_o(s)$ with cut at A.

$f_2 = 11.8125 \times 10^6 V_z/I_{yy}$, and again reversing the directions for application to the back face of the free-body diagram,

$$\sum M_o = V_z e_y - 150\left(6 \times 10^6 \frac{V_z}{I_{yy}}\right)$$

$$-200\left(11.8125 \times 10^6 \frac{V_z}{I_{yy}}\right) + 2(200)(150)\,27857\frac{V_z}{I_{yy}} = 0$$

which, with $I_{yy} = 13.5 \times 10^6$ mm^4, gives

$$V_z e_y = 1591.08 \times 10^6 \frac{V_z}{13.5 \times 10^6}$$

or $e_y = 117.9$ mm, as computed by the other approach.

Finally, the maximum shear stress occurs in the segment AD near D with the value

$$\tau = \frac{q}{t} = \frac{32413 V_z}{I_{yy} t} = \frac{32143(50000)}{13.5 \times 10^6(4)} \frac{\text{N}}{\text{mm}^2} = 29.8 \text{ MPa} \quad \blacklozenge$$

4-6 Thin-Walled Multitube Bending

The problem of the bending of a thin-walled multitube structure is treated in the same general manner as that of a single tube, the essential difference being that there is an unknown shear flow that exists in each of the tubes. The problem is statically indeterminate with the multiple unknown shear flows as redundants. Referring to Figure 4-53a showing two tubes, the coordinate system is centroidal with the loading applied through the shear center, the location of which is yet to be determined. The two unknown shear flows that are assumed to exist in tubes 1 and 2 at the locations of the two cuts are indicated by q_1 and q_2, respectively, shown in their assumed positive directions. The location of the cut in each tube is arbitrary. Additionally, as is always the case in a statics problem, the direction assumed for each of the unknown shear flows q_1 and q_2 indicated in Figure 4-53b is arbitrary. With tubes 1 and 2 cut, the shear flows in the two open sections are determinate and can be determined in the usual manner by evaluating

$$q_0(s) = \frac{V_y I_{yz} - V_z I_{zz}}{\Delta} A_s \bar{z} + \frac{V_z I_{yz} - V_y I_{yy}}{\Delta} A_s \bar{y}$$

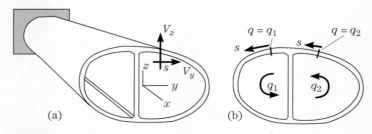

FIGURE 4-53 Two tubes in bending.

4-6 THIN-WALLED MULTITUBE BENDING

where I_{yy}, I_{zz} and I_{yz} are the centroidal second moments of, and $\bar{y}$ and $\bar{z}$ are measured with respect to, the centroidal axes of the entire cross section. Thus the resultant shear flows in tubes 1 and 2 are given, respectively, by

$$q_1(s) = q_1 + \frac{V_y I_{yz} - V_z I_{zz}}{\Delta} A_s \bar{z} + \frac{V_z I_{yz} - V_y I_{yy}}{\Delta} A_s \bar{y}$$

$$= q_1 + q_0(s)$$

and

$$q_2(s) = q_2 + \frac{V_y I_{yz} - V_z I_{zz}}{\Delta} A_s \bar{z} + \frac{V_z I_{yz} - V_y I_{yy}}{\Delta} A_s \bar{y}$$

$$= q_2 + q_0(s)$$

The function $q_0(s)$ is determined in the usual manner from the open section that results from introducing the cuts in tubes 1 and 2. Requiring the rates of twist in tubes 1 and 2 to be zero gives

$$\phi_1' = 0 = \frac{1}{2A_{01}G} \oint_{C_1} \frac{q_1(s)\,ds}{t}$$

and

$$\phi_2' = 0 = \frac{1}{2A_{02}G} \oint_{C_2} \frac{q_2(s)\,ds}{t}$$

or $\oint_{C_1} q_1(s)\,ds/t = \oint_{C_2} q_2(s)\,ds/t = 0$. Then, with the shear flows in each portion of the two tubes as shown in Figure 4-54, it follows that

$$\int_{ABD} \frac{(q_1 + q_0(s))\,ds}{t} + \int_{web} \frac{(q_1 - q_2 + q_0(s))\,ds}{t} = 0$$

and

$$\int_{web} \frac{(q_2 - q_1 + q_0(s))\,ds}{t} + \int_{DCA} \frac{(q_2 + q_0(s))\,ds}{t} = 0$$

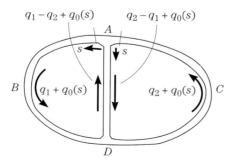

FIGURE 4-54 Directions of shear flows in a two-tube structure.

or, after simplification,

$$q_1\left(\oint_{C_1}\frac{ds}{t}\right) - q_2\left(\int_{web}\frac{ds}{t}\right) = -\oint_{C_1}\frac{q_0(s)ds}{t}$$

and

$$-q_1\left(\int_{web}\frac{ds}{t}\right) + q_2\left(\oint_{C_2}\frac{ds}{t}\right) = -\oint_{C_2}\frac{q_0(s)ds}{t}$$

a 2 by 2 set of linear algebraic equations for q_1 and q_2. Upon solving these equations, the shear flows and hence the shear stresses are completely determined. The location of the shear center is then determined by drawing the proper free-body diagram and requiring that the loads V_y and V_z together with the internal force resultants have no moment about an x axis. The specifics are more easily seen and understood in the example that follows.

EXAMPLE 4-11 Consider the dual tube shown in Figure 4-55 with $V_y = 0$ and V_z passing through the shear center. Take $t_1 = 0.1$ in, $t_2 = 0.2$ in and $t_3 = 0.3$ in. By virtue of the symmetry $I_{yz} = 0$ and Equation 4-24, the shear flow reduces to $q_0(s) = -V_z A_s \bar{z}/I_{yy}$. Denote $Q(s) = q_0(s) I_{yy}/V_z = -A_s \bar{z}$. As shown in Figure 4-55b, calculations for the shear flow in the left tube are begun at the cut at A and proceed in a counterclockwise direction to D. In the right tube integration is begun at the cut at E and proceeds in a clockwise direction to D, after which it is possible to proceed from D to A, completing the circuit. With lengths, areas and first moments of areas measured in in, in^2 and in^3, the function $Q(s)$ is computed according to the following equations.

On AB: $A_s = 0.2s, \bar{z} = 5, 0 \leq s \leq 10$ and $Q(s) = -s$ with $Q(10) = -10$.

On BC: $A_s = 0.1s, \bar{z} = 5 - 0.5s, 0 \leq s \leq 10$ and $Q(s) = -10 - 0.1s(5 - 0.5s)$ with $Q(10) = -10$.

On CD: $A_s = 0.2s, \bar{z} = -5, 0 \leq s \leq 10$ and $Q(s) = -10 + s$ with $Q(10) = 0$.

On EF: $A_s = 0.3s, \bar{z} = 5, 0 \leq s \leq 15$ and $Q(s) = -1.5s$ with $Q(15) = -22.5$.

On FG: $A_s = 0.3s, \bar{z} = 5 - 0.5s, 0 \leq s \leq 10$ and $Q(s) = -22.5 - 0.3s(5 - 0.5s)$ with $Q(10) = -22.5$.

On GD: $A_s = 0.3s, \bar{z} = -5, 0 \leq s \leq 15$ and $Q(s) = -22.5 + 1.5s$ with $Q(15) = 0$.

On DA: $A_s = 0.2s, \bar{z} = -5 + 0.5s, 0 \leq s \leq 10$ and $Q(s) = 0 + s(1 - 0.1s)$ with $Q(10) = 0$.

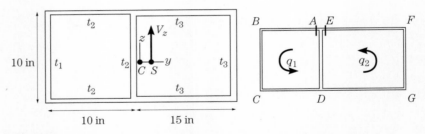

FIGURE 4-55 Cuts and positive directions for the unknown shear flows.

The function $Q(s)$ is also as shown in Figure 4-56. All values in Figure 4-56 must be multiplied by V_z/I_{yy}. The necessary integrals for evaluating q_1 and q_2 are

$$\oint_{C_1} \frac{q_0(s)\,ds}{t} = \left[-\frac{10(10)}{2(0.2)} - \frac{\left(10 + \frac{2}{3}(1.25)\right)10}{0.1} - \frac{10(10)}{2(0.2)} + \frac{\left(\frac{2(2.5)}{3}\right)10}{0.2} \right] \frac{V_z}{I_{yy}}$$

$$= -\frac{1500\,V_z}{I_{yy}}$$

$$\oint_{C_2} \frac{q_0(s)\,ds}{t} = \left[-\frac{\left(\frac{2(2.5)}{3}\right)10}{0.2} + \frac{22.5(15)}{2(0.3)} + \frac{\left(22.5 + \frac{2}{3}(3.75)\right)10}{0.3} + \frac{22.5(15)}{2(0.3)} \right] \frac{V_z}{I_{yy}}$$

$$= 1875\,\frac{V_z}{I_{yy}}$$

$$\oint_{C_1} \frac{ds}{t} = 250, \qquad \oint_{C_2} \frac{ds}{t} = 183.33 \quad \text{and} \quad \int_{\text{web}} \frac{ds}{t} = 50$$

where the integration in both tubes has been carried out in the counterclockwise direction. The equations are then

$$250 q_1 - 50 q_2 = \frac{1500 V_z}{I_{yy}}$$

and

$$-50 q_1 + 183.33 q_2 = -\frac{1875 V_z}{I_{yy}}$$

with solution $q_1 = 4.183 V_z/I_{yy}$ and $q_2 = -9.087 V_z/I_{yy}$. The resultant shear flows, all of which must be multiplied by V_z/I_{yy}, are shown in Figure 4-57. The resultant forces, also shown in Figure 4-57, are easily computed to be

$$[f_1 \;\; f_2 \;\; f_3 \;\; f_4 \;\; f_5 \;\; f_6 \;\; f_7] = [66.50 \;\; 16.94 \;\; 8.74 \;\; 149.4 \;\; 27.54 \;\; 59.94 \;\; 159.1] \frac{V_z}{I_{yy}}$$

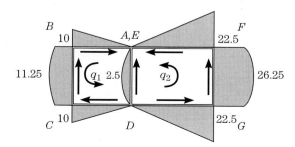

FIGURE 4-56 Shear flows in cut tubes.

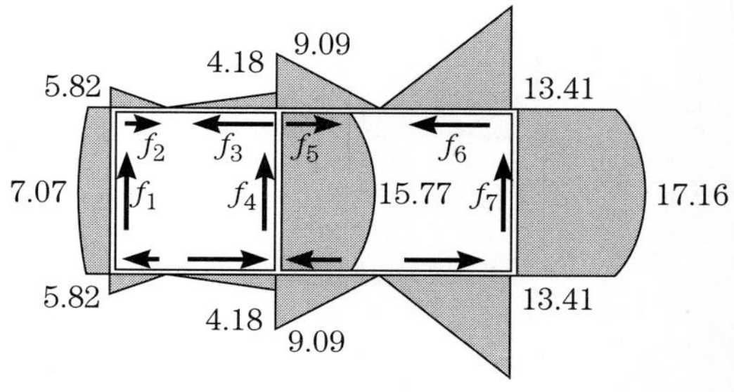

FIGURE 4-57 Resultant shear flows.

As a check,

$$\sum F_z = f_1 + f_4 + f_7 = (66.50 + 149.4 + 159.1)\frac{V_z}{I_{yy}}$$

$$= 375\frac{V_z}{I_{yy}} = V_z$$

since, as the reader should verify, $I_{yy} = 375$ in^4.

For determining the location of the shear center, the free-body diagram shown in Figure 4-58 is constructed. Consulting Figures 4-57 and 4-58

$$\sum M_o = V_z e_y + (f_2 - f_3 + f_5 - f_6)10 - f_4(10) - f_7(25) = 0$$

or

$$V_z e_y - (242 + 1494 + 3998)\frac{V_z}{I_{yy}} = 0$$

from which $e_y = 15.24$ in.

An alternate calculation for the shear center uses that fact that for a constant shear flow acting in a tube, the torque is $T = 2Aq$. Rather than the resultant shear flows pictured in Figure 4-57, those shown in Figure 4-56 can be used with the constant shear flows q_1 and q_2 included. These forces are shown in Figure 4-59. The values of the forces shown in Figure 4-59 are

$$[f_1 \ f_2 \ f_3 \ f_4] = [50.0 \ \ 16.67 \ \ 168.75 \ \ 250.00]\frac{V_z}{I_{yy}}$$

Keeping in mind that on the free-body diagram used to compute the location of the shear center these forces and shear flows must act opposite to the directions shown in Figure 4-59, it follows that

$$\sum M_o = V_z e_y + (f_1 - f_3)10 - f_2(10) - f_4(25) - 2A_1 q_1 - 2A_2 q_2 = 0$$

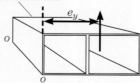

FIGURE 4-58 Free-body diagram for determining shear center.

Shear flow forces in opposite directions

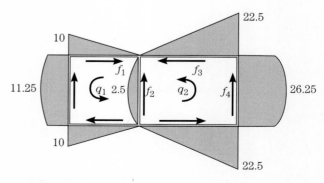

FIGURE 4-59 Forces in cut tubes.

or

$$V_z e_y = [(118.75 + 16.67)10 + 250(25) + 2(100)4.183 + 2(150)(-9.087)]\frac{V_z}{I_{yy}}$$

from which, with $I_{yy} = 375$ in^4, $e_y = 15.24$ in, as per the other approach. This approach is usually preferred from the standpoint that there are fewer forces involved in the calculation.

In any event, the maximum shear stress occurs in the middle web with

$$\tau_{max} = \frac{q_{max}}{t} = \frac{15.77}{0.2}\frac{V_z}{I_{yy}} = 0.210 V_z$$

◆

4-7 Combined Bending and Torsion

In Chapter 3 the problems of the torsion of a single tube and of the torsion of multitube sections were considered. In this chapter the corresponding bending problems were treated, where the transverse loads were assumed to act through the shear center. In this section the problem of the combined bending and torsional loading of multitube structures will be investigated using a straightforward application of the principle of superposition. It will be assumed that warping can be neglected.

Consider then a typical multitube structure acted upon by an arbitrary transverse load V in the yz plane, as shown in Figure 4-60a. As shown in Figure 4-60b, the line of action of the transverse force V is then considered to act through the shear center S with a compensating torque $T = V \cdot d$ applied so as to produce an equivalent force system.

In terms of these equivalent loads, split the original problem into two separate problems:

1. A pure torsional loading problem, to be handled per the theory of Chapter 3.
2. A transverse bending problem, to be treated per the theory of the present chapter.

The stresses will consist of (a) the bending stresses arising from the bending moments associated with the shear force V, (b) the shear stresses computed from the solution to the torsion problem and (c) the shear stresses computed from the solution to the bending problem. The solution of the original problem can then be obtained by superposition of the results from (1) and (2).

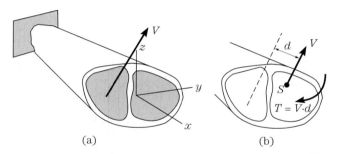

FIGURE 4-60 Shear loading of a thin-walled multitube structure.

EXAMPLE 4-12

The multitube beam is loaded as shown in Figure 4-61. Determine the bending and shear stresses. Take the uniform thickness to be $t = 0.1$ in.

Solution: Due to the horizontal plane of symmetry, the shear center will be located midway between the top and bottom surfaces. The two loads are moved so as to pass through the shear center with compensating couples introduced as shown in Figure 4-62. Thus the shear loads are $V_y = 100$ lbf and $V_z = 200$ lbf, and as shown in Figure 4-62, the torque is $(200X - 200)$ in-lbf where X is the horizontal distance, measured in inches, from the right edge of the section to the shear center. First consider the transverse bending problem and thus determine the distance X before solving the torsional problem.

Before embarking on the analysis of the stresses, consult Figures 4-61 and 4-63 to compute $\bar{z} = 2$ in (from symmetry)

$$\bar{y} = \frac{(4)(3)(0.1) + (4)(10)(0.1) + 2(10)(5)(0.1)}{32(0.1)} \text{ in} = 4.75 \text{ in}$$

$$I_{yy} = \left[\frac{3(0.1)4^3}{12} + 2(10)(0.1)2^2\right] \text{ in}^4 = 9.6 \text{ in}^4$$

$$I_{zz} = \left[4(.1)\{(4.75)^2 + (1.75)^2 + (5.25)^2\} \right.$$
$$\left. + 2\left\{\frac{(0.1)(10)^3}{12} + (0.1)(10)(0.25)^2\right\}\right] \text{ in}^4 = 38.067 \text{ in}^4$$

$I_{yz} = 0$ (from symmetry)

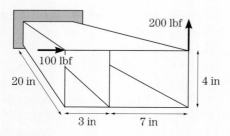

FIGURE 4-61 Combined loading of a multitube section.

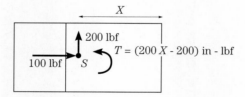

FIGURE 4-62 Equivalent force system.

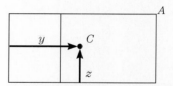

FIGURE 4-63 Area property calculations.

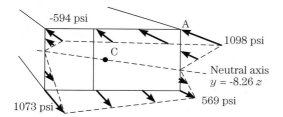

FIGURE 4-64 Bending distribution and location of the neutral axis.

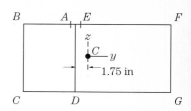

FIGURE 4-65 Locations of cuts in tubes.

Bending stresses: The maximum bending moments occur at the support. With $I_{yz} = 0$, $M_y = -4000$ in-lbf and $M_z = +2000$ in-lbf, it follows that

$$\sigma_x = \left[-y\frac{2000}{39.67} + z\frac{-4000}{9.6} \right] \text{ psi} = [-50.42y - 416.67z] \text{ psi}$$

with the maximum of $\sigma_x = -1098$ psi occurring at point A, the upper right corner. The stress distribution and location of the neutral axis are shown in Figure 4-64.

Transverse shear stresses: As outlined in section 4-6, it is necessary to introduce cuts in each of the tubes. These are shown as A and E in Figure 4-65. With $I_{yz} = 0$, the expression for evaluating shear flows is

$$\tau t = q = -\frac{V_z}{I_{yy}}A_s\overline{z} - \frac{V_y}{I_{zz}}A_s\overline{y} = -\frac{V_z}{I_{yy}}Q_y - \frac{V_y}{I_{zz}}Q_z$$

where distances, areas and first moments of areas are measured in in, in^2 and in^3, respectively, and $Q_y = A_s\overline{z}$ and $Q_z = A_s\overline{y}$.

On AB: $A_s = 0.1s$, $\overline{z} = 2$, $\overline{y} = -(3.5 + s)/2$, $0 \leq s \leq 3$ with $Q_y(s) = 0.2s$, $Q_z(s) = -0.05s(3.5 + s)$, $Q_y(3) = 0.6$ and $Q_z(3) = -0.975$.

On BC: $A_s = 0.1s$, $\overline{z} = 2 - 0.5s$, $\overline{y} = -4.75$, $0 \leq s \leq 4$ with $Q_y(s) = 0.6 + 0.1s(2 - 0.5s)$, $Q_z(s) = -0.975 - 0.475s$, $Q_y(4) = 0.6$ and $Q_z(4) = -2.875$.

On CD: $A_s = 0.1s$, $\overline{z} = -2$, $\overline{y} = -4.75 + 0.5s$, $0 \leq s \leq 3$ with $Q_y(s) = 0.6 - 0.2s$, $Q_z(s) = -2.875 - 0.1s(4.75 - 0.5s)$, $Q_y(3) = 0$ and $Q_z(3) = -3.85$.

On EF: $A_s = 0.1s$, $\overline{z} = 2$, $\overline{y} = -1.75 + 0.5s$, $0 \leq s \leq 7$ with $Q_y(s) = 0.2s$, $Q_z(s) = 0.1s(-1.75 + 0.5s)$, $Q_z(7) = 1.4$ and $Q_z(7) = 1.225$.

On FG: $A_s = 0.1s$, $\overline{z} = 2 - 0.5s$, $\overline{y} = 5.25$, $0 \leq s \leq 4$ with $Q_y(s) = 1.4 + 0.1s(2 - 0.5s)$, $Q_z(s) = 1.225 + 0.525s$, $Q_y(7) = 1.4$ and $Q_z(7) = 3.325$.

On GD: $A_s = 0.1s$, $\overline{z} = -2$, $\overline{y} = 5.25 - 0.5s$, $0 \leq s \leq 7$ with $Q_y(s) = 1.4 - 0.2s$, $Q_z(s) = 3.325 + 0.1s(5.25 - 0.5s)$, $Q_y(7) = 0$ and $Q_z(7) = 4.55$.

On DA: $A_s = 0.1s$, $\overline{z} = -2 + 0.5s$, $\overline{y} = -1.75$, $0 \leq s \leq 4$ with $Q_y(s) = 0.1s(-2 + 0.5s)$, $Q_z(s) = 0.7 - 0.1s(1.75)$, $Q_y(4) = 0$ and $Q_z(4) = 0$.

TABLE 4-1 Integrals of First Moments

Segment	$\int Q_y\, ds$	$\int Q_z\, ds$
AB	0.9000	−1.2375
BC	2.9333	−7.7000
CD	0.9000	−10.3125
EF	4.9000	1.4292
FG	6.1333	9.1000
GD	4.9000	30.4208
DA or DE	−0.5333	1.4000

The final values of zero for Q_z and Q_y on segment DA serve as a check on the calculations.

The integrals of Q_y and Q_z on each of the segments appear in Table 4-1. The corresponding distributions of shear flow and the forces associated with V_z and V_y, respectively, are shown in Figure 4-66 in each of the segments. The two shear flow distributions must of course be added to account for the simultaneous application of V_z and V_y.

Note that for the V_z part

$$\sum F_z = (61.11 + 11.11 + 127.78)\ \text{lbf} = 200\ \text{lbf}$$

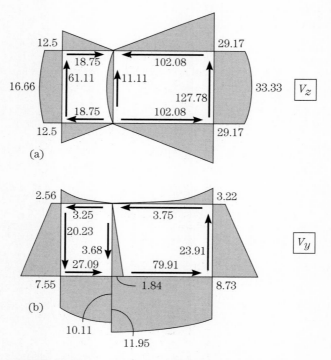

FIGURE 4-66 Shear flows and resultants (lbf) in individual segments.

4-7 COMBINED BENDING AND TORSION

and
$$\sum F_y = 0$$

and that for the V_y part
$$\sum F_y = (27.09 + 79.91 + 3.25 - 3.75) \text{ lbf} = 100 \text{ lbf}$$

with
$$\sum F_z = 0$$

both of which serve as checks on the calculations.

The equations that must be solved to determine the shear flows q_1 and q_2 in the two tubes are

$$q_1 \left(\oint_{C_1} \frac{ds}{t} \right) - q_2 \left(\int_{\text{web}} \frac{ds}{t} \right) = -\oint_{C_1} q_0(s) \frac{ds}{t}$$

and

$$-q_1 \left(\int_{\text{web}} \frac{ds}{t} \right) + q_2 \left(\oint_{C_2} \frac{ds}{t} \right) = -\oint_{C_2} q_0(s) \frac{ds}{t}$$

With

$$\oint_{C_1} \frac{ds}{t} = 140, \quad \int_{\text{web}} \frac{ds}{t} = 40, \quad \oint_{C_2} \frac{ds}{t} = 220$$

$$\oint_{C_1} q_0(s) \frac{ds}{t} = -405.9 \quad \text{and} \quad \oint_{C_2} q_0(s) \frac{ds}{t} = 4320.8$$

and the integrals around C_1 and C_2 taken in the counterclockwise direction, there results

$$140 q_1 - 40 q_2 = 405.9$$

and

$$-40 q_1 + 220 q_2 = -4320.8$$

with the solution $q_1 = -2.86$ lbf/in and $q_2 = -20.16$ lbf/in. These constant shear flows are equivalent to torques of $t_1 = 2A_1 q_1 = -68.7$ in-lbf and $t_2 = 2A_2 q_2 = -1129.0$ in-lbf. With all the resultants on the back face shown in the corresponding negative directions, the free-body diagram for locating the shear center is shown in Figure 4-67. Summing moments about axis OO yields

$$\sum M_{OO} = [100(2) - 200 \cdot Y - 68.7 - 1129.0 + 7.43(3) + 151.69(10) + (105.83 - 15.5)(4)] \text{ in-lbf} = 0$$

from which $Y = 4.514$ in. The final bending shear flow distribution resulting from applying the loads at the shear center can be obtained by superposing $q_1 = -2.861$ lbf/in and $q_2 = -20.160$ lbf/in upon those from the addition of V_y and V_z pictured in Figure 4-66.

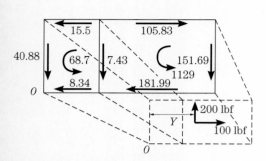

FIGURE 4-67 Free-body diagram for locating shear center.

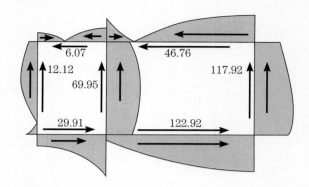

FIGURE 4-68 Final shear flow distributions and resultant forces.

Torsional stresses: In addition to the bending shear stresses, those due to the torque must also be superposed. With $Y = 4.514$ in, the torque that must be applied is

$$T = [200(10 - 4.514) - 200] \text{ in-lbf} = 897.2 \text{ in-lbf}$$

From Chapter 3, the equations that must be solved are

$$2A_1 G\phi' = \left(\oint_{C_1} \frac{ds}{t}\right) q_1 - \left(\int_W \frac{ds}{t}\right) q_2$$

and

$$2A_2 G\phi' = -\left(\int_W \frac{ds}{t}\right) q_1 + \left(\oint_{C_2} \frac{ds}{t}\right) q_2$$

with the total torque then given by

$$T = 2A_1 q_1 + 2A_2 q_2$$

These equations become

$$140 q_1 - 40 q_2 = 2(12) G\phi'$$

and

$$-40 q_1 + 220 q_2 = 2(28) G\phi'$$

with solution $q_2 = 0.301 G\phi'$ and $q_1 = 0.258 G\phi'$ and

$$897.2 \text{ in-lbf} = 2(12)(0.258 G\phi') + 2(28)(0.301 G\phi') = 23.037 G\phi'$$

from which $G\phi' = 38.95$ with $q_1 = 10.05$ lbf/in and $q_2 = 11.72$ lbf/in. These constant shear flows must be superposed on those obtained from solving the bending problem. The final distributions along with the corresponding resultants are shown in Figure 4-68. The reader should check that the corresponding resultant force and moment associated with these individual forces are correct.

4-8 Summary

In general a bar can be subjected to loads that produce axial, bending, shear and torsional resultants, that is, three internal force resultants and three internal moment resultants. For the unsymmetrical bending problem only four resultants, that is, two transverse shear forces and two bending moments, are transmitted. On the basis of the theory developed for unsymmetrical sections, external loads must be applied through the shear center, that is, so that there is no resultant torque about the axis of the beam. The corresponding bending stress is then a linear function of the in-plane coordinates given by

$$\sigma_x = -y\frac{M_z I_{yy} + M_y I_{yz}}{\Delta} + z\frac{M_y I_{zz} + M_z I_{yz}}{\Delta}$$

Based on this result for the bending stress, the shear flow in thin-walled open sections can be determined on the basis of an equilibrium argument as

$$\tau t = q = \frac{V_y I_{yz} - V_z I_{zz}}{\Delta} A_s \bar{z} + \frac{V_z I_{yz} - V_y I_{yy}}{\Delta} A_s \bar{y}$$

Stresses in single tubes and multitube cross sections subjected to loading passing through the shear center constitute statically indeterminate problems that are essentially handled by considering shear flows to be internal redundants that are then determined on the basis of requiring no twist of the beam. The resulting equations, one for each tube, can then be used to determine the redundant internal shear flows and, ultimately, the location of the shear center for the cross section.

The problem of combined bending and torsion of multitube cross sections is treated using superposition, wherein the loading is decomposed into a shear passing through the shear center and a compensating torque. Each of the pure torsion and pure bending problems are handled separately using results from Chapters 3 and 4, respectively, with the shear stresses from the bending problem superposed upon those for the torsion problem.

REFERENCES

1. Oden, J. T., and Ripperger, E. A. *Mechanics of Elastic Structures*. McGraw-Hill, New York, 1981.
2. Timoshenko, S. P. *Strength of Materials: Part I, Elementary Theory and Problems*, 3rd ed. Krieger, New York, 1955.

EXERCISES

Section 4-2

1. Determine the location of the neutral axis and the maximum bending stress for the rectangular cross-section beam loaded as shown. Take $b = 2$ in, $h = 6$ in, $L = 6$ ft, $P = 1000$ lbf, and $\theta = 20°$. By what factor is the maximum bending stress increased compared to the case where $\theta = 0$?

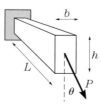

2. Determine the location of the neutral axis and the maximum bending stress for the T-section beam loaded as shown. Take $b = 3$ in, $h = 6$ in, $t = 0.188$ in, $L = 6$ ft, $P = 1000$ lbf, and $\theta = 7°$. Assume that the beam is simply supported about the y and z axes at both ends and that the load P is applied midspan. By what factor is the maximum bending stress increased compared to the case where $\theta = 0$?

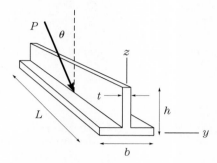

3. Repeat problem 1 with $b = 45$ mm, $h = 130$ mm, $L = 1.6$ m, $P = 5.3$ kN and $\theta = 15°$.

4. Repeat problem 2 with $b = 70$ mm, $h = 140$ mm, $t = 4$ mm, $L = 1.6$ m, $P = 2.3$ kN and $\theta = 6°$.

5. For the section subjected to a bending moment as shown below, determine the location of the neutral axis and the maximum bending stress. Take $M = 6000$ in-lbf, $b = 12$ in, $h = 6$ in and $t = 0.375$ in.

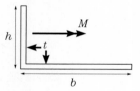

6. Repeat problem 5 with $M = 610$ N-m, $b = 0.29$ m, $h = 0.15$ m and $t = 0.01$ m.

7. For the cantilever beam loaded as shown below, determine the location of the neutral axis and the maximum bending stress. Take $P = 120$ lbf, $b = 8$ in, $h = 5$ in, $t = 0.5$ in, $L = 10$ ft and $\theta = 10°$.

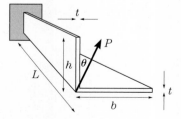

8. Repeat problem 7 with $\theta = 80°$.

9. Repeat problem 7 with $P = 600$ N, $b = 0.19$ m, $h = 0.1$ m, $t = 0.015$ m, $L = 3$ m and $\theta = 10°$.

10. Repeat problem 9 with $\theta = 80°$.

11. Determine the magnitude and direction of the displacement at the end of the cantilever beam described in problem 1. Take $E = 3 \times 10^7$ psi.

EXERCISES

12. Determine the magnitude and direction of the displacement at the middle of the simply supported beam described in problem 2. Take $E = 10^7$ psi.

13. Determine the magnitude and direction of the displacement at the end of the cantilever beam described in problem 7. Take $E = 10^7$ psi.

14. For the uniform-thickness cantilever beam shown below, determine the location of the neutral axis and the maximum bending stress. Determine the magnitude and direction of the displacement at the point where the load P is applied. Take $P = 120$ lbf, $b = 9$ in, $h = 5$ in, $t = 0.5$ in, $L = 10$ ft, $w_0 = 15$ lbf/ft and $E = 3 \times 10^7$ psi.

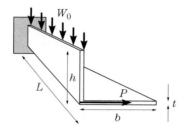

15. Repeat problem 14 with $P = 600$ N, $b = 0.19$ m, $h = 0.1$ m, $t = 0.018$ m, $L = 3$ m, $w_0 = 200$ N/m and $E = 205$ GPa.

16. For the uniform-thickness beam shown below, determine the location of the neutral axis and the maximum bending stress (a) at the load P and (b) at a point midway between the support and the load P. Assume that the beam is simply supported in both directions at its ends. Take $b = 4$ in, $h = 8$ in, $t = 0.3$ in, $L = 9$ ft, $P = 1000$ lbf and $w_0 = 100$ lbf/ft. Take P to be applied at the middle of the beam where the flange and web intersect.

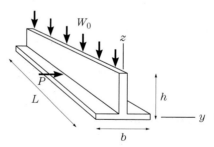

17. Repeat problem 16 with $b = 90$ mm, $h = 190$ mm, $t = 10$ mm, $L = 2.7$ m, $P = 4.5$ kN and $w_0 = 1000$ N/m.

Section 4-3

18. For the angle section shown, where the shear forces are applied through the shear center S, determine and sketch the shear stress distribution. What is the maximum shear stress? Take $b = 4$ in, $h = 4$ in, $t = 0.125$ in, $V_y = 5000$ lbf and $V_z = 0$.

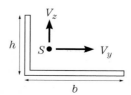

19. Repeat problem 18 with $b = 4$ in, $h = 4$ in, $t = 0.125$ in, $V_y = 5000$ lbf and $V_z = 5000$ lbf.
20. Repeat problem 18 with $b = 4$ in, $h = 4$ in, $t = 0.125$ in, $V_y = 5000$ lbf and $V_z = -5000$ lbf.
21. Repeat problem 18 with $b = 8$ in, $h = 4$ in, $t = 0.125$ in, $V_y = 5000$ lbf and $V_z = 0$.
22. Repeat problem 18 with $b = 8$ in, $h = 4$ in, $t = 0.125$ in, $V_y = 5000$ lbf and $V_z = 5000$ lbf.
23. Repeat problem 18 with $b = 8$ in, $h = 4$ in, $t = 0.125$ in, $V_y = 5000$ lbf and $V_z = -5000$ lbf.
24. Repeat problem 18 with $b = 8$ in, $h = 4$ in, $t = 0.125$ in, $V_y = 0$ and $V_z = 5000$.
25. For the section shown, where the shear forces are applied through the shear center S, determine and sketch the shear stress distribution. What is the maximum shear stress? Take $a = 60$ mm, $h = 60$ mm, $b = 60$ mm, $t = 6$ mm, $V_y = 60$ kN and $V_z = 0$.

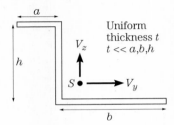

26. Repeat problem 25 with $a = 60$ mm, $h = 60$ mm, $b = 60$ mm, $t = 6$ mm, $V_z = 40$ kN and $V_y = 0$.
27. Repeat problem 25 with $a = 60$ mm, $h = 60$ mm, $b = 120$ mm, $t = 6$ mm, $V_y = 80$ kN and $V_z = 0$.
28. Repeat problem 25 with $a = 60$ mm, $h = 90$ mm, $b = 120$ mm, $t = 6$ mm, $V_z = 50$ kN and $V_y = 50$ kN.
29. Repeat problem 25 with $a = 60$ mm, $h = 90$ mm, $b = 60$ mm, $t = 6$ mm, $V_z = 60$ kN and $V_y = 0$.
30. Repeat problem 25 with $a = 120$ mm, $h = 60$ mm, $b = 120$ mm, $t = 6$ mm, $V_z = 30$ kN and $V_y = 60$ kN.
31. For the section shown, where the shear forces are applied through the shear center S, determine and sketch the shear stress distribution. What is the maximum shear stress? Take $a = 2$ in, $b = 2$ in, $h = 2$ in, $t = 0.125$ in, $V_y = 2000$ lbf and $V_z = 0$.

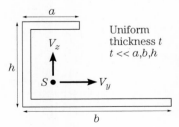

32. Repeat problem 31 with $a = 2$ in, $b = 2$ in, $h = 2$ in, $t = 0.125$ in, $V_y = 0$ and $V_z = 2000$ lbf.
33. Repeat problem 31 with $a = 2$ in, $h = 4$ in, $b = 4$ in, $t = 0.125$ in, $V_y = 2000$ lbf and $V_z = 0$.
34. Repeat problem 31 with $a = 2$ in, $h = 4$ in, $b = 4$ in, $t = 0.125$ in, $V_y = 2000$ lbf and $V_z = 4000$ lbf.

EXERCISES

35. Repeat problem 31 with $a = 2$ in, $h = 2$ in, $b = 4$ in, $t = 0.125$ in, $V_y = 2000$ lbf and $V_z = 2000$ lbf.

36. Repeat problem 31 with $a = 2$ in, $h = 2$ in, $b = 4$ in, $t = 0.125$ in, $V_y = 2000$ lbf and $V_z = -2000$ lbf.

37. For the section shown, where the shear forces are applied through the shear center S, determine and sketch the shear stress distribution. What is the maximum shear stress? Take $b = h$, $t \ll b$, $V_y = V$ and $V_z = 0$.

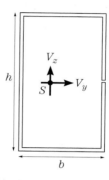

38. Repeat problem 37 with $V_y = 0$ and $V_z = V$.

39. For the section shown, where the shear forces are applied through the shear center S, determine and sketch the shear stress distribution. What is the maximum shear stress? Take $h = 2b$, $h = 4a$, $t \ll a$, $V_y = V$ and $V_z = 0$.

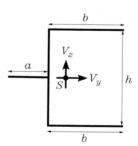

40. Repeat problem 39 with $V_y = 0$ and $V_z = V$.

41. For the section shown, where the shear forces are applied through the shear center S, determine and sketch the shear stress distribution. What is the maximum shear stress? Take $t \ll R$, $V_y = V$ and $V_z = 0$.

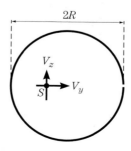

42. Repeat problem 41 with $V_y = 0$ and $V_z = V$.

43. For the section shown, where the shear forces are applied through the shear center S, determine and sketch the shear stress distribution. What is the maximum shear stress? Take $t \ll R$, $V_y = V$ and $V_z = 0$.

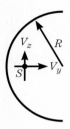

44. Repeat problem 43 with $V_y = 0$ and $V_z = V$.

45. For the section shown, where the shear forces are applied through the shear center S, determine and sketch the shear stress distribution. What is the maximum shear stress? Take $b = h$, $\alpha = 30°$, $t \ll h$, $V_y = V$ and $V_z = 0$. For $h = 3$ in and $t = 0.1875$ in, what is the allowable V if $\tau_{max} = 8$ ksi?

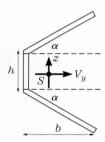

46. Repeat problem 45 with $V_y = 0$ and $V_z = V$. For $h = 3$ in and $t = 0.1875$ in, what is the allowable V if $\tau_{max} = 8$ ksi?

Section 4-4

47. Determine the location of the shear center for the section shown. Take $b = a = h$.

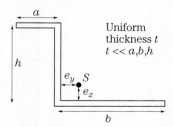

48. Repeat problem 47 with $a = h$ and $b = 2h$.

49. Repeat problem 47 with $b = h$ and $a = h/2$.

EXERCISES

50. Determine the location of the shear center for the section shown. Take $b = h$ and $a = h/2$.

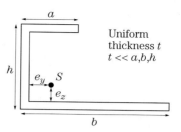

51. Repeat problem 50 with $b = 2h$ and $a = h$.

52. Repeat problem 50 with $b = 3h$ and $a = h$.

53. Determine the location of the shear center for the uniform-thickness cross section shown. Take $b = h$ and $\alpha = 30°$.

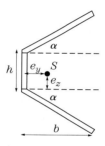

54. Repeat problem 53 with $b = 0.5h$ and $\alpha = 45°$.

55. Repeat problem 53 with $b = h$ and $\alpha = 60°$.

56. Determine the location of the shear center for the uniform-thickness cross section shown. Take $b = h$ and $t \ll b$.

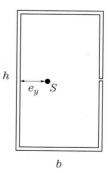

57. Repeat problem 56 with $h = 2b$.

58. Repeat problem 56 with $b = 2h$.

59. Determine the location of the shear center for the uniform-thickness cross section shown. Take $b = h = 2a$.

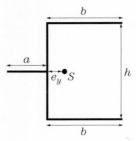

60. Repeat problem 59 with $h = 2b$, $a = b/2$.

61. Determine the location of the shear center for the uniform-thickness cross section shown.

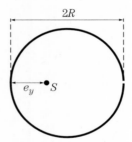

62. Determine the location of the shear center for the uniform-thickness cross section shown.

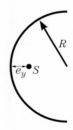

Section 4-5

63. Determine the maximum shear stress and the location of the shear center for the section shown. Take $b = 0.3$ m, $h = 0.2$ m, $t_1 = 0.01$ m, $t_2 = 0.02$ m, $V_y = 30$ kN and $V_z = 20$ kN.

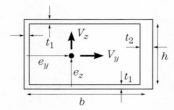

EXERCISES

64. Repeat problem 63 with $b = 0.3$ m, $h = 0.2$ m, $t_1 = 0.01$ m, $t_2 = 0.04$ m, $V_y = 30$ kN and $V_z = 20$ kN.

65. Repeat problem 63 with $b = 15$ in, $h = 20$ in, $t_1 = 0.125$ in, $t_2 = 0.0625$ in, $V_y = 10$ kips and $V_z = 20$ kips.

66. Determine the maximum shear stress and the location of the shear center for the section shown. Take $b = 0.3$ m, $h = 0.2$ m, $t_1 = 0.01$ m, $t_2 = 0.02$ m, $V_y = 30$ kN and $V_z = 20$ kN.

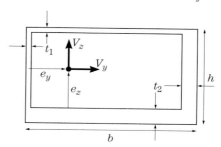

67. Repeat problem 66 with $b = 0.3$ m, $h = 0.2$ m, $t_1 = 0.01$ m, $t_2 = 0.04$ m, $V_y = 25$ kN and $V_z = 30$ kN.

68. Repeat problem 66 with $b = 18$ in, $h = 15$ in, $t_1 = 0.125$ in, $t_2 = 0.094$ in, $V_y = 12$ kips and $V_z = 18$ kips.

69. Determine the maximum shear stress for the section shown. Take $b = R$, the uniform thickness $t \ll R$ and $V_y = V_z = V$. Also determine the distance e_y to the location of the shear center.

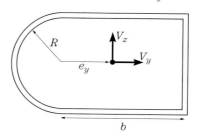

70. Repeat problem 69 with $b = 0$.

71. Repeat problem 69 with $b = 2R$.

Section 4-6

72. For the uniform 0.2 in thick section shown, determine the maximum shear stress. Take $R = 6$ in, $b = 10$ in and $V_z = 15000$ lbf. Also determine the distance e_y to the shear center.

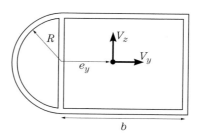

73. Determine the maximum shear stress for the section shown. Take all thicknesses as 0.1 in. Also determine the location of the shear center.

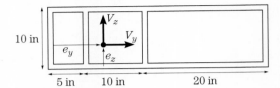

74. Determine the maximum shear stress for the section shown. All thicknesses are 0.1 in except along AB, where the thickness is 0.2 in. Also determine the location of the shear center. Take $V_y = 3500$ lbf and $V_z = 4000$ lbf.

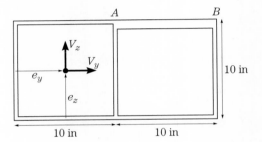

Section 4-7

75. Show that the resultant shear stresses obtained in Example 4-12 are equivalent to the two loads applied at the shear center plus the compensating torque.

76. For the section shown, determine the bending and shear stresses. Take $t_1 = 0.06$ in, $t_2 = 0.18$ in, $a = 6$ in, $b = 12$ in, $L = 60$ in, $P = 400$ lbf and $Q = 0$.

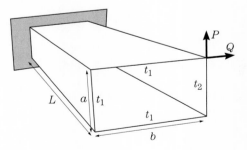

77. Repeat problem 76 with $t_1 = 0.06$ in, $t_2 = 0.18$ in, $a = 6$ in, $b = 12$ in, $L = 60$ in, $P = 0$ and $Q = 800$ lbf.

78. Repeat problem 76 with $t_1 = 2$ mm, $t_2 = 6$ mm, $a = 0.15$ m, $b = 0.3$ m, $L = 1.5$ m, $P = 1.8$ kN and $Q = 0$.

79. Repeat problem 76 with $t_1 = 2$ mm, $t_2 = 6$ mm, $a = 0.15$ m, $b = 0.3$ m, $L = 1.5$ m, $P = 0$ and $Q = 4$ kN.

EXERCISES

80. For the section loaded as shown, determine the bending and shear stresses. Take $t_1 = 0.1$ in, $t_2 = 0.3$ in, $t_3 = 0.2$ in, $a = 4$ in, $b = 4$ in, $L = 120$ in and $P = 400$ lbf.

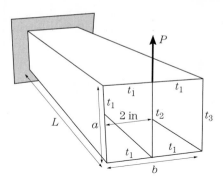

81. Repeat problem 80 with $t_1 = 0.1$ in, $t_2 = 0.2$ in, $t_3 = 0.3$ in, $a = 4$ in, $b = 8$ in, $L = 100$ in and $P = 600$ lbf.

82. Repeat problem 80 with $t_2 = 4$ mm, $t_2 = 12$ mm, $t_3 = 8$ mm, $a = 100$ mm, $b = 100$ mm, $L = 2.5$ m and $P = 2$ kN.

83. Repeat problem 80 with $t_2 = 4$ mm, $t_2 = 8$ mm, $t_3 = 12$ mm, $a = 100$ mm, $b = 200$ mm, $L = 2.0$ m and $P = 3$ kN.

84. For the uniform 0.2 in thick section shown, determine the bending and shear stresses. Take $h = 6$ in, $a = 6$ in, $b = 8$ in, $L = 98$ in and $P = 1000$ lbf.

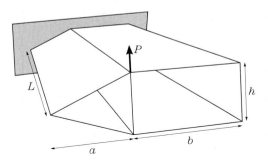

85. Repeat problem 84 with $h = 0.15$ m, $a = 0.15$ m, $b = 0.20$ m, $L = 2.0$ m and $P = 4.25$ kN.

5 Plane-Curved Beams

5-1 Introduction
5-2 Development of the Governing Equations
5-3 Bending Stresses
5-4 Transverse Shear Stresses
5-5 Displacements
5-6 Summary
References
Exercises

5-1 Introduction

In Chapter 4 the theory for straight bars was developed. The results, borne out by experiments, are that the bending stress in a straight bar can be computed using the flexure formula $\sigma = -My/I$. Frequently there are applications where the axis or centerline of the beam is curved rather than straight. Curved pipes and mechanical presses, both indicated in Figure 5-1, are instances where stresses and displacements must be determined on the basis of a theory that accounts for the nonstraight geometry of the structure.

In situations where the h/R ratio of a curved beam is small, as indicated in Figure 5-2a, the flexure formula $\sigma = -My/I$ continues to give acceptable accuracy. How-

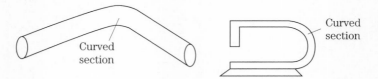

FIGURE 5-1 Examples of curved beams.

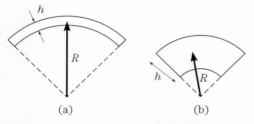

FIGURE 5-2 Thin and thick curved beams.

ever, when the h/R ratio is large, as indicated in Figure 5-2b, the flexure formula can no longer predict the bending stress accurately and the curvature must be taken into account. In this chapter a general theory of plane-curved, untwisted bars subjected to loadings that result in deformations in the plane of curvature, will be developed.

5-2 Development of the Governing Equations

The equations that govern the behavior of curved beams are developed in the usual manner, namely, by consideration of the geometry of the region, loading, equilibrium, deformation and material behavior. The equations are developed for a general untwisted cross section that has a plane of symmetry.

Geometry. Consider Figure 5-3, which shows the geometry of such a bar. The size and orientation of the cross section must be unchanging along the axis of the bar, taken to be the x axis passing everywhere through the centroid of the section. The x axis is everywhere tangent to a circular arc of radius R, that is, $dx = R\, d\theta$, where θ is the angular variable associated with a change in location along the bar. The y axis lies along the local direction of the radius R with the z axis such that a right-handed system is formed. The xy plane is a plane of symmetry for the cross section. As will be seen presently, this assumption is necessary to prevent out-of-plane effects.

Loading. The loadings that will be considered are shown in Figure 5-4a, which indicates distributed loadings q_x and q_y acting per unit of length along the centroidal axis. Figure 5.4b indicates that concentrated loads X applied in the local x direction through the centroidal axis and Y applied along the y direction, and concentrated moments M applied about the z axis, are also permissible.

Equilibrium. Figure 5-5 shows a free-body diagram of a differential element of the beam acted upon by external distributed loadings and internal force resultants N and V and internal moment M. Summing forces in the radial direction yields

$$-N \sin \frac{d\theta}{2} - V \cos \frac{d\theta}{2} + q_y^* R\, d\theta + (V + dV)\cos \frac{d\theta}{2} - (N + dN)\sin \frac{d\theta}{2} = 0$$

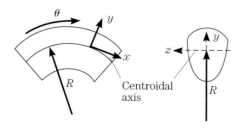

FIGURE 5-3 Geometry for a plane-curved beam.

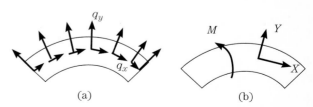

FIGURE 5-4 In-plane and out-of-plane external loadings.

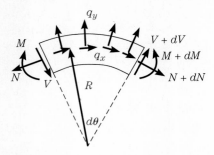

FIGURE 5-5 Free-body diagram of a curved beam segment.

where q_y^* represents an average value of q_y on the element. Passing to the limit yields

$$\frac{dV}{d\theta} - N + Rq_y = 0 \qquad (5\text{-}1a)$$

The student is asked to show in the exercises that summing forces in the tangential direction yields

$$\frac{dN}{d\theta} + V + Rq_x = 0 \qquad (5\text{-}1b)$$

and that moment equilibrium about the z axis yields

$$\frac{dM}{d\theta} + RV = 0 \qquad (5\text{-}1c)$$

In terms of the arc length $s = R\theta$ along the centroidal axis of the beam, these three equations can be expressed as

$$\frac{dV}{ds} - \frac{N}{R} + q_y = 0, \qquad \frac{dN}{ds} + \frac{V}{R} + q_x = 0 \quad \text{and} \quad \frac{dM}{ds} + V = 0$$

Consulting Figure 5-6, it is seen that the relations between the stresses σ_x and τ_{xy} and the force resultants N, V and M are

$$N = \int_{\text{Area}} \sigma_x \, dA, \qquad M = -\int_{\text{Area}} y\sigma_x \, dA \quad \text{and} \quad V = \int_{\text{Area}} \tau_{xy} \, dA \qquad (5\text{-}2)$$

It is left for the student to show that unless the cross section is symmetric with respect to the xy plane, moments about the y axis result, producing bending out of the plane.

Deformations. Referring to Figure 5-7, the displacements are taken as

$$U = u(x) + y\theta(x) \qquad (5\text{-}3a)$$
$$V = v(x) \qquad (5\text{-}3b)$$

where $u(x)$ and $v(x)$ represent the components of the displacements of the centerline of the beam, and $-\theta(x)$ represents the rotation of the cross section about the z axis.

5-2 DEVELOPMENT OF THE GOVERNING EQUATIONS

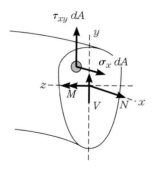

FIGURE 5-6 Stresses and force resultants.

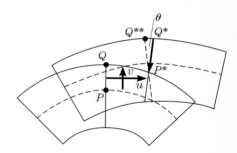

FIGURE 5-7 Displacements and rotations.

The fact that U is assumed to depend linearly on y means that *plane sections remain plane*, that is, that lines PQ and P^*Q^* are straight.

The student is asked to show in the exercises that by specializing the cylindrical coordinate strain-displacement relations of problem 36 in Chapter 1 to the present situation, the strain-displacement relations are

$$\epsilon_x = \frac{1}{g}\left(\frac{\partial U}{\partial x} + kV\right) \tag{5-4a}$$

$$\epsilon_y = \frac{\partial V}{\partial y} \tag{5-4b}$$

and

$$\gamma_{xy} = \frac{1}{g}\left(g\frac{\partial U}{\partial y} - kU + \frac{\partial V}{\partial x}\right) \tag{5-4c}$$

where $g = 1 + ky$ and where $k = 1/R$ is the curvature. Substituting Equations (5-3a–b) into Equations (5-4a–c) yields $\epsilon_y = 0$ and

$$g\epsilon_x = u' + y\theta' + kv \tag{5-4d}$$

$$g\gamma_{xy} = v' - ku + \theta \tag{5-4e}$$

At this point in the development of the theory, the eight equations (5-1), (5-2), (5-4d–e) contain the ten unknowns M, N, V, σ_x, τ_{xy}, ϵ_x, γ_{xy}, u, v and θ. Further, Equations (5-1) and (5-2) contain the kinetic variables M, N, V, σ_x, and τ_{xy}, whereas Equations (5-4d–e) contain the kinematical variables ϵ_x, γ_{xy}, u, v and θ, so that there is no way to relate the forces to the displacements.

The additional equations that are necessary to bridge the gap come from the material behavior.

Stress-Strain. Assume that the beam is composed of a one-dimensional elastic isotropic solid behaving according to

$$\sigma_x = E\epsilon_x \qquad \tau_{xy} = G\gamma_{xy} \tag{5-5}$$

Combination. In the expression for γ_{xy}, v' is the slope due to bending and $ku = u/R$ is the rotation of the cross section due to the tangential displacement u, as indicated in Figure 5-7. Requiring that $\theta = -(v' - ku)$ forces the point Q^* to coincide with the point Q^{**}, that is, the line P^*Q^* remains perpendicular to the axis of the beam, with the result that from Equation (5-4d), $\gamma_{xy} = 0$, resulting in no shear deformation. The remaining strain-displacement relation can then be expressed as

$$\epsilon_x = u' + kv - \frac{y}{1+ky}(v'' + k^2 v) \tag{5-6}$$

Note that due to the presence of $g = 1 + ky$ in the denominator of ϵ_x, the strains are not a linear function of the y coordinate as in the case of a straight beam. This is because although it has been assumed that plane sections remain plane, the original lengths of fibers above and below the centroidal axis are not the same, resulting in the nonlinear dependence of the strains on the coordinate y.

Combining Equations (5-2), (5-5) and (5-6) results in the force displacement relations

$$N = AE(u' + kv) - EI_1(v'' + k^2 v) \tag{5-7a}$$
$$V = 0 \tag{5-7b}$$
$$M = EI_2(v'' + k^2 v) \tag{5-7c}$$

where

$$I_1 = \int_{\text{Area}} \frac{y}{g}\, dA \quad \text{and} \quad I_2 = \int_{\text{Area}} \frac{y^2}{g}\, dA$$

In Equation (5-7c) the integral

$$I_2 = \int_{\text{Area}} \frac{y^2}{g}\, dA = \int_{\text{Area}} \frac{y^2}{1+ky}\, dA$$

is the counterpart of

$$I_z = \int_{\text{Area}} y^2\, dA$$

which is the second moment of the area for the straight bar. As the radius $R \to \infty$, $k \to 0$, and

$$\lim_{k \to 0} I_2 = I_z$$

It is left for the student to show in the exercises that the two area properties I_1 and I_2 are related by $kI_2 = -I_1$. Using this fact, Equations (5-7a) and (5-7c) can be combined to yield

$$v'' + k^2 v = \frac{M}{EI_2} \tag{5-8a}$$

and

$$u' + kv = \frac{N - kM}{AE} \tag{5-8b}$$

so that with Equations (5-5a) and (5-6) the normal stress is given by

$$\sigma_x = \frac{N - kM}{A} - \frac{My}{(1 + ky)I_2} \qquad (5\text{-}9)$$

The remaining equations are the equilibrium equations, namely,

$$\frac{dV}{ds} - \frac{N}{R} + q_y = 0, \qquad \frac{dN}{ds} + \frac{V}{R} + q_x = 0 \quad \text{and} \quad \frac{dM}{ds} + V = 0$$

On the basis of Equation (5-7b), that is, $V = 0$, the shear force V is usually eliminated from the equilibrium equations to yield

$$\frac{d^2M}{ds^2} - \frac{N}{R} + q_y = 0$$

and

$$\frac{dN}{ds} - \frac{1}{R}\frac{dM}{ds} + q_x = 0$$

which can be integrated to determine N and M for the purposes of computing the bending stresses from Equation (5-9). It is also frequently possible to determine N, M and V on the basis of drawing free-body diagrams. In either case, once N, M and V are known, the displacements can then be determined by integrating Equations (5-8). All these issues will be considered in the sections that follow.

5-3 Bending Stresses

As pointed out in the section above, it is clear from Equation (5-9) that the stresses are not distributed in a linear fashion across the cross section. Although it was assumed that plane sections remain plane by virtue of Equations (5-3), the fact that the original lengths of elements $ds_0 = R(1 + ky)\,d\theta$ depend on the distance y from the centroidal axis dictates that the strains and hence the stresses will depend on the original length. This is reflected in the occurrence of the $1 + ky$ factor in the denominator of the second term in Equation (5-9). The stresses appear generally as shown in Figure 5-8.

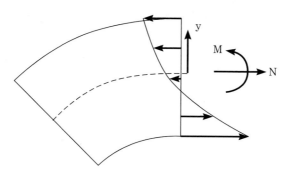

FIGURE 5-8 Normal stresses in a curved beam.

Note that even for the case of pure bending the stresses are given by

$$\sigma_x = -\frac{kM}{A} - \frac{My}{(1+ky)I_2}$$

and that the location at which the stresses vanish, that is, the neutral axis, is

$$y = -\frac{RI_2}{I_2 + AR^2}$$

showing that for pure bending the neutral and centroidal axes do not coincide. In general, the maximum bending stress occurs at the inside radius due to the presence of the $1 + ky$ term. Depending on the loading and the dimensions of the cross section in relation to the location of the center of curvature, the ratio of the stress at the inside radius to that at the outside radius can be quite large.

EXAMPLE 5-1 Consider the problem of the combined loading of a curved beam with a rectangular cross section, as shown in Figure 5-9. Consulting the free-body diagram, it is easily concluded that $N = P \sin \theta$, $M_z = PR \sin \theta$ and $V = -P \cos \theta$ so that $N - kM = 0$ and

$$\sigma_x = -\frac{Ry}{R+y}\frac{PR \sin \theta}{I_2}$$

with a maximum value at the support ($\theta = \pi/2$) of

$$\sigma_x = -\frac{Ry}{R+y}\frac{PR}{I_2}$$

The area property I_2 is given by

$$I_2 = \int_{Area} \frac{y^2\, dA}{1+ky} = \int_{-h/2}^{h/2} \frac{by^2\, dy}{1+ky} = bR^3 \left[\ln\left(\frac{1+\frac{kh}{2}}{1-\frac{kh}{2}}\right) - kh \right]$$

Expanding the ln term and simplifying yields

$$I_2 = \frac{bh^3}{12}\left[1 + \frac{3}{20}(kh)^2 + \frac{3}{112}(kh)^4 + \cdots\right]$$

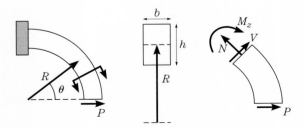

FIGURE 5-9 Bending of a rectangular cross section curved bar.

5-3 BENDING STRESSES

showing clearly the effect of the curvature compared to the corresponding straight beam.

For a specific example, let $h = 0.5R$ so that with $I_2 = 1.0392bh^3$ the combined stresses become

$$\sigma_x \bigg|_{y=-h/2} = \frac{3.849PR^2}{bh^3}$$

$$\sigma_x \bigg|_{y=+h/2} = -\frac{2.309PR^2}{bh^3}$$

The geometry is shown in Figure 5-10a along with the stress distribution as a function of the coordinate y. Shown for comparison in Figure 5-10b are the geometry and the corresponding distribution for $h = R$. In both instances the dotted line represents the linear distribution for the corresponding straight beam of the same cross-sectional dimensions. The maximum bending stress clearly occurs at the inside surface. For this particular loading and cross section, the ratio of the stresses at the inside and outside surfaces is

$$\left|\frac{\sigma_{inner}}{\sigma_{outer}}\right| = \frac{1 + \dfrac{h}{2R}}{1 - \dfrac{h}{2R}}$$

with the values 5/3 and 3, respectively, for the cases shown in Figures 5-10a and 5-10b.

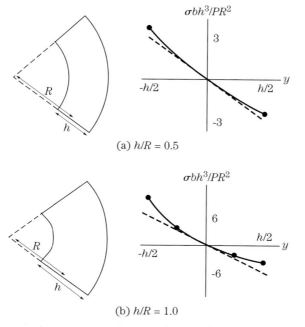

(a) $h/R = 0.5$

(b) $h/R = 1.0$

FIGURE 5-10 Curved beam stresses for various values of h/R.

The larger the h/R ratio, that is, the sharper the curvature, the more the stress distribution deviates from the corresponding linear distribution. ◆

EXAMPLE 5-2 A press is to be constructed from a T-section, as shown in Figure 5-11. Investigate the relative merits of the configurations shown, that is, where the top of the T-section is toward (a) the outside and (b) the inside. Take $P = 3000$ lbf. In either case the centroid is located as shown in Figure 5-11b. In general the second moment I_2 can be expressed as

$$I_2 = \int_{\text{Area}} \frac{y^2 \, dA}{1 + ky} = \int_{\text{Area}} \frac{Ry^2 \, dA}{R + y} = -\int_{\text{Area}} \frac{R^2 y \, dA}{R + y}$$

the last equality following from $y^2/(R + y) = y - Ry/(R + y)$ with y measured from the centroid of the section.

For the geometry of case (a) with $dA = b\,dy$

$$k^2 I_2 = -\int_{-1.9\text{in}}^{0.1\text{in}} \frac{(1 \text{ in})y \, dy}{R + y} - \int_{0.1\text{in}}^{1.1\text{in}} \frac{(3 \text{ in})y \, dy}{R + y}$$

which, with $R = 3.9$ in, can be evaluated to yield $I_2 = 4.777$ in^4. The force resultants at the horizontal section A-A are $N = P$ and $M = P(4 \text{ in} + 3.9 \text{ in})$, resulting in

$$\sigma = \frac{P - kP(4 \text{ in} + 3.9 \text{ in})}{5 \text{ in}^2} - \frac{(7.9 \text{ in})P}{4.777 \text{ in}^4} \frac{y}{1 + ky}$$

Inside:

$$\frac{\sigma}{P} = \frac{1 - 1 - \dfrac{4}{3.9}}{5 \text{ in}^2} - \frac{(7.9 \text{ in})}{4.777 \text{ in}^4}\left(\frac{-1.9 \text{ in}}{1 - \dfrac{1.9}{3.9}}\right)$$

$$= \frac{(-0.205 + 6.127)}{\text{in}^2}$$

$$= \frac{5.922}{\text{in}^2}$$

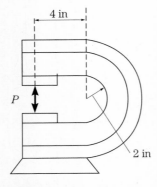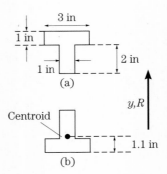

FIGURE 5-11 T-section press.

5-3 BENDING STRESSES

Outside:

$$\frac{\sigma}{P} = \frac{1 - 1 - \frac{4}{3.9}}{5 \text{ in}^2} - \frac{(7.9 \text{ in})}{4.777 \text{ in}^4}\left(\frac{1.1 \text{ in}}{1 + \frac{1.1}{3.9}}\right)$$

$$= \frac{(-0.205 - 1.419)}{\text{in}^2}$$

$$= \frac{-1.624}{\text{in}^2}$$

so that the maximum stress at the inner radius is $\sigma_{max} = 17.8$ ksi.
For the geometry of case (b) with $dA = b\,dy$

$$k^2 I_2 = -\int_{-1.1 \text{ in}}^{-0.1 \text{ in}} \frac{(3 \text{ in})y\,dy}{R + y} - \int_{-0.1 \text{ in}}^{1.9 \text{ in}} \frac{(1 \text{ in})y\,dy}{R + y}$$

which, with $R = 3.1$ in, can be evaluated to yield $I_{20} = 3.406$ in^4. The force resultants at the horizontal section A-A are $N = P$ and $M = P(4 \text{ in} + 3.1 \text{ in})$, resulting in

$$\sigma = \frac{P - kP(4 \text{ in} + 3.1 \text{ in})}{5 \text{ in}^2} - \frac{(7.1 \text{ in})P}{3.406 \text{ in}^4}\frac{y}{1 + ky}$$

Inside:

$$\frac{\sigma}{P} = \frac{1 - 1 - \frac{4}{3.1}}{5 \text{ in}^2} - \frac{(7.1 \text{ in})}{3.406 \text{ in}^4}\left(\frac{-1.1 \text{ in}}{1 - \frac{1.1}{3.1}}\right)$$

$$= \frac{(-0.258 + 3.554)}{\text{in}^2}$$

$$= \frac{3.296}{\text{in}^2}$$

Outside:

$$\frac{\sigma}{P} = \frac{1 - 1 - \frac{4}{3.1}}{5 \text{ in}^2} - \frac{(7.1 \text{ in})}{3.406 \text{ in}^4}\left(\frac{1.9 \text{ in}}{1 + \frac{1.9}{3.1}}\right)$$

$$= \frac{(-0.258 - 2.456)}{\text{in}^2}$$

$$= \frac{-2.714}{\text{in}^2}$$

so that the maximum stress at the inner radius is 9.89 ksi. Thus configuration (b), with the top of the T toward the inside of the radius, results in a smaller maximum stress. ◆

5-4 Transverse Shear Stresses

In the course of the development of the equations for the in-plane deformation problem, it was assumed that the shear strain was zero. This translated into the fact that the transverse shear stress resultant V was zero, obviously in contradiction to the equilibrium requirements for a general loading situation. In this section a means for estimating the shear stresses based solely on equilibrium requirements will be developed. The development parallels the steps taken to derive the result $\tau = VQ/Ib$ for estimating shear stresses for the classical straight beam considered in Chapter 4.

Consider in Figure 5-12 the free-body diagram of a portion of a curved bar showing the shear stresses τ and normal stresses σ at an interior radial location. Acting on a face a distance y from the centroidal axis, the resultant force associated with the shear stresses is approximately $\tau(R + y)b\Delta\theta$, where τ indicates the average shear stress over the thickness b, as shown in Figure 5-12b. In addition, the resultant forces V_y^* and F associated with the shear and normal stresses, respectively, acting on the faces corresponding to the area A', are shown.

Summing forces in the x direction yields

$$\sum F_x = \Delta F \cos \frac{\Delta\theta}{2} + (2V_y^* + \Delta V_y^*)\sin \frac{\Delta\theta}{2} - \tau(R + y)b\Delta\theta = 0$$

Passing to the limit with $\Delta x = R\Delta\theta$ yields

$$\tau b(1 + ky) = \frac{\partial F}{\partial x} + kV_y^*$$

The force F is given by

$$F = \int_{A'} \sigma_x \, dA = \int_{A'} \left(\frac{NR - M}{AR} - \frac{y}{1 + ky}\frac{M}{I_2} \right) dA$$

with

$$\frac{\partial F}{\partial x} = \int_{A'} \left(\frac{\frac{dN}{dx}R - \frac{dM}{dx}}{AR} - \frac{y}{1 + ky}\frac{\frac{dM}{dx}}{I_2} \right) dA$$

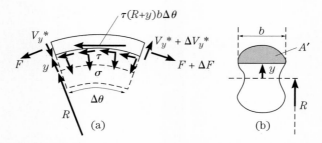

FIGURE 5-12 Free-body diagram for determining transverse shear stresses.

5-4 TRANSVERSE SHEAR STRESSES

where, as indicated in Figure 5-12b, A' is the area outside of the radial position y. From the equilibrium equations $dN/dx = -kV_y$ and $dM/dx = -V_y$, with the result that

$$\frac{\partial F}{\partial x} = \int_{A'} \frac{y}{1 + ky} \frac{V_y}{I_2} \, dA$$

The quantity V_y^* can be represented as

$$V_y^* = \int_{A'} \tau b \, dy$$

so that

$$\tau b(1 + ky) = \int_{A'} \left(\frac{Ry}{R + y} \frac{V_y}{I_2} \right) dA + k \int_{A'} \tau b \, dy$$

giving an equation from which the distribution of the average shear stress τ as a function of the radial position y within the cross section can be computed.

◆**EXAMPLE 5-3** Consider the shear stresses acting on the rectangular cross section shown in Figure 5-13. In this case b is a constant, and the expression for τ can be written as

$$\tau(1 + ky) = \frac{V_y}{I_2} \int_y^{h/2} \frac{y}{1 + ky} \, dy + k \int_y^{h/2} \tau \, dy$$

Computing the derivative using Leibnitz rule for differentiation of an integral yields

$$\tau k + (1 + ky)\tau' = -\frac{V_y}{I_2} \frac{y}{1 + ky} - k\tau$$

or

$$(1 + ky)^2 \tau' + 2k\tau(1 + ky) = -y\frac{V_y}{I_2}$$

or

$$[(1 + ky)^2 \tau]' = -y\frac{V_y}{I_2}$$

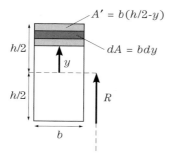

FIGURE 5-13 Rectangular cross section.

Integrating yields

$$(1 + ky)^2 \tau = -\frac{y^2}{2}\frac{V_y}{I_2} + C_1$$

C_1 is evaluated on the basis that the shear stress must vanish at $y = \pm h/2$, leading to $C_1 = h^2 V_y / 8 I_2$ and then to

$$(1 + ky)^2 \tau = \frac{V_y}{2I_2}\left(\frac{h^2}{4} - y^2\right) \tag{5-10}$$

where I_2 is given by

$$I_2 = \int_{-h/2}^{h/2} \frac{Ry^2}{R+y} b\, dy = bR^3 \left(\log\left(\frac{1+\alpha}{1-\alpha}\right) - 2\alpha\right)$$

where $\alpha = h/2R$.

As presented in [1], the shear stress given by the corresponding elasticity solution can be represented as

$$\frac{\tau b h}{V_y} = \frac{\alpha\left(1 + ky + \dfrac{(1-\alpha^2)}{(1+ky)^3} - \dfrac{2(1+\alpha^2)}{1+ky}\right)}{(1+\alpha^2)\log\left(\dfrac{1+\alpha}{1-\alpha}\right) - 2\alpha}$$

For $h/R = 1$, Figure 5-14 shows a plot of $\tau(y)$ as given by the curved-beam theory and the elasticity theory, together with the parabolic distribution corresponding to a straight beam. The results from the approximate curved-beam theory show excellent agreement

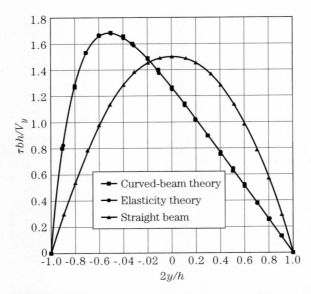

FIGURE 5-14 Transverse shear stresses in a rectangular cross section curved beam.

compared to the solution from elasticity theory. The maximum shear stress of approximately 1.69 times V_y/bh occurs at the location $y = -0.25h$ for both the curved-beam and elasticity solutions, with the maximum value exceeding the corresponding value for the straight beam by approximately 13%. For values of h/R less than approximately 0.2, the curved-beam shear stresses are essentially given by the straight-beam results. ◆

EXAMPLE 5-4 Consider the problem of the transverse shear stresses in an I-section, as shown in Figure 5-15. Assume that $t \ll b$ and $b \approx h$, that is, that it is a thin-walled cross section. The general expression developed for the shear stresses is given by Equation (5-10) as

$$\tau b(1 + ky) = \int_{A'} \left(\frac{Ry}{R + y} \frac{V_y}{I_2} \right) dA + \frac{A'V_y}{AR}$$

The second term, $A'V_y/AR$, is intended to account for the portion of the total force acting on the portion A'. For the I-section the shear stresses in the flanges must act in a horizontal direction and hence contribute nothing to the summation of forces along the axis of the beam, that is, in the flanges the shear stresses will be computed according to

$$\tau b(1 + ky) = \int_{A'} \left(\frac{Ry}{R + y} \frac{V_y}{I_2} \right) dA$$

It also then follows that the entire shear force V_y will be carried by the web.

Consider Figure 5-16, which shows a free-body diagram of a portion of the outer flange. With the shear flow acting in the positive s direction, as shown, it follows that

$$\tau t \left(1 + k\frac{h}{2}\right) = -\frac{\partial F}{\partial x} = -\int_{A'} \left(\frac{R\frac{h}{2}}{R + \frac{h}{2}} \frac{V_y}{I_2} \right) dA$$

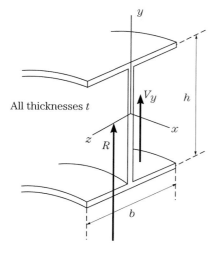

FIGURE 5-15 Transverse shear stresses in an I-section

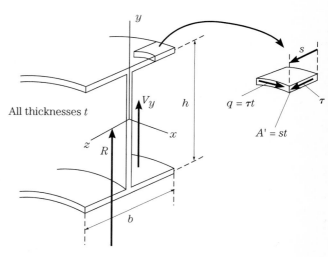

FIGURE 5-16 Free-body diagram for the shear stress in the flange.

Then, with $t \ll h$ and $dA = t\, ds$,

$$\tau t\left(1 + k\frac{h}{2}\right) = -\int_{A'} \left(\frac{\dfrac{Rh}{2}}{R + \dfrac{h}{2}} \frac{V_y}{I_2}\right) t\, ds$$

$$= -\frac{Rhts}{R + \dfrac{h}{2}} \frac{V_y}{I_{20}}$$

or

$$q = \tau t = -\frac{hts}{(1+\alpha)^2} \frac{V_y}{I_2}$$

where $\alpha = h/2R$. The shear flow in the other portion of the top flange is of the same form and is shown in Figure 5-18, below.

For the shear flows in the web, consider the free-body diagram shown in Figure 5-17. The quantity q_U is twice the shear flow in the flange at the junction between the flanges and the web. Summing forces in the x direction yields

$$(2V_y^* + \Delta V_y^*)\sin\frac{\Delta\theta}{2} - q(y)(R+y)\Delta\theta + q_U\left(R + \frac{h}{2}\right)\Delta\theta + \Delta F \cos\frac{\Delta\theta}{2} = 0$$

or

$$q(y)\left(1 + \frac{y}{R}\right) = q_U(1+\alpha) + \frac{V_y^*}{R} + \frac{\partial F}{\partial x}$$

Represent V_y^*, the portion of the total shear V_y as

$$V_y^* = \int_{A'} q\, dy = \int_y^{h/2} q\, dy$$

and as before

$$\frac{\partial F}{\partial x} = \int_{A'}\left(\frac{Ry}{R+y}\frac{V_y}{I_2}\right) dA = \int_y^{h/2}\left(\frac{Ry}{R+y}\frac{V_y}{I_2}\right) t\, dy$$

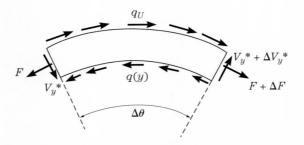

FIGURE 5-17 Free-body diagram for shear flows in the web.

5-4 TRANSVERSE SHEAR STRESSES

from which

$$q(y)(1 + ky) = q_U(1 + \alpha) + k \int_y^{h/2} q \, dy + \int_y^{h/2} \left(\frac{y}{1 + ky} \frac{V_y}{I_2} \right) t \, dy$$

The appearance of q inside the first integral on the right requires that we compute the derivative of this to obtain

$$kq(y) + (1 + ky)q'(y) = -kq(y) - \frac{y}{1 + ky} \frac{V_y}{I_2} t$$

or

$$(1 + ky)q'(y) + 2kq(y) = -\frac{y}{1 + ky} \frac{V_y}{I_2} t$$

Multiplying by $(1 + ky)$, collecting and integrating yield

$$q(1 + ky)^2 = -\frac{y^2}{2} \frac{V_y}{I_2} t + C_1$$

The constant C_1 is evaluated on the basis that the shear flow at $y = h/2$ is known to be q_U so that

$$q_U(1 + \alpha)^2 = -\frac{h^2}{8} \frac{V_y}{I_2} t + C_1$$

from which $C_1 = q_U(1 + \alpha)^2 + h^2 V_y \, t/8I_{20}$ with the shear flow in the web then given as

$$q(1 + ky)^2 = q_U(1 + \alpha)^2 + \frac{V_y}{2I_2} \left(\frac{h^2}{4} - y^2 \right) t$$

or

$$q = \frac{V_y t}{2I_2(1 + ky)^2} \left[bh + \left(\frac{h^2}{4} - y^2 \right) \right] \tag{5-11}$$

The maximum value occurs at $y^* = -(bh + h^2/4)/R$. Note that the corresponding result for the straight beam is

$$q = \frac{V_y t}{2I_z} \left[bh + \left(\frac{h^2}{4} - y^2 \right) \right]$$

which Equation (5-11) reduces to $h/R \to 0$. It is left for the student to show that the shear stress in the bottom flange at the intersection with the web is

$$q_B = -\frac{htb}{2(1 - \alpha)^2} \frac{V_y}{I_2}$$

and decays linearly to zero at the outside edges of the bottom flange.

For the specific geometry $h/R = 0.5$ and $b/h = 0.5$, the shear flow distribution for the curved beam is shown as shaded in Figure 5-18, where the maximum value occurs at $y^* = -0.375h$. The corresponding results for the straight beam are shown as dotted

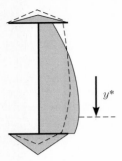

FIGURE 5-18 Shear flows in an I-section.

lines in Figure 5-18. For the curved beam the maximum value of $1.305V_y/bh$ is approximately 16% larger than the corresponding maximum value of $1.125V_y/bh$ for the straight beam. The student is asked to show in the exercises that the integral of the shear flow in the web is equal to the applied shear load. ◆

5-5 Displacements

The basic equations that are to be integrated for determining the displacement components u and v are

$$v'' + k^2 v = \frac{M}{EI_2} \tag{5-8a}$$

and

$$u' + kv = \frac{N - kM}{AE} \tag{5-8b}$$

Given M, Equation (5-8a) can be integrated to determine v, after which Equation (5-8b) can be integrated to determine u. The resulting expressions for u and v will contain three constants of integration.

For a statically determinate problem, it will have been possible to draw free-body diagrams and determine N and M completely. There will then be three boundary conditions on the displacement that must be satisfied and that will give three equations for the determination of the three constants of integration. Several common types of supports, together with the boundary conditions (both on the displacements and on the force resultants) that must be satisfied, are indicated in Figure 5-19.

Boundary conditions on displacement components u, v and v' are referred to as *geometric* boundary conditions or *essential* boundary conditions. Boundary conditions on the force resultants M, N and V are referred to as *force* boundary conditions or *nonessential* boundary conditions. For any sufficiently supported statically determinate beam, it will turn out that there will be precisely three geometric boundary conditions and three force boundary conditions. The three geometric boundary conditions will be used to determine the three constants of integration, and the three force boundary conditions will be satisfied identically by the resulting solution.

5-5 DISPLACEMENTS

(a) Clamped end

$u = 0$
$v = 0$
$v' = 0$

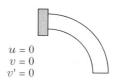

(b) Free end

$u' + kv = 0$
$v'' + k^2 v = 0$
$v''' + k^2 v' = 0$

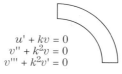

(c) Simple support #1

$u = 0$
$v'' + k^2 v = 0$
$v''' + k^2 v' = 0$

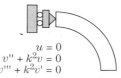

(d) Simple support #2

$v = 0$
$u' + kv = 0$
$v'' + k^2 v = 0$

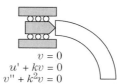

FIGURE 5-19 Boundary conditions for various common supports.

EXAMPLE 5-5 Consider the problem of determining the displacements for the problem of Example 5-1, shown again here in Figure 5-20a. The boundary conditions that must be imposed are shown in Figure 5-20b. From the free-body diagram of Figure 5-20 it is easily shown that $M = PR \cos \theta, N = P \cos \theta$ and $V = P \sin \theta$. Equation 5-9 becomes

$$v'' + k^2 v = \frac{PR \cos \theta}{EI_2}$$

Converting $v'' = d^2v/ds^2$ to $v'' = k^2 d^2v/d\theta^2$, there results

$$\frac{d^2 v}{d\theta^2} + v = \frac{PR^3 \cos \theta}{EI_2}$$

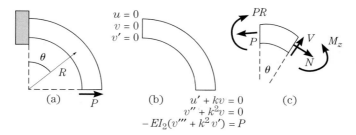

FIGURE 5-20 Boundary conditions.

The solution consists of the homogeneous solution
$$v_h = C_1 \cos\theta + C_2 \sin\theta$$
plus the particular solution
$$v_p = \frac{PR^3 \theta \sin\theta}{2EI_2}$$
so that
$$v = C_1 \cos\theta + C_2 \sin\theta + \frac{PR^3 \theta \sin\theta}{2EI_2}$$
Satisfying the boundary conditions $v(0) = 0$ and $v'(0) = 0$ yields $C_1 = C_2 = 0$ so that
$$v(\theta) = \frac{PR^3 \theta \sin\theta}{2EI_2}$$
The remaining equation is
$$u' + kv = \frac{N - kM}{AE}$$
which, for the M and N above, can be expressed as
$$\frac{du}{d\theta} = -v = -\frac{PR^3 \theta \sin\theta}{2EI_2}$$
Integrating yields
$$u(\theta) = \frac{PR^3}{2EI_2}(\theta \cos\theta - \sin\theta) + C_3$$
which, with the remaining boundary condition $u(0) = 0$, gives $C_3 = 0$. Thus there results finally
$$u(\theta) = \frac{PR^3}{2EI_2}(\theta \cos\theta - \sin\theta)$$
and
$$v(\theta) = \frac{PR^3 \theta \sin\theta}{2EI_2}$$
The reader should verify that the remaining force boundary conditions at $\theta = \pi/2$ are satisfied identically. ◆

The energy approach of Chapter 2 can also be used to determine displacements. For the curved beam the energy principle is based on evaluating the expression for the bending energy, namely,
$$U^* = \int_{Vol} \frac{\sigma^2}{2E} dVol = \int_{Vol} \frac{1}{2E}\left(\frac{NR - M}{AR} - \frac{y}{1 + ky}\frac{M}{I_2}\right)^2 dVol$$

5-5 DISPLACEMENTS

It is left for the student to show in the exercises that with $dVol = (1 + ky)\, dA\, ds$, this can be expressed as

$$U^* = \int_0^L \left[\frac{(NR - M)^2}{2EAR^2} + \frac{M^2}{2EI_2} \right] ds$$

where L is the centroidal length of the beam. It is frequently preferable to express this in terms of the subtended angle Θ as

$$U^* = \int_0^\Theta \left[\frac{(NR - M)^2}{2EAR^2} + \frac{M^2}{2EI_2} \right] R\, d\theta$$

where N and M are now expressed as functions of θ. Engesser's first theorem can then be used to determine displacements and rotations according to $\Delta = \partial U^*/\partial p$ and $\phi = \partial U^*/\partial m$, respectively, where p and m are real or dummy loads.

EXAMPLE 5-6 Consider the same problem as described in Examples 5-1 and 5-5. The displacements at the lower end of the beam shown in Figure 5-21a are desired. A dummy load is placed at the point in question in order to be able to calculate a displacement in the vertical direction. Referring to the free-body diagram of Figure 5-21b, it is easily seen that $N = P \sin\theta + q \cos\theta$, $V = -P \cos\theta + q \sin\theta$ and $M = PR \sin\theta - qR(1 - \cos\theta)$. The complementary energy can then be expressed as

$$U^* = \int_0^{\pi/2} \left[\frac{(qR)^2}{2EAR^2} + \frac{(PR \sin\theta - qR(1 - \cos\theta))^2}{2EI_2} \right] R\, d\theta$$

The displacements at the loaded end can then be computed as

$$u\left(\frac{\pi}{2}\right) = \left.\frac{\partial U^*}{\partial q}\right|_{q=0} = \int_0^{\pi/2} \frac{PR \sin\theta(-R(1 - \cos\theta))R\, d\theta}{EI_2} = -\frac{PR^3}{2EI_2}$$

$$v\left(\frac{\pi}{2}\right) = \left.\frac{\partial U^*}{\partial P}\right|_{q=0} = \int_0^{\pi/2} \frac{PR \sin\theta(R \sin\theta)R\, d\theta}{EI_2} = \frac{PR^3\pi}{4EI_2}$$

coinciding with the values of the displacements of Example 5-5 at $\theta = \pi/2$. It is left to the homework exercises to show that by placing dummy loads p and q at arbitrary locations along the beam, as shown in Figure 5-22, it is possible to generate the functions

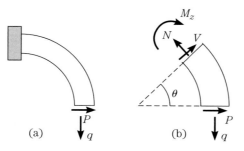

FIGURE 5-21 Free-body diagram.

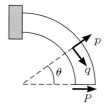

FIGURE 5-22 Dummy loads for determining displacement functions.

$u(\theta)$ and $v(\theta)$ of Example 5-5 according to

$$u(\theta) = \left.\frac{\partial U^*}{\partial q}\right|_{p=q=0} \quad \text{and} \quad v(\theta) = \left.\frac{\partial U^*}{\partial p}\right|_{p=q=0}$$

that is, the displacement functions can be determined using an energy approach. ◆

For statically indeterminate problems, it is still possible to integrate Equations (5-9) and (5-10) to determine the displacements. As a result of solving the equilibrium equations there will be one or more redundants. As a result of satisfying equilibrium the expressions for M and N will contain these redundants. However, there will also be one or more additional geometric boundary conditions that can be used to solve for the redundants. This technique is demonstrated in the following example.

◆**EXAMPLE 5-7** Consider the statically indeterminate problem shown in Figure 5-23. It is easily determined that the reactions at the upper support are $M = (P - Y)R$, $H = P$ and $V = Y$, where Y will be considered to be the redundant. The internal force resultants are then given by

$$M(\theta) = M + YR \sin \theta - PR(1 - \cos \theta)$$

and

$$N(\theta) = P \cos \theta + Y \sin \theta$$

so that the equations to be integrated are

$$v'' + k^2 v = \frac{PR \cos \theta - YR(1 - \sin \theta)}{EI_2}$$

and

$$u' + kv = \frac{Y}{AE}$$

or, in terms of θ,

$$\frac{d^2 v}{d\theta^2} + v = \frac{PR^3 \cos \theta - YR^3(1 - \sin \theta)}{EI_2}$$

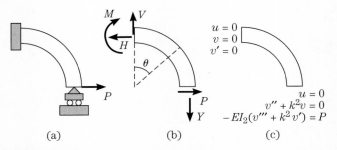

FIGURE 5-23 Statically indeterminate problem.

and

$$\frac{du}{d\theta} + v = \frac{YR}{AE}$$

Integrating the first equation yields

$$v = C_1 \cos\theta + C_2 \sin\theta + \frac{PR^3\theta \sin\theta}{2EI_2} - \frac{YR^3}{EI_2}\left(1 + \frac{\theta \cos\theta}{2}\right)$$

Satisfying $v(0) = 0$ and $v'(0) = 0$ gives $C_1 = YR^3/EI_2$ and $C_2 = YR^3/2EI_2$, so that

$$v(\theta) = \frac{PR^3\theta \sin\theta}{2EI_2} + \frac{YR^3}{EI_2}\left(\cos\theta + \frac{\sin\theta}{2} - 1 - \frac{\theta \cos\theta}{2}\right)$$

The second equation then becomes

$$\frac{du}{d\theta} = \frac{YR}{AE} - \frac{PR^3\theta \sin\theta}{2EI_2} - \frac{YR^3}{EI_2}\left(\cos\theta + \frac{\sin\theta}{2} - 1 - \frac{\theta \cos\theta}{2}\right)$$

from which

$$u(\theta) = \frac{YR\theta}{AE} - \frac{PR^3(\sin\theta - \theta \cos\theta)}{2EI_2}$$
$$- \frac{YR^3}{EI_2}\left(\sin\theta - \frac{\cos\theta}{2} - \theta - \frac{\theta \sin\theta + \cos\theta}{2}\right) + C_3$$

Using the boundary condition $u(0) = 0$ gives $C_3 = -YR^3/EI_2$. At this point the redundant Y is still unknown. However, the boundary condition or constraint (as it is sometimes called), namely, $u(\pi/2) = 0$, must still be enforced, leading to

$$u\left(\frac{\pi}{2}\right) = 0 = \frac{YR\pi}{2AE} - \frac{PR^3}{2EI_2} - \frac{YR^3}{EI_2}\left(1 - \frac{3\pi}{4}\right) - \frac{YR^3}{EI_2}$$

from which, after a bit of algebra,

$$Y = \frac{\dfrac{P}{2}}{\dfrac{3\pi}{4} - 2 + \dfrac{\pi}{2}\dfrac{I_2}{AR^2}} = \frac{P}{2\Psi}$$

with the obvious definition for Ψ. Finally, then,

$$u(\theta) = \frac{PR^3}{2EI_2\Psi}\left[\Psi(\theta \cos\theta - \sin\theta) + \frac{I_2\theta}{AR^2} - \sin\theta + \cos\theta + \theta - 1 + \frac{\theta \sin\theta}{2}\right]$$

If the curvature is small, $I_2 \approx I_z$ and $I_2/AR^2 \approx 0$ with some resulting simplification in the expression for $u(\theta)$. ◆

The energy method can also be used in connection with the principle of least work to treat the statically indeterminate problem. The equilibrium equations are solved just as in Example 5-5 to express the internal force resultants in terms of the applied loads

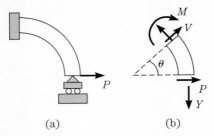

FIGURE 5-24 Statically indeterminate problem using the principle of least work.

and the redundants, after which the principle of least work is used to generate the required number of additional equations.

EXAMPLE 5-8 Consider the problem of Example 5-7, shown again here in Figure 5-24. Requiring equilibrium for the free-body diagram of Figure 5-24b, it follows that

$$N = Y \cos \theta + P \sin \theta$$

and

$$M = PR \sin \theta - YR(1 - \cos \theta)$$

where the reaction Y is considered the redundant. The internal complementary energy is

$$U^* = \int_0^{\pi/2} \left(\frac{Y^2}{2AE} + \frac{(PR \sin \theta - YR(1 - \cos \theta))^2}{2EI_2} \right) R \, d\theta$$

The principle of least work then leads to

$$\frac{\partial U^*}{\partial Y} = 0 = \int_0^{\pi/2} \frac{YR \, d\theta}{AER} - \int_0^{\pi/2} \frac{(PR \sin \theta - YR(1 - \cos \theta))R(1 - \cos \theta)R \, d\theta}{EI_2}$$

or, after evaluating the integrals and simplifying, to

$$\left(\frac{3\pi}{4} - 2 + \frac{\pi}{2} \frac{I_2}{AR^2} \right) Y = \frac{P}{2}$$

as per the integration approach of Example 5-7. All the internal force resultants are now known, and it is possible to evaluate the stresses. Displacements can now be determined using Engesser's first theorem, if desired. ◆

5-6 Summary

A theory for determining stresses and displacements for the in-plane deformation of plane-curved beams was developed. This theory predicts that the normal stresses are given by

$$\sigma_x = \frac{N - kM}{A} - \frac{My}{(1 + ky)I_2}$$

where $k = 1/R$ is the curvature of the centroidal axis, $I_2 = \int y^2\, dA/(1 + ky)$ and N and M_z are the axial and bending force resultants. For either statically determinate or statically indeterminate problems, the axial displacement u and the transverse displacement v can be determined by integrating

$$v'' + k^2 v = \frac{M}{EI_2}$$

and

$$u' + kv = \frac{N - kM}{AE}$$

or by using an energy method.

REFERENCES

1. Timoshenko, S. P., and Goodier, J. N. *Theory of Elasticity*, 2nd ed. McGraw-Hill, New York, 1951.
2. Langhaar, H. L. *Energy Methods in Applied Mechanics*. Wiley, New York, 1962.
3. Oden, J. T., and Ripperger, E. A. *Mechanics of Elastic Structures*. McGraw-Hill, New York, 1981.
4. Washizu, K. "Some Considerations on a Naturally Curved and Twisted Slender Beam." *Journal of Mathematics and Physics*, vol. 43, no. 2, pp. 111–116, June 1964.
5. Roark, R. J., and Young, W. C. *Formulas for Stress and Strain*, 5th ed. McGraw-Hill, New York, 1975.

EXERCISES

Sections 5-1 & 5-2

1. Show that in connection with Figure 5-5, equilibrium of forces in the tangential directions yields $dN/d\theta + V + Rq_x = 0$ or $dN/ds + V/R + q_x = 0$.

2. Show that in connection with Figure 5-5, equilibrium of moments yields $dM/d\theta + RV = 0$ or $dM/ds + V = 0$.

   For each of the problems shown below, integrate Equations 5-1 to determine the internal force resultants N, M and V. Also draw and use free-body diagrams to verify your results.

3.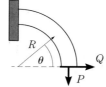

4.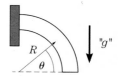

5.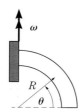

6.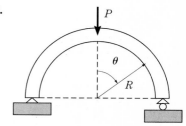

7. 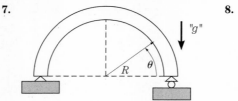   8.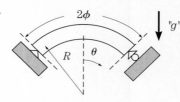

9. Retaining the linear portions of the strain-displacement relations of problem 36 in Chapter 1, let $u = V$, $v = U$, $r = R + y$ and $x = R\theta$ and hence verify Equations 5-4.

10. Show that $kI_2 = -I_1$ and that $I_2 = R^2 [\int dA/g - A]$.

11. Carry through the development of the governing equations if shear deformation is not neglected, that is, if $v' - ku + \theta \neq 0$.

Section 5-3

12. Verify that for a circular cross section bar with radius a and area A
$$I_2 = AR^2 \left[2\left(\frac{R}{a}\right)^2 - 1 - 2\left(\frac{R}{a}\right)^2 \sqrt{1 - \left(\frac{a}{R}\right)^2} \right]$$

13. Using the results of problem 12, show that if $a/R \to 0$, then $I_2 \to \pi a^4/4$, which is the correct result for the straight circular cross section bar.

14. Verify that for a hollow circular cross section bar with outer radius a and inner radius b
$$I_2 = 2\pi R^4 \left[\sqrt{1 - \left(\frac{b}{R}\right)^2} - \sqrt{1 - \left(\frac{a}{R}\right)^2} \right] - \pi(a^2 - b^2)R^2$$

15. Using the result of problem 14, show that as $a/R \to 0$ and $b/R \to 0$, $I_2 \to \pi(a^4 - b^4)/4$, which is the correct result for the straight hollow circular cross section bar.

16. For the section shown below, determine the ratio of the bending stress at the inner and outer radii as a function of h/R. Take $h = 2b$ and $t = h/20$.

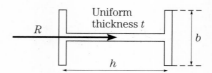

17. For the T-section oriented as shown below, determine the ratio of the bending stress at the inner and outer radii as a function of h/R. Take $h = 2b$ and $t = h/20$.

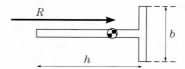

18. For the T-section oriented as shown below, determine the ratio of the bending stress at the inner and outer radii as a function of h/R. Take $h = 2b$ and $t = h/20$.

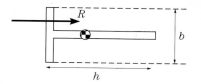

19. Show that for the trapezoidal section shown below,

$$I_2 = R^2 \left[R \frac{b_1 r_2 - b_2 r_1}{r_2 r_1} - \log\left(\frac{1 + \frac{h - y_b}{R}}{1 - \frac{y_b}{R}} \right) + (b_2 - b_1)R - \frac{b_1 + b_2}{2}(r_2 - r_1) \right]$$

where

$$R = \frac{r_1(2b_1 + b_2) + r_2(b_1 + 2b_2)}{3(b_1 + b_2)}$$

and

$$y_b = R - r_1$$

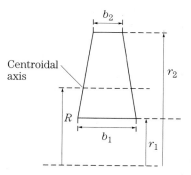

20. Show that the results of problem 19 for the trapezoidal section reduce to those for the rectangular cross section in Example 5-1 if $b_1 = b_2$.

21. The curved beam shown below has a square cross section 2.5 in on a side with an inner diameter of 5 in. If $P = 3000$ lbf, determine the circumferential stress distribution on the section AB and hence the location of the neutral axis and the maximum circumferential stress.

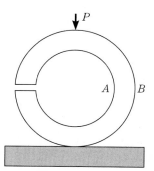

22. For the circular cross section bar shown below, determine the maximum circumferential stress at the support. Take $P = 6$ kN, $a = 40$ mm, $b = 50$ mm and $c = 75$ mm.

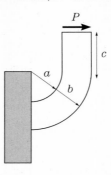

23. Repeat problem 22 with $P = 1600$ lbf, $a = 2$ in, $b = 3$ in and $c = 4$ in.

24. For the crane hook shown below, determine the maximum allowable value of P if the maximum circumferential stress at section AB is not to exceed 150 MPa. Compare the results with those obtained by assuming a straight beam. Take $b_1 = 80$ mm, $b_2 = 50$ mm, $H = 90$ mm and $a = 70$ mm.

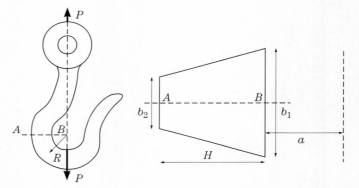

25. For the crane hook shown in problem 24, determine the maximum allowable value of P if the maximum circumferential stress at section AB is not to exceed 20 ksi. Compare the results with those obtained by assuming a straight beam. Take $b_1 = 3$ in, $b_2 = 2$ in, $H = 3.5$ in and $a = 3$ in.

26. The press shown below is to be designed to handle loads up to 25 tons. If $a = 3$ ft, $L = 4$ ft and $t/b = 1/3$, and the maximum allowable circumferential stress is 15 ksi, determine the minimum allowable value of b.

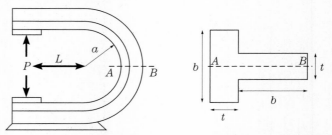

27. The press shown in problem 26 is to be designed to handle loads up to 200 kN. If $a = 0.75$ m, $L = 1.0$ m and $t/b = 1/3$, and the maximum allowable circumferential stress is 100 MPa, determine the minimum allowable value of b.

EXERCISES

28. For the beam shown below, determine the circumferential stress distribution at section AB. Take $a = 20$ in, $t = 2$ in, $b = 10$ in, $L = 40$ in, $P = 4000$ lbf, and $\theta = 45°$.

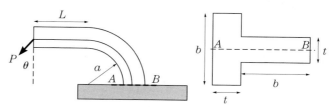

29. Repeat problem 28 with $a = 0.5$ m, $t = 0.05$ m, $b = 0.25$ m, $L = 1.0$ m, $P = 15$ kN, and $\theta = 45°$.

30. The triangular cross section beam is loaded as shown below. Specialize the results for I_2 for the trapezoidal section described in problem 19 and hence determine the circumferential stress distribution across section AB. Take $a = 16$ in, $h = 10$ in, $b = 6$ in, $L = 20$ in and $P = 9000$ lbf.

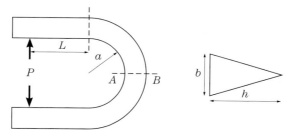

31. Repeat problem 30 with $a = 0.4$ m, $h = 0.25$ m, $b = 0.1875$ m, $L = 0.5$ m and $P = 40$ kN.

32. For nonstandard shapes such as those shown in the figure below, it is possible to proceed as follows for determining the section properties. Subdivide the area into rectangles as shown. The area, centroidal radius and second moment I_2 are then given, respectively, by

$$A \approx \sum b_i t_i, \qquad R \approx \frac{\sum b_i t_i r_i}{A}, \qquad \text{and} \qquad I_2 = R^2 \left[R \sum \frac{b_i t_i}{R_i} - A \right]$$

For calibration as to the number of segments required, take two, four and eight equal-thickness segments and apply these formulae to the case of the rectangular cross section shown with $h = R$. For this case the student should verify that A and R are determined exactly and that $I_2 = 0.0986\, bh^3$ is the exact value against which the results for that property should be compared.

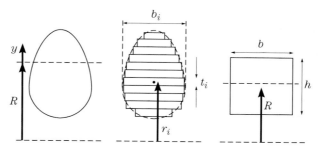

33. A curved beam consisting of an inner trapezoid and two hemispherical pieces as shown below, is to be used in constructing a press. With $L = 36$ in, $a = 20$ in, $b_1 = 8$ in, $b_2 = 4$ in, $h = 10$ in and $P = 22$ kips, determine (a) I_2 for the section and (b) the circumferential stress distribution.

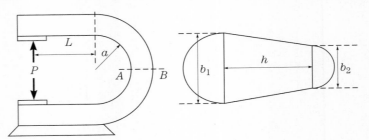

34. Repeat problem 33 with $L = 0.9$ m, $a = 0.5$ m, $b_1 = 0.2$ m, $b_2 = 0.1$ m, $h = 0.25$ m and $P = 100$ kN.

35. A large trailer is pin-connected to a truck by an I-section curved-beam, as shown below. The trailer carries a load W positioned as shown. If the maximum circumferential stress is not to exceed 10 ksi, determine the dimensions of the cross section. Take $h = 2b$, $t/b = 1/10$, $R = 3$ ft, $B = 16$ ft, $H_1 = 8$ ft, $H_2 = 4$ ft, $d = 6$ ft and $W = 40$ tons. Neglect the weight of the trailer.

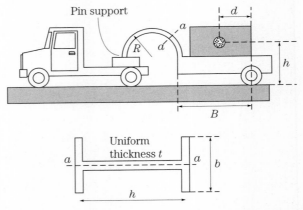

36. Repeat problem 35 if the truck decelerates at the rate of 10 ft/sec^2. Assume that the brakes on the trailer are not functioning and neglect the mass of the trailer.

Section 5-4

37. Verify by integrating Equation 5-10 for the shear stresses in a rectangular section that $\int \tau \, dA = V$, that is, that the distribution of shear stresses is in fact equivalent to the resultant force V that is transmitted.

38. Verify by integrating Equation 5-11 for the shear stresses in the web of an I-section that $\int \tau \, dA = V$, that is, that the distribution of shear stresses in the web is in fact equivalent to the resultant force V that is transmitted.

39. Determine the shear stress distribution in a typical T-section as indicated in the figure below. Take $h = 1.5b$, $h = R$ and $t = b/8$.

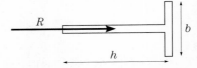

EXERCISES

Section 5-5

40. For the semicircular arch shown below, integrate Equations 5-8 to determine the displacements u and v.

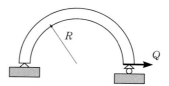

41. Integrate Equations 5-8 to determine the displacements u and v for the quartercircle beam loaded under its own weight as shown below.

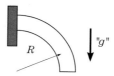

42. For the rotating quartercircle beam shown below, integrate Equations 5-8 to determine the displacements u and v.

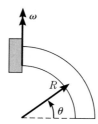

43. For the semicircular arch loaded under its own weight as shown below, integrate Equations 5-8 to determine the displacements u and v.

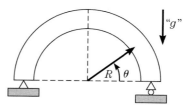

44. For the symmetrically oriented arch loaded under its own weight as shown below, integrate Equations 5-8 to determine the displacements u and v.

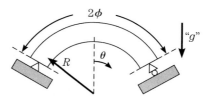

45. Integrate Equations 5-8 to determine the displacements u and v for the quartercircle beam subjected to a uniform radial pressure as shown below.

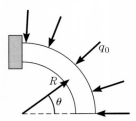

46. Integrate Equations 5-8 to determine the displacements u and v for the quartercircle beam subjected to a uniform tangential pressure as shown below.

47. Use the principle of least work to determine the redundants and then determine the decrease in the vertical diameter for the complete ring shown below.

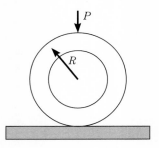

48. Use the principle of least work to determine the redundants and then determine the decrease in the vertical radius for the half ring shown below.

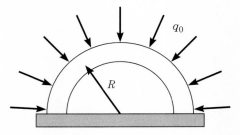

49. A ring rests on a horizontal surface under the action of its own weight, as shown below. Use the principle of least work to determine the redundants and then determine the decrease in the vertical diameter and the increase in the horizontal diameter.

EXERCISES

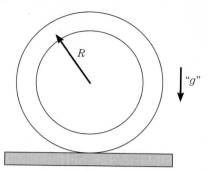

50. A ring rotates about a diameter, as shown below. Use the principle of least work to determine the redundants and then determine the decrease in the vertical diameter and the increase in the horizontal diameter.

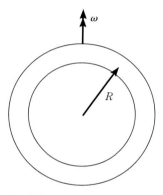

Variational formulation exercises

51. Show that the strain energy of bending for the curved beam shown below can be expressed as
$$U = \int_{Vol} \frac{E\varepsilon_x^2}{2} \, dVol = \int_0^L \left(\frac{AE(u' + kv)^2}{2} + \frac{EI_2(v'' + k^2v)^2}{2} \right) ds$$
Take the external potential energy to be
$$\Omega = -\int_0^L (q_x u + q_y v) \, ds - V_0 v(L) - N_0 u(L) - M_0(v'(L) - ku(L))$$
and use the calculus of variations to determine the conditions that define the stationary value for the functional $U + \Omega$. Interpret all your results in terms of the corresponding force resultants.

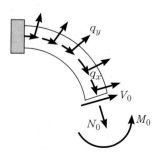

6 Thick-Walled Cylinders

6-1 Introduction
6-2 Development of the Theory
6-3 Internal and External Pressures
    6-3-1 Shrink-Fit Problems
6-4 Long Cylinders
6-5 Thermal Problems
6-6 Rotating Disks
6-7 Variable-Thickness Rotating Disks
6-8 Finite-Element Solutions
6-9 Summary
References
Exercises

6-1 Introduction

The student is familiar with the problem of the thin-walled cylindrical pressure vessel, shown in Figure 6-1. The geometry of a thin-walled pressure vessel is defined by its thickness t and average radius R. At locations away from the ends of the pressure vessel, the small t/R ratio makes it possible to argue that the radial displacement is essentially constant over the thickness, resulting in a hoop stress σ_θ that is also constant, that is, $\sigma_\theta = pR/t$. The corresponding axial stress σ_x is given by $\sigma_x = pR/2t$.

For geometries where the thin-walled assumption is not valid, as indicated in Figure 6-2a, a more general analysis must be carried out to account for the variations of displacement and stress over the thickness. As shown in Figure 6-2b, the loadings or inputs that will be considered are uniform external and internal pressures p_o and p_i,

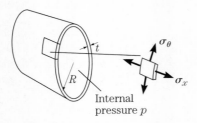

FIGURE 6-1 Thin-walled pressure vessel.

6-2 DEVELOPMENT OF THE THEORY

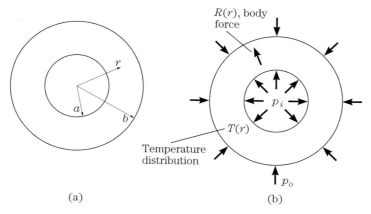

FIGURE 6-2 Thick-walled geometry.

respectively, symmetric body forces $R(r)$ and symmetric temperature distributions $T(r)$. For an axisymmetric region, and for inputs and material properties that do not depend on the circumferential coordinate θ, assume that none of the displacements or stresses depends on θ.

6-2 Development of the Theory

Equilibrium. A free-body diagram for a differential volume $t(\Delta r)(r\Delta\theta)$ is shown in Figure 6-3b. The dimension of the region [perpendicular to the paper] is t. Shown are the normal stresses σ_r and σ_θ and the body force $R(r)$ per unit volume. It is left to the exercises for the student to investigate whether or not there are any axisymmetric problems involving the shear stress $\tau_{r\theta}$.

From the free-body diagram,

$$\sum F_r = t\left(r\sigma_r + \frac{d(r\sigma_r)}{dr}\Delta r\right)\Delta\theta - t(r\sigma_r)\Delta\theta - 2t\sigma_\theta \Delta r \sin\left(\frac{\Delta\theta}{2}\right) + tRr\Delta r\Delta\theta = 0$$

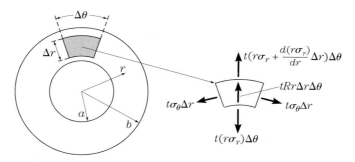

FIGURE 6-3 Free-body diagram for considering equilibrium.

Passing to the limit as Δr and $\Delta\theta$ approach zero yields the equilibrium equations

$$\frac{d(r\sigma_r)}{dr} - \sigma_\theta + rR = 0$$

or

$$\frac{d\sigma_r}{dr} + \frac{\sigma_r - \sigma_\theta}{r} + R = 0 \tag{6-1}$$

which must be satisfied by σ_r and σ_θ for $a \leq r \leq b$.

Deformation. Within the context of axisymmetric problems involving pressures, body forces and temperature, the tangential component of displacement vanishes with the only in-plane component of displacement being the radial displacement $u(r)$. The student is asked to investigate other possibilities in the exercises. Shown in Figure 6-4 are initial radial and circumferential line segments, AB and CD, respectively, along with the deformed positions, A^*B^* and C^*D^*. Consulting Figure 6-4b, the extensional strain in the radial direction is given by

$$\epsilon_r = \lim_{\Delta r \to 0} \frac{A^*B^* - AB}{AB} = \lim_{\Delta r \to 0} \frac{[(r + \Delta r + u + \Delta u) - (r + u)] - \Delta r}{\Delta r}$$
$$= \frac{du}{dr} = u'(r) \tag{6-2a}$$

Similarly, with reference to Figure 6-4c, the extensional strain in the circumferential direction is given by

$$\epsilon_\theta = \lim_{\Delta\theta \to 0} \frac{C^*D^* - CD}{CD} = \lim_{\Delta\theta \to 0} \frac{(r + u)\Delta\theta - r\Delta\theta}{r\Delta\theta} = \frac{u}{r} \tag{6-2b}$$

Note that there are five unknowns, namely, σ_r, σ_θ, ϵ_r, ϵ_θ and u, appearing in only three equations for equilibrium and deformation. Not only is the number of equations defi-

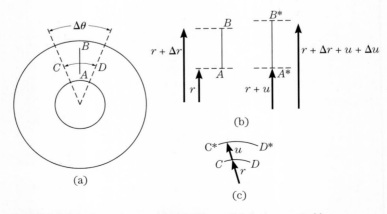

FIGURE 6-4 Deformations associated with a radial displacement $u(r)$.

cient, but there is also no way to relate the kinetic variables σ_r and σ_θ contained in the equilibrium equations to the kinematic variables ϵ_r, ϵ_θ and u appearing in the strain-displacement equations. The additional equations that address these points come from the material behavior, which is discussed next.

Material Behavior. Assume that the disk is constructed of a linearly elastic isotropic material with the stress-strain relations given by

$$\epsilon_r = \frac{\sigma_r - \nu(\sigma_\theta + \sigma_z)}{E} + \alpha T \tag{6-3a}$$

$$\epsilon_\theta = \frac{\sigma_\theta - \nu(\sigma_z + \sigma_r)}{E} + \alpha T \tag{6-3b}$$

$$\epsilon_z = \frac{\sigma_z - \nu(\sigma_r + \sigma_\theta)}{E} + \alpha T \tag{6-3c}$$

where E is Young's modulus, ν is Poisson's ratio and α is the coefficient of thermal expansion. The third of these three equations is included by virtue of the fact that the two stresses of interest, namely, σ_r and σ_θ, appear. These stress-strain relations will be specialized to two technically important classes of physical problems that are useful in applications.

Plane Stress. Plane stress refers to a geometry, frequently referred to as a disk, where the z dimension of region t is small compared to the outer radius, as shown in Figure 6-5, that is, $t \ll b$. The $z = $ constant planes are stress-free, that is, σ_z is zero on both faces. We assume that this precludes the possibility that any appreciable σ_z can exist in the region between the faces and thus take $\sigma_z \equiv 0$, termed *plane stress*. The first two of the stress-strain relations can thus be written as

$$\epsilon_r = \frac{\sigma_r - \nu\sigma_\theta}{E} + \alpha T \tag{6-4a}$$

$$\epsilon_\theta = \frac{\sigma_\theta - \nu\sigma_r}{E} + \alpha T \tag{6-4b}$$

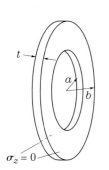

FIGURE 6-5 Geometry for plane stress.

Solving these for σ_r and σ_θ gives

$$\sigma_r = \frac{E}{1-\nu^2}(\epsilon_r + \nu\epsilon_\theta) - \frac{E\alpha T}{1-\nu} \tag{6-5a}$$

and

$$\sigma_\theta = \frac{E}{1-\nu^2}(\epsilon_\theta + \nu\epsilon_r) - \frac{E\alpha T}{1-\nu} \tag{6-5b}$$

There are now five equations for the five unknowns identified previously.

Combination. The governing equations are obtained as follows. Substitute the strain-displacement relations into the stress-strain relations to obtain

$$\sigma_r = \frac{E}{1-\nu^2}\left(u' + \nu\frac{u}{r}\right) - \frac{E\alpha T}{1-\nu} \tag{6-6a}$$

and

$$\sigma_\theta = \frac{E}{1-\nu^2}\left(\frac{u}{r} + \nu u'\right) - \frac{E\alpha T}{1-\nu} \quad (6\text{-}6b)$$

Substitute these in turn into the equilibrium equation to give, after a bit of algebra,

$$\frac{d^2u}{dr^2} + \frac{1}{r}\frac{du}{dr} - \frac{u}{r^2} = (1+\nu)\alpha T' - \frac{1-\nu^2}{E}R \quad (6\text{-}7)$$

This differential equation must be integrated subjected to appropriate boundary conditions. These boundary conditions are

At $r = a$:   Either $u(a) = U_a$

or $\sigma_r\Big|_{r=a} = \frac{E}{1-\nu^2}\left(u' + \nu\frac{u}{r}\right)\Big|_{r=a} - \frac{E\alpha T(a)}{1-\nu} = \Sigma(a)$

At $r = b$:   Either $u(b) = U_b$

or $\sigma_r\Big|_{r=b} = \frac{E}{1-\nu^2}\left(u' + \nu\frac{u}{r}\right)\Big|_{r=b} - \frac{E\alpha T(b)}{1-\nu} = \Sigma(b)$

where U_a and U_b are prescribed displacements at $r = a$ and $r = b$, respectively, and $\Sigma(a)$ and $\Sigma(b)$ are prescribed radial stresses at $r = a$ and $r = b$, respectively.

Although the stress σ_z has been assumed to be zero for plane stress, the corresponding strain ϵ_z is not zero and is given by the third of the stress-strain equations as

$$\epsilon_z = -\frac{\nu(\sigma_r + \sigma_\theta)}{E} + \alpha T$$

The strain-displacement relation $\epsilon_z = \partial w/\partial z$ can be used in connection with this equation to estimate the displacements in the axial direction.

Plane Strain. Plane strain refers to a situation where, as shown in Figure 6-6, constraints in the form of radially dependent normal stresses $\sigma_z(r)$ prevent any displace-

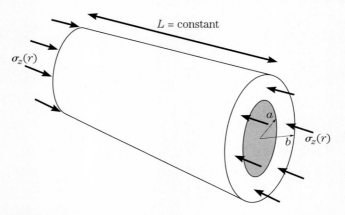

FIGURE 6-6 Geometry for plane strain.

ment w in the direction normal to the z axis. With $w = 0$ it follows that $\epsilon_z = 0$, referred to as *plane strain*. The third of the stress-strain relations then yields

$$\sigma_z = \nu(\sigma_r + \sigma_\theta) - E\alpha T$$

which, when substituted into the first two stress-strain equations, gives

$$\epsilon_r = \frac{1-\nu^2}{E}\left(\sigma_r - \frac{\nu}{1-\nu}\sigma_\theta\right) + \alpha(1+\nu)T \tag{6-8a}$$

$$\epsilon_\theta = \frac{1-\nu^2}{E}\left(\sigma_\theta - \frac{\nu}{1-\nu}\sigma_r\right) + \alpha(1+\nu)T \tag{6-8b}$$

Observe that if in the corresponding equations for plane stress, namely Equations (6-4a) and (6-4b), we replace E with $E/(1-\nu^2)$, ν with $\nu/(1-\nu)$ and α with $(1+\nu)\alpha$, Equations (6-8a) and (6-8b) result. Solving Equations (6-8) for σ_r and σ_θ gives

$$\sigma_r = \frac{E}{(1+\nu)(1-2\nu)}((1-\nu)\epsilon_r + \nu\epsilon_\theta) - \frac{E\alpha T}{1-2\nu} \tag{6-9a}$$

and

$$\sigma_\theta = \frac{E}{(1+\nu)(1-2\nu)}((1-\nu)\epsilon_\theta + \nu\epsilon_r) - \frac{E\alpha T}{1-2\nu} \tag{6-9b}$$

Combination. To obtain the governing equations for plane strain, proceed as in the case of plane stress by first substituting for the strains to obtain

$$\sigma_r = \frac{E}{(1+\nu)(1-2\nu)}\left((1-\nu)u' + \nu\frac{u}{r}\right) - \frac{E\alpha T}{1-2\nu}$$

and

$$\sigma_\theta = \frac{E}{(1+\nu)(1-2\nu)}\left((1-\nu)\frac{u}{r} + \nu u'\right) - \frac{E\alpha T}{1-2\nu}$$

Substituting into the equilibrium equations then yields

$$\frac{d^2u}{dr^2} + \frac{1}{r}\frac{du}{dr} - \frac{u}{r^2} = \frac{1+\nu}{1-\nu}\alpha T' - \frac{(1+\nu)(1-2\nu)}{E(1-\nu)}R \tag{6-10}$$

The appropriate boundary conditions are

At $r = a$: Either $u(a) = U_a$

or $\sigma_r\Big|_{r=a} = \frac{E}{(1+\nu)(1-2\nu)}\left((1-\nu)u' + \nu\frac{u}{r}\right)\Big|_{r=a}$

$$-\frac{E\alpha T(a)}{1-2\nu} = \Sigma(a)$$

At $r = b$: Either $u(b) = U_b$

or $\sigma_r\Big|_{r=b} = \frac{E}{(1+\nu)(1-2\nu)}\left((1-\nu)u' + \nu\frac{u}{r}\right)\Big|_{r=a}$

$$-\frac{E\alpha T(b)}{1-2\nu} = \Sigma(b)$$

where U_a and U_b are prescribed displacements at $r = a$ and $r = b$, respectively, and $\Sigma(a)$ and $\Sigma(b)$ are prescribed radial stresses at $r = a$ and $r = b$, respectively.

For the case of plane strain, it must be remembered that axial stresses in the amount

$$\sigma_z = \nu(\sigma_r + \sigma_\theta) - E\alpha T$$
$$= \frac{E}{(1+\nu)(1-2\nu)}\left(u' + \frac{u}{r}\right) - E\alpha T$$

must be imposed at the ends of the cylinder to maintain the plane-strain condition.

Homogeneous Solutions. When there are no temperature or body force inputs, the equilibrium equation for either the case of plane stress or plane strain reduces to

$$\frac{d^2 u}{dr^2} + \frac{1}{r}\frac{du}{dr} - \frac{u}{r^2} = 0 = \frac{d}{dr}\left(\frac{1}{r}\frac{d}{dr}(ru)\right)$$

which can be integrated to yield

$$u = C_1 r + \frac{C_2}{r}$$

where C_1 and C_2 are arbitrary constants. For plane stress, the stresses can then be computed using Equations (6-5a and b) as

$$\sigma_r = \frac{E}{1-\nu}C_1 - \frac{E}{1+\nu}\frac{C_2}{r^2} = A - \frac{B}{r^2} \tag{6-11a}$$

$$\sigma_\theta = \frac{E}{1-\nu}C_1 + \frac{E}{1+\nu}\frac{C_2}{r^2} = A + \frac{B}{r^2} \tag{6-11b}$$

where A and B are arbitrary constants. For plane strain, using Equations (6-9a and b) the stresses are

$$\sigma_r = \frac{E}{(1+\nu)(1-2\nu)}C_1 - \frac{E}{1+\nu}\frac{C_2}{r^2} = A^* - \frac{B^*}{r^2} \tag{6-12a}$$

$$\sigma_\theta = \frac{E}{(1+\nu)(1-2\nu)}C_1 + \frac{E}{1+\nu}\frac{C_2}{r^2} = A^* + \frac{B^*}{r^2} \tag{6-12b}$$

where again A^* and B^* are arbitrary constants. These equations are useful in situations where stresses are prescribed at the boundaries as in the case of internal and external pressures, which are treated in the next section.

6-3 Internal and External Pressures

One of the simplest and most useful applications of the theory developed in section 6-2 is the problem of an annular region subjected to internal and external pressures. It is assumed that there are no thermal inputs nor any other inputs resulting in body forces. The details of the problem are presented in the example that follows.

EXAMPLE 6-1

For the problem shown in Figure 6-7, determine the stresses and displacements for (a) the plane-stress case and (b) the plane-strain case.

Solution: With temperature $T(r)$ and the body force $R(r)$ both zero, the governing equation of equilibrium is

$$\frac{d^2u}{dr^2} + \frac{1}{r}\frac{du}{dr} - \frac{u}{r^2} = 0$$

with the solution $u = C_1 r + C_2/r$. For either the plane-stress or the plane-strain case, the boundary conditions can be represented as

$$\sigma_r\bigg|_{r=a} = \left(A - \frac{B}{r^2}\right)\bigg|_{r=a} = A - \frac{B}{a^2} = -p_i$$

$$\sigma_r\bigg|_{r=b} = \left(A - \frac{B}{r^2}\right)\bigg|_{r=b} = A - \frac{B}{b^2} = -p_o$$

Solving for A and B yields

$$A = \frac{a^2 p_i - b^2 p_o}{b^2 - a^2} \qquad B = \frac{a^2 b^2 (p_i - p_o)}{b^2 - a^2}$$

with

$$\sigma_r = \frac{p_i a^2}{b^2 - a^2}\left(1 - \frac{b^2}{r^2}\right) - \frac{p_o b^2}{b^2 - a^2}\left(1 - \frac{a^2}{r^2}\right)$$

$$\sigma_\theta = \frac{p_i a^2}{b^2 - a^2}\left(1 + \frac{b^2}{r^2}\right) - \frac{p_o b^2}{b^2 - a^2}\left(1 + \frac{a^2}{r^2}\right)$$

For the plane-stress case Equations (6-8) give $C_1 = (1-\nu)A/E$ and $C_2 = (1+\nu)B/E$, and the displacement $u = C_1 r + C_2/r$ can be expressed as

$$u(r) = \frac{(1-\nu)}{E}\frac{a^2 p_i - b^2 p_o}{b^2 - a^2}r + \frac{(1+\nu)}{E}\frac{a^2 b^2 (p_i - p_o)}{b^2 - a^2}\frac{1}{r}$$

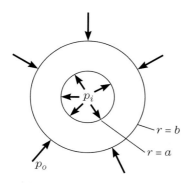

FIGURE 6-7 Internal and external pressures.

For the plane-strain case, Equations (6-9) give $C_1 = (1+\nu)(1-2\nu)A/E$ and $C_2 = (1+\nu)B/E$, with the displacement given by

$$u(r) = \frac{(1-2\nu)(1+\nu)}{E}\frac{a^2 p_i - b^2 p_o}{b^2 - a^2}r + \frac{(1+\nu)}{E}\frac{a^2 b^2 (p_i - p_o)}{b^2 - a^2}\frac{1}{r}$$

For the technically important case where there is only an internal pressure $p_i = p$, the stresses reduce to

$$\sigma_r = -\frac{pa^2}{b^2 - a^2}\left(\frac{b^2}{r^2} - 1\right)$$

$$\sigma_\theta = \frac{pa^2}{b^2 - a^2}\left(\frac{b^2}{r^2} + 1\right)$$

and the displacements to

$$u = \frac{a^2 p}{E(b^2 - a^2)}\left[(1-\nu)r + (1+\nu)\frac{b^2}{r}\right]$$

and

$$u = \frac{1+\nu}{E}\frac{a^2 p}{(b^2 - a^2)}\left[(1-2\nu)r + \frac{b^2}{r}\right]$$

for the plane-stress and plane-strain cases, respectively. The radial stress, which vanishes at the outer radius and has the value $\sigma_r(a) = -p$ at the inner radius, is shown in Figure 6-8 for several b/a ratios. The hoop stress σ_θ is positive at all points $a \leq r \leq b$ with the values at the inside radius given by

$$\left.\sigma_\theta\right|_{r=a} = \frac{p(b^2 + a^2)}{b^2 - a^2}$$

$$\left.\sigma_\theta\right|_{r=b} = \frac{2pa^2}{b^2 - a^2}$$

which is also shown in Figure 6-8.

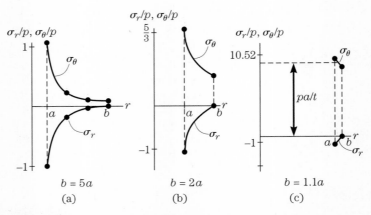

FIGURE 6-8 Radial and hoop stresses for several geometries.

6-3 INTERNAL AND EXTERNAL PRESSURES

Clearly, as the wall thickness decreases, that is, as $a \to b$, the hoop stress is dominant with the maximum and minimum values approximately the same. In this regard, it is left as an exercise for the student to show that the average hoop stress over the interval $a \le r \le b$ is $(\sigma_\theta)_{\text{avg}} = pa/(b-a) = pa/t$, as shown in Figure 6-8c. This is essentially the thin-walled result mentioned at the beginning of this chapter, showing that the thin-walled theory is correct as a limiting case of the thick-wall theory. Also note that for $b \approx a$, the magnitude of the maximum radial stress is approximately

$$\left|(\sigma_r)_{\max}\right| = \frac{t}{a}(\sigma_\theta)_{\max}$$

that is, the radial stresses are much smaller than the hoop stresses.

Only for a small t/a can the approximate thin-walled expressions $\sigma_\theta = pa/t$ and $\sigma_r \approx 0$ be used.

Note that when thermal and body force inputs are absent, the stresses do not depend on the elastic constants but only on the values of the stresses prescribed at the inner and outer radii. The corresponding displacements do, however, depend on the elastic constants.

Finally, for the plane-strain case the axial stress σ_z is given by

$$\sigma_z = \nu(\sigma_r + \sigma_\theta) = \frac{a^2 p_i - b^2 p_o}{b^2 - a^2}$$

which has a constant value over the $z = $ constant ends of the region. ◆

6-3-1 Shrink-Fit Problems

Another important technical problem involves shrinking one tube over another. This is usually accomplished by taking two tubes, a smaller tube whose outer diameter is slightly larger than the inner diameter of the larger tube, as shown in Figure 6-9. The amount δ by which the outer diameter of the inner tube exceeds the inner diameter of the outer tube is frequently referred to as the *interference*. The larger tube is then heated until the smaller tube can be inserted, with the assembly then allowed to cool to room temperature. The details of the analysis are carried out in the example that follows.

EXAMPLE 6-2 Two annular regions with an interference of magnitude δ as shown in Figure 6-9 are to be combined by a shrink-fit process. Determine the stresses for both plane-stress and plane-strain situations.

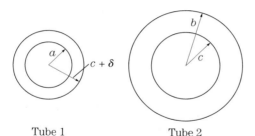

Tube 1   Tube 2

FIGURE 6-9 Shrink-fit geometries.

Solution: We formulate this problem as follows. Let u_1 be the displacement in the inner tube and u_2 that in the outer tube. At the final state, $T = R = 0$ and both displacements u_1 in the inner tube and u_2 in the outer tube satisfy the same equation, namely,

$$u_1'' + \frac{u_1'}{r} - \frac{u_1}{r^2} = 0 \qquad a \leq r \leq c$$

$$u_2'' + \frac{u_2'}{r} - \frac{u_2}{r^2} = 0 \qquad c \leq r \leq b$$

With σ_{r1} and σ_{r2} as the radial stresses in the two disks, the four boundary conditions are

Inner boundary: $\qquad \sigma_{r1}(a) = 0$
Interface: $\qquad \sigma_{r1}(c) = \sigma_{r2}(c) \quad \text{and} \quad u_2(c) - u_1(c) = \delta$
Outer boundary: $\qquad \sigma_{r2}(b) = 0$

The solutions to the equations of equilibrium are

$$u_1 = C_1 r + \frac{C_2}{r}$$

and

$$u_2 = C_3 r + \frac{C_4}{r}$$

Assuming both tubes are composed of the same material, the radial stresses for the two tubes for the plane-stress case are

$$\sigma_{r1} = \frac{E}{1-\nu} C_1 - \frac{E}{1+\nu} \frac{C_2}{r^2}$$

and

$$\sigma_{r2} = \frac{E}{1-\nu} C_3 - \frac{E}{1+\nu} \frac{C_4}{r^2}$$

Satisfying the boundary conditions leads to the four simultaneous equations

$$\frac{E}{1-\nu} C_1 - \frac{E}{1+\nu} \frac{C_2}{a^2} = 0$$

$$\frac{E}{1-\nu} C_1 - \frac{E}{1+\nu} \frac{C_2}{c^2} = \frac{E}{1-\nu} C_3 - \frac{E}{1+\nu} \frac{C_4}{c^2}$$

$$C_3 c + \frac{C_4}{c} - C_1 c - \frac{C_2}{c} = \delta$$

$$\frac{E}{1-\nu} C_3 - \frac{E}{1+\nu} \frac{C_4}{b^2} = 0$$

6-3 INTERNAL AND EXTERNAL PRESSURES

which can be solved to give

$$C_1 = -\frac{b^2 - c^2}{b^2 - a^2}\frac{\phi}{1+\phi}\frac{\delta}{c} \qquad C_2 = -a^2\frac{b^2 - c^2}{b^2 - a^2}\frac{1}{1+\phi}\frac{\delta}{c}$$

$$C_3 = \frac{c^2 - a^2}{b^2 - a^2}\frac{\phi}{1+\phi}\frac{\delta}{c} \qquad C_4 = b^2\frac{c^2 - a^2}{b^2 - a^2}\frac{1}{1+\phi}\frac{\delta}{c}$$

where $\phi = (1-\nu)/(1+\nu)$. The radial and hoop stresses are given by

$$\sigma_{r1} = \frac{E}{1-\nu}C_1 - \frac{E}{1+\nu}\frac{C_2}{r^2} = -\frac{E\,\delta}{2\,c}\frac{b^2 - c^2}{b^2 - a^2}\left(1 - \frac{a^2}{r^2}\right)$$

$$\sigma_{\theta 1} = \frac{E}{1-\nu}C_1 + \frac{E}{1+\nu}\frac{C_2}{r^2} = -\frac{E\,\delta}{2\,c}\frac{b^2 - c^2}{b^2 - a^2}\left(1 + \frac{a^2}{r^2}\right)$$

and

$$\sigma_{r2} = \frac{E}{1-\nu}C_3 - \frac{E}{1+\nu}\frac{C_4}{r^2} = \frac{E\,\delta}{2\,c}\frac{c^2 - a^2}{b^2 - a^2}\left(1 - \frac{b^2}{r^2}\right)$$

$$\sigma_{\theta 2} = \frac{E}{1-\nu}C_3 + \frac{E}{1+\nu}\frac{C_4}{r^2} = \frac{E\,\delta}{2\,c}\frac{c^2 - a^2}{b^2 - a^2}\left(1 + \frac{b^2}{r^2}\right)$$

These are shown in Figure 6-10 for the particular case where $b = 3a$ and $c = 2a$. In that figure, $\Sigma = 2c\sigma/E\delta$. The maximum hoop stress occurs at the inner radius a with the value $\sigma_{\theta 1} = -40E\delta/64c$. The corresponding positive value in the outer tube at the interface is $\sigma_{\theta 2} = 39E\delta/64c$.

As outlined above, the corresponding stresses for the plane-strain case are obtained by replacing E by $E/(1-\nu^2)$, leading to

$$\sigma_{r1} = -\frac{E}{2(1-\nu^2)}\frac{\delta}{c}\frac{b^2 - c^2}{b^2 - a^2}\left(1 - \frac{a^2}{r^2}\right)$$

$$\sigma_{\theta 1} = -\frac{E}{2(1-\nu^2)}\frac{\delta}{c}\frac{b^2 - c^2}{b^2 - a^2}\left(1 + \frac{a^2}{r^2}\right)$$

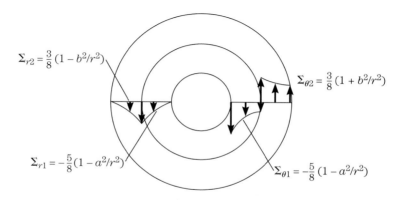

$\Sigma_{r2} = \frac{3}{8}(1 - b^2/r^2)$

$\Sigma_{\theta 2} = \frac{3}{8}(1 + b^2/r^2)$

$\Sigma_{r1} = -\frac{5}{8}(1 - a^2/r^2)$

$\Sigma_{\theta 1} = -\frac{5}{8}(1 - a^2/r^2)$

FIGURE 6-10 Radial and hoop stresses for the interference problem.

and

$$\sigma_{r2} = \frac{E}{2(1-\nu^2)} \frac{\delta}{c} \frac{c^2 - a^2}{b^2 - a^2}\left(1 - \frac{b^2}{r^2}\right)$$

$$\sigma_{\theta 2} = \frac{E}{2(1-\nu^2)} \frac{\delta}{c} \frac{c^2 - a^2}{b^2 - a^2}\left(1 + \frac{b^2}{r^2}\right)$$

Each of the stresses for the plane-strain case is clearly larger by the factor $1/(1-\nu^2)$ compared to its plane-stress counterpart. ◆

6-4 Long Cylinders

Many applications involve long cylinders or tubes that are either open (as in Figure 6-11a) or closed (as in Figure 6-11b). For the open tube, it is no longer valid to use plane stress on the basis of the argument that $\sigma_z = 0$ at the ends and that the thickness in the axial direction is small. For the closed tube, there is clearly the possibility that the length will change with a resulting nonzero axial strain ϵ_z. Both of these cases will be treated in this section by making suitable assumptions regarding the stresses away from the ends and ultimately superposing whatever state(s) of stress are necessary in order to satisfy, at least approximately, the conditions at the ends.

Open Cylinder. For the case of the open cylinder, a section perpendicular to the axis of the tube through any point shows that there is no resultant force transmitted by the axial stresses in the tube. This in turn implies that the average axial stress must be zero. With this in mind, the displacement u and the stresses σ_r and σ_θ are determined using a plane-strain formulation in the usual manner by solving the equations of equilibrium and satisfying the boundary conditions at the inner and outer radii. Then, according to the plane-strain theory, the axial stress necessary to accomplish $w = \epsilon_z = 0$ is given by

$$\sigma_z = \sigma_z(r) = \nu(\sigma_r + \sigma_\theta) - E\alpha T$$

The resultant force associated with this axial stress distribution is

$$F_z = \int_{\text{Area}} (\nu(\sigma_r + \sigma_\theta) - E\alpha T)\, dA$$

An average stress computed from F_z is then superposed at the ends of the tube in the opposite direction to produce a stress distribution that has no resultant force. The corresponding average strain in the axial direction is then given by $\epsilon_z = F_z/AE$ with the elongation given approximately by $e = L\epsilon_z$, where L is the length of the tube. The stresses σ_r and σ_θ are not affected by the addition of the force P. It is left for the

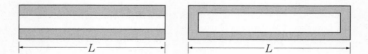

FIGURE 6-11 Open and closed thick-walled cylinders.

student to show that the displacement u is altered from the plane-strain solution by the amount

$$u = -\frac{\nu F_z}{AE} r$$

EXAMPLE 6-3 An open tube is subjected to an internal pressure p, as shown in Figure 6-12. Investigate the state of stress and the displacements.

Solution: As mentioned above, the argument of the thin disk can no longer be used to assume that $\sigma_z = 0$. It is true, however, that the average of σ_z must be zero for equilibrium. In order to satisfy this condition, formulate the problem as a plane-strain problem and then superpose the appropriate stress σ_z to obtain the zero average. With $T = R = 0$, the equation of equilibrium is given from Equation (6-7) as

$$\frac{d^2 u}{dr^2} + \frac{1}{r}\frac{du}{dr} - \frac{u}{r^2} = 0$$

The boundary conditions are $\sigma_r(a) = -p$ and $\sigma_r(b) = 0$, leading to

$$\sigma_r = -\frac{pa^2}{b^2 - a^2}\left(\frac{b^2}{r^2} - 1\right)$$

$$\sigma_\theta = \frac{pa^2}{b^2 - a^2}\left(\frac{b^2}{r^2} + 1\right)$$

and

$$u = (1+\nu)\frac{pa^2}{E(b^2-a^2)}\left[(1-2\nu)r + \frac{b^2}{r}\right]$$

According to the plane-strain theory, the stress σ_z corresponding to the strain $\epsilon_z = 0$ is $\sigma_z = \nu(\sigma_r + \sigma_\theta) = 2\nu pa^2/(b^2 - a^2)$. The resultant associated with this axial stress is the tensile force

$$F_z = \sigma_z \pi (b^2 - a^2) = 2\nu p \pi a^2$$

The stress-free state is then obtained by applying a compressive force of magnitude $2\nu p \pi a^2$, resulting in the stress $\sigma_z = -2\nu pa^2/(b^2 - a^2)$. The corresponding strain is $\epsilon_z = -2\nu pa^2/E(b^2 - a^2)$ with the change in length given by

$$e = L\epsilon_z = -\frac{2\nu pa^2 L}{E(b^2 - a^2)}$$

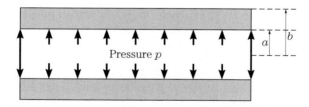

FIGURE 6-12 Open tube subjected to pressure.

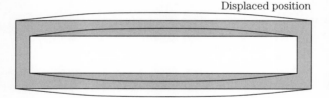

FIGURE 6-13 Radial displacements in a closed tube.

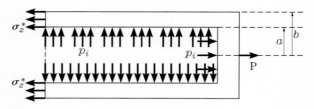

FIGURE 6-14 Closed cylinder.

Closed Cylinder. Shown exaggerated in magnitude, the radial deformations in a closed tube would appear as indicated in Figure 6-13. There will be a region near the ends where the deformations and stresses would clearly depend on the axial coordinate but generally also a sizable region away from the ends where it is reasonable to assume that the displacement u depends only on r. In this middle portion, neither plane stress nor plane strain is appropriate; proceed as follows. Draw a free-body diagram exposing the internal axial stress in the middle portion, as shown in Figure 6-14. Shown are the internal pressure p_i, a possible external resultant force P and an *average* axial stress σ_z^*. Equilibrium requires that σ_z^* satisfy

$$P + p_i A_i = \sigma_z^* A^*$$

where $A_i = \pi a^2$ and $A^* = \pi(b^2 - a^2)$. The stress-strain relations are taken as

$$\epsilon_r = \frac{\sigma_r - \nu \sigma_\theta}{E} - \frac{\nu \sigma_z^*}{E} + \alpha T$$

$$\epsilon_\theta = \frac{\sigma_\theta - \nu \sigma_r}{E} - \frac{\nu \sigma_z^*}{E} + \alpha T$$

$$\epsilon_z = \frac{\sigma_z^* - \nu(\sigma_r + \sigma_\theta)}{E} + \alpha T$$

These are precisely the plane-stress equations with αT replaced by $\alpha T - \nu \sigma_z^*/E$, with the result that solving for the stresses yields

$$\sigma_r = \frac{E}{1 - \nu^2}(\epsilon_r + \nu \epsilon_\theta) - \frac{E \alpha T}{1 - \nu} + \frac{\nu \sigma_z^*}{1 - \nu}$$

$$\sigma_\theta = \frac{E}{1 - \nu^2}(\epsilon_\theta + \nu \epsilon_r) - \frac{E \alpha T}{1 - \nu} + \frac{\nu \sigma_z^*}{1 - \nu}$$

6-4 LONG CYLINDERS

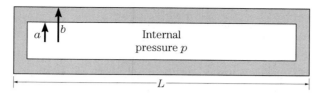

FIGURE 6-15 Thick-walled tube subjected to an internal pressure p.

With $\nu\sigma_z^*/E$ a constant, the equation of equilibrium is unchanged from the usual plane-stress result, as are the stresses σ_r and σ_θ. The displacement u is changed by the amount

$$u = -\frac{\nu\sigma_z^*}{E}r$$

The strain in the axial direction is then given by

$$\epsilon_z = \frac{\sigma_z^* - \nu(\sigma_r + \sigma_\theta)}{E} + \alpha T$$

where σ_r and σ_θ are replaced by the solutions to the plane-stress boundary value problem. The average axial strain is then computed as

$$\epsilon_z^* = \frac{1}{A^*}\int_{A^*}\epsilon_z\, dA$$

and the elongation approximated by $e = L\epsilon_z^*$, where L is the dimension shown in Figure 6-15.

EXAMPLE 6-4 A thick-walled tube with dimensions as shown in Figure 6-15 is subjected to an internal pressure P. Determine the stresses in and the elongation of the middle portion taken to have length L.

Solution: The strain-displacement equations are written as

$$\epsilon_r = \frac{\sigma_r - \nu\sigma_\theta}{E} - \frac{\nu\sigma_z^*}{E}$$

$$\epsilon_\theta = \frac{\sigma_\theta - \nu\sigma_r}{E} - \frac{\nu\sigma_z^*}{E}$$

$$\epsilon_z = \frac{\sigma_z^* - \nu(\sigma_r + \sigma_\theta)}{E}$$

where $\sigma_z^* = pa^2/(b^2 - a^2)$. Solving for σ_r and σ_θ yields

$$\sigma_r = \frac{E}{1-\nu^2}(\epsilon_r + \nu\epsilon_\theta) + \frac{\nu\sigma_z^*}{1-\nu}$$

$$\sigma_\theta = \frac{E}{1-\nu^2}(\epsilon_\theta + \nu\epsilon_r) + \frac{\nu\sigma_z^*}{1-\nu}$$

Substituting these into the equilibrium equation gives

$$\frac{d^2u}{dr^2} + \frac{1}{r}\frac{du}{dr} - \frac{u}{r^2} = 0$$

with boundary conditions

$$\sigma_r(a) = \left[\frac{E}{1-\nu^2}\left(u' + \frac{\nu u}{r}\right) + \frac{\nu \sigma_z^*}{1-\nu}\right]\bigg|_{r=a} = -p$$

and

$$\sigma_r(b) = \left[\frac{E}{1-\nu^2}\left(u' + \frac{\nu u}{r}\right) + \frac{\nu \sigma_z^*}{1-\nu}\right]\bigg|_{r=b} = 0$$

The solution is

$$\sigma_r = -\frac{pa^2}{b^2 - a^2}\left(\frac{b^2}{r^2} - 1\right)$$

$$\sigma_\theta = \frac{pa^2}{b^2 - a^2}\left(\frac{b^2}{r^2} + 1\right)$$

and

$$u = \frac{pa^2}{E(b^2 - a^2)}\left[(1-\nu)r + \frac{(1+\nu)b^2}{r}\right] - \frac{\nu \sigma_z^* r}{E}$$

$$= \frac{pa^2}{E(b^2 - a^2)}\left[(1-2\nu)r + \frac{(1+\nu)b^2}{r}\right]$$

Clearly, the stresses σ_r and σ_θ in the middle of the tube are unaffected by the closed ends, whereas the radial displacement is decreased by the amount $\nu \sigma_z^* r/E$.

The strain ϵ_z is given by

$$\epsilon_z = \frac{\sigma_z^* - \nu(\sigma_r + \sigma_\theta)}{E} = \frac{pa^2}{E(b^2 - a^2)}(1 - 2\nu)$$

which is a constant so that

$$e \approx L\epsilon_z = \frac{pa^2 L}{E(b^2 - a^2)}(1 - 2\nu) \qquad \blacklozenge$$

These same ideas are applicable irrespective of the type of loading, that is, for thermal and body force loadings. Several applications will be investigated in detail in the following sections.

6-5 Thermal Problems

One of the most important applications of the theory developed above occurs in situations where a thick-walled cylinder is subjected to axisymmetric thermal loads. Examples include pipes that carry a heated liquid, fins intended to serve as devices for dissipating heat and turbine disks. In each of these applications, it is necessary to combine the stresses arising from the thermal loads with those arising from the displacement or stress-type boundary conditions.

Recall that for a plane-stress situation, the governing equation of equilibrium in the absence of body forces is

$$\frac{d}{dr}\left(\frac{1}{r}\frac{d}{dr}(ru)\right) = (1+\nu)\alpha T'$$

which can be integrated to yield

$$u = \frac{1}{r}\int_a^r (1+\nu)\alpha rT\, dr + C_1 r + \frac{C_2}{r}$$

The two constants of integration are determined from the boundary conditions as follows:

At $r = a$: Either $u(a) = U_a$

or $\left.\sigma_r\right|_{r=a} = \left.\frac{E}{1-\nu^2}\left(u' + \nu\frac{u}{r}\right)\right|_{r=a} - \frac{E\alpha T(a)}{1-\nu} = \Sigma(a)$

At $r = b$: Either $u(b) = U_b$

or $\left.\sigma_r\right|_{r=b} = \left.\frac{E}{1-\nu^2}\left(u' + \nu\frac{u}{r}\right)\right|_{r=b} - \frac{E\alpha T(b)}{1-\nu} = \Sigma(b)$

where U_a and U_b are prescribed displacements at $r = a$ and $r = b$, respectively, and $\Sigma(a)$ and $\Sigma(b)$ are prescribed radial stresses at $r = a$ and $r = b$, respectively. The corresponding strain in the z or axial direction is

$$\epsilon_z = -\frac{\nu(\sigma_r + \sigma_\theta)}{E} + \alpha T$$

from which the elongation can be determined.

For plane strain, the governing equilibrium equation is

$$\frac{d}{dr}\left(\frac{1}{r}\frac{d}{dr}(ru)\right) = \frac{1+\nu}{1-\nu}\alpha T'$$

which can be integrated to yield

$$u = \frac{1}{r}\int_a^r \frac{1+\nu}{1-\nu}\alpha rT\, dr + C_1 r + \frac{C_2}{r}$$

with boundary conditions

At $r = a$: Either $u(a) = U_a$

or $\left.\frac{E}{(1+\nu)(1-2\nu)}\left((1-\nu)u' + \nu\frac{u}{r}\right)\right|_{r=a} - \frac{E\alpha T(a)}{1-2\nu} = \Sigma(a)$

At $r = b$: Either $u(b) = U_b$

or $\left.\frac{E}{(1+\nu)(1-2\nu)}\left((1-\nu)u' + \nu\frac{u}{r}\right)\right|_{r=a} - \frac{E\alpha T(b)}{1-2\nu} = \Sigma(b)$

where U_a and U_b are prescribed displacements at $r = a$ and $r = b$, respectively, and $\Sigma(a)$ and $\Sigma(b)$ are prescribed radial stresses at $r = a$ and $r = b$, respectively. For the case of plane strain, it must be remembered that axial stresses in the amount

$$\sigma_z = \nu(\sigma_r + \sigma_\theta) - E\alpha T$$
$$= \frac{E}{(1+\nu)(1-2\nu)}\left(u' + \frac{u}{r}\right) - E\alpha T$$

must be imposed at the ends of the cylinder to maintain the plane-strain condition. In the remainder of this section, several examples will be presented for hollow and solid disks.

Hollow Disks. Consider the problem of a thin, hollow disk whose temperature distribution is a function of r, as indicated in Figure 6-16, and whose inner and outer radii are free from stress. For a thin disk or a long cylinder unconstrained in the axial direction, the plane-stress case is appropriate with the equilibrium equation and boundary conditions as

$$\frac{d}{dr}\left(\frac{1}{r}\frac{d}{dr}(ru)\right) = (1+\nu)\alpha T'$$

and

$$\sigma_r\bigg|_{r=a} = \sigma_r\bigg|_{r=b} = 0$$

Successive integrations of the equilibrium equation yield

$$u = \left(\frac{1}{r}\right)\int_a^r r(1+\nu)\alpha T\, dr + C_1 r + \frac{C_2}{r}$$

The corresponding stresses are

$$\sigma_r = -\frac{\alpha E}{r^2}\int_a^r rT\, dr + \frac{E}{1-\nu^2}\left[C_1(1+\nu) - \frac{C_2(1-\nu)}{r^2}\right]$$

$$\sigma_\theta = \frac{\alpha E}{r^2}\int_a^r rT\, dr - \alpha ET + \frac{E}{1-\nu^2}\left[C_1(1+\nu) + \frac{C_2(1-\nu)}{r^2}\right]$$

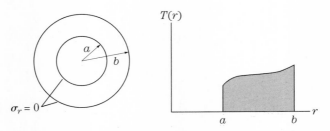

FIGURE 6-16 Radial temperature distribution.

6-5 THERMAL PROBLEMS

Satisfying the boundary conditions leads to the equations

$$A - \frac{B}{a^2} = 0$$

and

$$A - \frac{B}{b^2} = \left(\frac{\alpha E}{b^2}\right) \int_a^b rT \, dr$$

where $A = EC_1/(1 - \nu)$ and $B = EC_2/(1 + \nu)$. The solutions are

$$A = \frac{\alpha E}{b^2 - a^2} \int_a^b rT \, dr \quad \text{and} \quad B = \frac{\alpha E a^2}{b^2 - a^2} \int_a^b rT \, dr$$

with

$$\sigma_r = -\frac{\alpha E}{r^2} \int_a^r rT \, dr + \frac{\alpha E}{b^2 - a^2}\left(1 - \frac{a^2}{r^2}\right) \int_a^b rT \, dr \qquad (6\text{-}13\text{a})$$

$$\sigma_\theta = \frac{\alpha E}{r^2} \int_a^r rT \, dr + \frac{\alpha E}{b^2 - a^2}\left(1 + \frac{a^2}{r^2}\right) \int_a^b rT \, dr - E\alpha T \qquad (6\text{-}13\text{b})$$

and

$$u = \frac{1}{r} \int_a^r r(1 + \nu)\alpha T \, dr + \left[(1 - \nu)r + (1 + \nu)\frac{a^2}{r}\right]\frac{\alpha}{b^2 - a^2} \int_a^b rT \, dr$$

Given a specific temperature distribution, the integrals can be evaluated to determine the stresses and displacements. ◆

EXAMPLE 6-5 An important practical problem occurs when a heated liquid flows through a pipe. Assuming a steady condition, the temperature in the pipe satisfies Laplace's equation $\nabla^2 T(r) = 0$, the solution of which the student should verify is given by

$$T(r) = T_a \frac{\ln\left(\frac{r}{b}\right)}{\ln\left(\frac{a}{b}\right)}$$

where T_a is the temperature at the inside radius a with the temperature at the outside radius b taken to be zero. Assuming the pressure in the liquid to be small, that is, the inside pressure $p_i \approx 0$ with the outside pressure $p_o = 0$, the solution for the stresses is obtained by substituting the temperature into Equations (6-13a) and (6-13b). The results are

$$\sigma_r = \frac{\alpha E T_a}{2 \ln\left(\frac{b}{a}\right)} \left[\ln\left(\frac{r}{b}\right) + \frac{(b^2 - r^2)a^2 \ln\left(\frac{b}{a}\right)}{r^2(b^2 - a^2)}\right] \qquad (6\text{-}14\text{a})$$

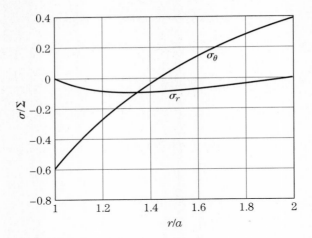

FIGURE 6-17 Radial and hoop stresses.

and

$$\sigma_\theta = \frac{\alpha E T_a}{2 \ln\left(\frac{b}{a}\right)} \left[1 + \ln\left(\frac{r}{b}\right) - \frac{(b^2 + r^2)a^2 \ln\left(\frac{b}{a}\right)}{r^2(b^2 - a^2)} \right] \quad \text{(6-14b)}$$

and are shown in graphical form in Figure 6-17 as σ_r/Σ and σ_θ/Σ, where $\Sigma = \alpha E T_a$. The maximum normal stress, which is compressive, is the hoop stress at the inside surface. ◆

For the corresponding plane-strain case, the stresses and displacement are

$$\sigma_r = -\frac{\alpha E}{(1-\nu)r^2} \int_a^r rT\, dr + \frac{\alpha E}{(1-\nu)(b^2 - a^2)}\left(1 - \frac{a^2}{r^2}\right) \int_a^b rT\, dr \quad \text{(6-15a)}$$

$$\sigma_\theta = \frac{\alpha E}{(1-\nu)r^2} \int_a^r rT\, dr + \frac{\alpha E}{(1-\nu)(b^2 - a^2)}\left(1 + \frac{a^2}{r^2}\right) \int_a^b rT\, dr - \frac{E\alpha T}{1-\nu} \quad \text{(6-15b)}$$

and

$$u = \frac{1+\nu}{1-\nu}\left[\frac{1}{r}\int_a^r r\alpha T\, dr + \left((1-2\nu)r + \frac{a^2}{r}\right)\frac{\alpha}{b^2 - a^2}\int_a^b rT\, dr\right] \quad \text{(6-15c)}$$

with

$$\sigma_z = \nu(\sigma_r + \sigma_\theta) - E\alpha T$$

that must be imposed at the ends of the cylinder to maintain the plane-strain condition.

EXAMPLE 6-6

Consider the problem where the temperature distribution is the same as in Example 6-5 but where the ends are constrained against axial motion. The corresponding plane-strain results are obtained from Equations (6-14a and b) in Example 6-5 by replacing E with $E/(1-\nu^2)$, ν with $\nu/(1-\nu)$ and α with $\alpha(1+\nu)$, resulting in

$$\sigma_r = \frac{\alpha E T_a}{2(1-\nu)\ln\left(\frac{b}{a}\right)}\left[\ln\left(\frac{r}{b}\right) + \frac{(b^2-r^2)a^2\ln\left(\frac{b}{a}\right)}{r^2(b^2-a^2)}\right]$$

and

$$\sigma_\theta = \frac{\alpha E T_a}{2(1-\nu)\ln\left(\frac{b}{a}\right)}\left[1 + \ln\left(\frac{r}{b}\right) - \frac{(b^2+r^2)a^2\ln\left(\frac{b}{a}\right)}{r^2(b^2-a^2)}\right]$$

The axial stress σ_z is given by $\sigma_z = \nu(\sigma_r + \sigma_\theta) - E\alpha T$ or

$$\sigma_z = \frac{\nu\alpha E T_a}{2(1-\nu)\ln\left(\frac{b}{a}\right)}\left[1 + 2\ln\left(\frac{r}{b}\right) - \frac{2a^2\ln\left(\frac{b}{a}\right)}{b^2-a^2}\right] - \alpha E T_a \frac{\ln\left(\frac{r}{b}\right)}{\ln\left(\frac{a}{b}\right)}$$

For $b=2a$ these three stresses are shown in Figure 6-18 as σ_r/Σ, σ_θ/Σ and σ_z/Σ, where $\Sigma = \alpha E T_a/(1-\nu)$.

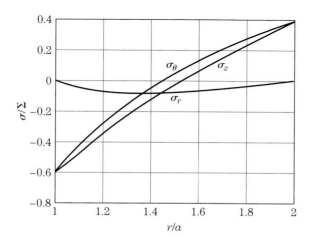

FIGURE 6-18 Radial, hoop and axial stresses.

Solid Disks. For a solid disk, $r = 0$ is a point at which the displacements and stresses are of interest so that the displacement must be written as

$$u = \left(\frac{1}{r}\right) \int_0^r r(1 + \nu)\alpha T \, dr + C_1 r$$

with

$$\sigma_r = -\frac{\alpha E}{r^2} \int_0^r rT \, dr + \frac{E}{1-\nu} C_1$$

$$\sigma_\theta = \frac{\alpha E}{r^2} \int_0^r rT \, dr - \alpha ET + \frac{E}{1-\nu} C_1$$

Assuming that the outer radius $r = b$ is stress-free, it follows that

$$0 = -\frac{\alpha E}{b^2} \int_0^b rT \, dr + \frac{E}{1-\nu} C_1$$

and that

$$u = (1 + \nu)\alpha \left[\frac{r}{b^2} \int_0^b rT \, dr + \frac{1}{r} \int_0^r rT \, dr \right] \quad (6\text{-}16a)$$

$$\sigma_r = \frac{\alpha E}{b^2} \int_0^b rT \, dr - \frac{\alpha E}{r^2} \int_0^r rT \, dr \quad (6\text{-}16b)$$

and

$$\sigma_\theta = \frac{\alpha E}{b^2} \int_0^b rT \, dr + \frac{\alpha E}{r^2} \int_0^r rT \, dr - E\alpha T \quad (6\text{-}16c)$$

It is left for the student to show in the homework exercises that u, σ_r and σ_θ are finite as r approaches zero.

▶ **EXAMPLE 6-7** A solid cylinder initially at a temperature T_0 is immersed in a fluid at a lower temperature T_1. A thermal analysis of the problem indicates that the temperature profiles near the center of the cylinder appear as shown in Figure 6-19. Approximate a typical profile by a suitable polynomial and hence investigate the resulting stresses and displacements.

Solution: Figure 6-20 shows the comparison of one of the curves from Figure 6-19 with the assumption that $T(r) = (T_0 - T_1)(1 - (r/b)^3)$, that is, that by representing the temperature profile as a cubic polynomial, good agreement is obtained. With $\Delta T = T_0 - T_1$, Equations (6-16b and c) yield

$$\sigma_r = \frac{E\alpha \Delta T}{5} \left[\left(\frac{r}{b}\right)^3 - 1 \right]$$

and

$$\sigma_\theta = \frac{E\alpha \Delta T}{5} \left[4\left(\frac{r}{b}\right)^3 - 1 \right]$$

and are shown in Figure 6-21 as σ_r/Σ and σ_θ/Σ where $\Sigma = E\alpha \Delta T$. The largest value is clearly the tensile hoop stress at the outer surface.

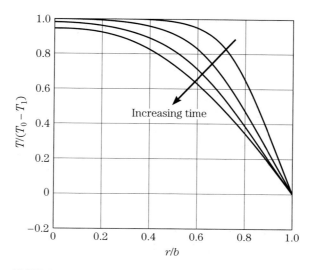

FIGURE 6-19 Temperature profiles for a cylinder immersed in a liquid at a lower temperature.

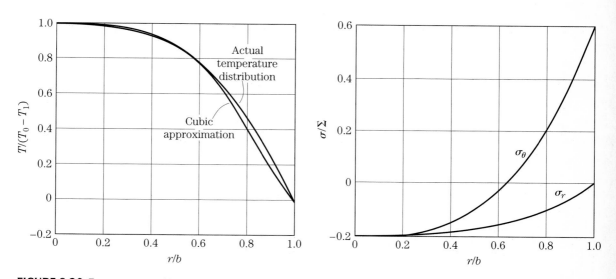

FIGURE 6-20 Temperature profile approximated by a cubic.

FIGURE 6-21 Radial and hoop stresses in immersed cylinder.

6-6 Rotating Disks

Rotating disks of various cross-sectional shapes constitute a part of essentially any rotating assembly so that the analysis of the corresponding deformations and stresses is of great practical importance. In this section the problems of the constant-thickness solid and hollow disks will be investigated.

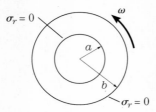

FIGURE 6-22 Constant-thickness rotating thin, hollow disk.

Consider the general problem of a thin hollow disk rotating at a constant angular velocity ω, as shown in Figure 6-22. The body force R is given by $R = \rho r \omega^2$ where ρ is the mass density of the material. For the thin disk, the plane-stress formulation is appropriate. Using Equation (6-7) with $T = 0$, the governing equation for the radial displacement is

$$\frac{d^2 u}{dr^2} + \frac{1}{r}\frac{du}{dr} - \frac{u}{r^2} = \frac{d}{dr}\left(\frac{1}{r}\frac{d}{dr}(ru)\right) = -\frac{1-\nu^2}{E}\rho\omega^2 r$$

Two integrations yield

$$u = C_1 r + \frac{C_2}{r} - \frac{1-\nu^2}{8E}\rho\omega^2 r^3$$

with the stresses then given by

$$\sigma_r = \frac{E}{1-\nu^2}\left(u' + \nu\frac{u}{r}\right) = \frac{E}{1-\nu}C_1 - \frac{E}{1+\nu}\left(\frac{C_2}{r^2}\right) - \frac{3+\nu}{8}\rho\omega^2 r^2$$

$$\sigma_\theta = \frac{E}{1-\nu^2}\left(\frac{u}{r} + \nu u'\right) = \frac{E}{1-\nu}C_1 - \frac{E}{1+\nu}\left(\frac{C_2}{r^2}\right) - \frac{1+3\nu}{8}\rho\omega^2 r^2$$

Satisfying the boundary conditions on the radial stress σ_r at the inner and outer radii leads to

$$\frac{E}{1-\nu}C_1 - \frac{E}{1+\nu}\left(\frac{C_2}{a^2}\right) = \frac{3+\nu}{8}\rho\omega^2 a^2$$

and

$$\frac{E}{1-\nu}C_1 - \frac{E}{1+\nu}\left(\frac{C_2}{b^2}\right) = \frac{3+\nu}{8}\rho\omega^2 b^2$$

from which

$$\frac{EC_1}{1-\nu} = \frac{(3+\nu)\rho\omega^2(a^2+b^2)}{8} \quad \text{and} \quad \frac{EC_2}{1+\nu} = \frac{(3+\nu)\rho\omega^2 a^2 b^2}{8}$$

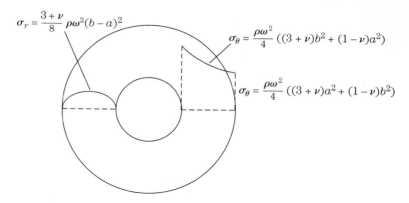

FIGURE 6-23 Stresses in a hollow rotating disk.

with the stresses and displacements as

$$\sigma_r = \frac{3+\nu}{8}\rho\omega^2\left(a^2 + b^2 - \frac{a^2b^2}{r^2} - r^2\right)$$

$$\sigma_\theta = \frac{3+\nu}{8}\rho\omega^2\left(a^2 + b^2 + \frac{a^2b^2}{r^2} - \frac{1+3\nu}{3+\nu}r^2\right)$$

$$u = \frac{\rho\omega^2}{8E}\left[(1-\nu)(3+\nu)(a^2+b^2)r + \frac{(1+\nu)(3+\nu)a^2b^2}{r} - (1-\nu^2)r^3\right]$$

A plot of the stresses is shown in Figure 6-23, where it is indicated that the maximum stress is the hoop stress at the inner radius, that is,

$$(\sigma_\theta)_{\max} = \frac{\rho\omega^2}{4}((3+\nu)b^2 + (1-\nu)a^2)$$

The maximum radial stress that occurs at $r = \sqrt{ab}$ is

$$(\sigma_r)_{\max} = \frac{3+\nu}{8}\rho\omega^2(b-a)^2$$

which can be shown to be less than $(\sigma_\theta)_{\max}$.

Solid Disk. For the solid disk, C_2 must be taken as zero in order that the stresses not become infinite at the center of the disk. The boundary condition on the stress σ_r at $r = b$ then gives

$$\frac{E}{1+\nu}C_1 - \frac{3+\nu}{8}\rho b^2\omega^2 = 0$$

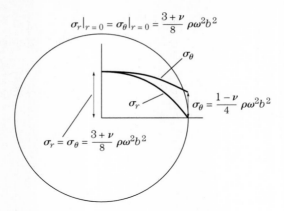

FIGURE 6-24 Stresses in a solid rotating disk.

after which the stresses and displacement are

$$\sigma_r = \frac{3+\nu}{8}\rho\omega^2(b^2 - r^2)$$

$$\sigma_\theta = \frac{3+\nu}{8}\rho\omega^2 b^2 - \frac{1+3\nu}{8}\rho\omega^2 r^2$$

$$u = \frac{(1+\nu)(3+\nu)}{8E}\rho b^2\omega^2 r - \frac{1-\nu^2}{8E}\rho\omega^2 r^3$$

As shown in Figure 6-24, the maximum stress occurs at the center of the disk with

$$\sigma_r\bigg|_{r=0} = \sigma_\theta\bigg|_{r=0} = \frac{3+\nu}{8}\rho\omega^2 b^2$$

◆**EXAMPLE 6-8** Rotating disks are sometimes an integral part of a shaft, as indicated in Figure 6-25a. In other applications, the disk is shrunk-fit onto the shaft, as indicated in Figure 6-25b. As a result of the shrink fit, the actual distribution of pressure between the shaft and disk

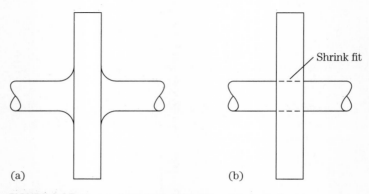

FIGURE 6-25 Modes of attachment of rotating disks to shafts.

6-6 ROTATING DISKS

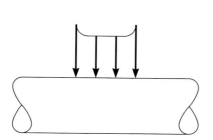

FIGURE 6-26 Actual radial stress distribution between the shaft and disk.

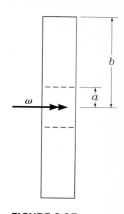

FIGURE 6-27 Geometry of shaft and disk.

is as shown in Figure 6-26. An approximate analysis consists of assuming that in the case of the shrink fit, the shaft acts as an inner disk with a uniform pressure resisting the shrink, so that the ideas developed in this chapter can be used to approximate the displacements and stresses in the shaft and disk under the combined action of the shrink fit and the rotation. Assume that the shaft and disk are made of the same material. The geometry is as shown in Figure 6-27.

With subscripts 1 and 2 denoting the inner and outer disks, respectively, the solutions for the displacement and radial stresses in the two disks can be expressed as

$$u_1 = C_1 r - A r^3$$

$$\sigma_{r1} = \frac{E}{1+\nu} C_1 - B r^2$$

$$u_2 = C_3 r + \frac{C_4}{r} - A r^3$$

$$\sigma_{r2} = \frac{E}{1+\nu} C_3 - \frac{E}{1-\nu} \frac{C_4}{r^2} - B r^2$$

where $A = (1-\nu^2)\rho\omega^2/8E$ and $B = (3+\nu)\rho\omega^2/8$. Note that the C_2/r term in the displacement for the inner disk has been discarded in order that the displacements be finite. The boundary conditions are

$$u_2\big|_{r=a} - u_1\big|_{r=a} = \delta$$

$$\sigma_{r1}\big|_{r=a} = \sigma_{r2}\big|_{r=a}$$

$$\sigma_{r2}\big|_{r=b} = 0$$

leading to the three equations

$$C_3 + \frac{C_4}{a^2} - C_1 = \frac{\delta}{a}$$

$$C_1 = C_3 - \phi\frac{C_4}{a^2}$$

and

$$C_3 - \phi\frac{C_4}{b^2} = \frac{(1-\nu)Bb^2}{E}$$

where $\phi = (1+\nu)/(1-\nu)$. Solving for the three constants yields

$$C_1 = \frac{(1+\nu)Bb^2}{E} + \frac{\phi}{1+\phi}\frac{\delta}{a}\frac{a^2-b^2}{b^2}$$

$$C_3 = \frac{(1+\nu)Bb^2}{E} + \frac{\phi}{1+\phi}\frac{\delta}{a}\frac{a^2}{b^2}$$

$$C_4 = \frac{a^2}{1+\phi}\frac{\delta}{a}$$

The stresses can then be expressed as

$$\sigma_{r1} = \frac{3+\nu}{8}\rho\omega^2(b^2 - r^2) - \frac{E}{2}\frac{\delta}{a}\frac{b^2-a^2}{b^2}$$

$$\sigma_{\theta 1} = \frac{3+\nu}{8}\rho\omega^2\left(b^2 - \frac{1+3\nu}{3+\nu}r^2\right) - \frac{E}{2}\frac{\delta}{a}\frac{b^2-a^2}{b^2}$$

$$\sigma_{r2} = \frac{3+\nu}{8}\rho\omega^2(b^2 - r^2) + \frac{E}{2}\frac{\delta}{a}\frac{a^2}{b^2}\left(1 - \frac{b^2}{r^2}\right)$$

$$\sigma_{\theta 2} = \frac{3+\nu}{8}\rho\omega^2\left(b^2 - \frac{1+3\nu}{3+\nu}r^2\right) + \frac{E}{2}\frac{\delta}{a}\frac{a^2}{b^2}\left(1 + \frac{b^2}{r^2}\right)$$

Of particular interest is the angular velocity at which the compressive stress at the interface $r = a$ is reduced to zero, that is,

$$\frac{3+\nu}{8}\rho\omega^2(b^2 - a^2) - \frac{E}{2}\frac{\delta}{a}\frac{b^2-a^2}{b^2} = 0$$

from which

$$\omega^2 = \frac{4E}{(3+\nu)\rho b^2}\frac{\delta}{a}$$

giving the critical angular velocity. At this speed, the maximum hoop stress occurs in the outer disk with the value

$$\sigma_{\theta 2}\bigg|_{r=a} = \frac{E}{2}\frac{\delta}{a}\left[2 + \frac{b^2}{a^2} - \frac{1+3\nu}{3+\nu}\frac{a^2}{b^2}\right]$$

$$= \frac{3+\nu}{8}\rho\omega^2 b^2\left[2 + \frac{b^2}{a^2} - \frac{1+3\nu}{3+\nu}\frac{a^2}{b^2}\right]$$

6-7 Variable-Thickness Rotating Disks

Generally for a rotating disk the stresses are larger near the center than at the outer portions. The desire to design disks with stresses somewhat more uniform throughout the volume suggests that we place more of the material near the center of the disk where the stresses are larger. Several possibilities along these lines are indicated in Figure 6-28. As long as the thickness doesn't change rapidly with respect to the radius, that is, it is a gradually changing thickness, it is possible to develop a plane-stress theory that accounts for the variable thickness in an approximate manner. It is left to the student to show that rather than the equilibrium equation

$$\frac{d(r\sigma_r)}{dr} - \sigma_\theta + rR = 0$$

developed in section 6-2, the analogous equation that accounts for the variable thickness is

$$\frac{d(tr\sigma_r)}{dr} - t\sigma_\theta + trR = 0$$

or, in the case of the rotating disk,

$$\frac{d(tr\sigma_r)}{dr} - t\sigma_\theta + t\rho\omega^2 r^2 = 0$$

For $t = $ constant, this clearly reduces to the appropriate equation. For variable t, the equation can be rewritten as

$$\frac{d(r\sigma_r)}{dr} - \sigma_\theta + \rho\omega^2 r^2 + \frac{1}{t}\frac{dt}{dr} r\sigma_r = 0$$

If the stresses are then eliminated using Equations (6-6) with $T = 0$, there results

$$u'' + \frac{u'}{r} - \frac{u}{r^2} + \frac{1}{t}\frac{dt}{dr}\left(u' + \nu\frac{u}{r}\right) = -(1-\nu^2)\rho\omega^2 r/E$$

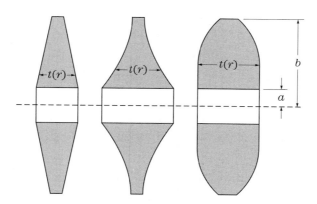

FIGURE 6-28 Variable-thickness disks.

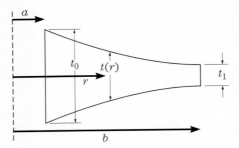

FIGURE 6-29 Power-law variable thickness.

The last term on the left clearly contains the effect of the variable thickness. In general this term makes it somewhat more difficult to obtain solutions to the equilibrium equation. Consider, however, the choice $t/t_0 = (a/r)^s$, as pictured in Figure 6-29. Given the values t_0 and a, the parameter s can be chosen on the basis that the thickness at $r = b$ is t_1, that is,

$$\frac{t_1}{t_0} = \left(\frac{a}{b}\right)^s$$

from which s can be determined. In situations where the radius does not vary strictly according to a power law, it may be possible to choose the parameters involved so as to closely approximate the actual dependence of t on r.

For this form of the dependence of the thickness on the radius, the equilibrium equation can be written as

$$r^2 u'' + r(1 - s)u' - (1 + \nu s)u = -\frac{1 - \nu^2}{E}\rho\omega^2 r^3$$

This is an equi-dimensional equation with solutions of the form $u = Ar^\lambda$, leading to the indicial equation

$$\lambda^2 - s\lambda + (1 + \nu s) = 0$$

with roots

$$\lambda_1 = \frac{s}{2} - \sqrt{\frac{s^2}{4} + 1 + \nu s} \quad \text{and} \quad \lambda_2 = \frac{s}{2} + \sqrt{\frac{s^2}{4} + 1 + \nu s}$$

The particular solution is easily verified to be

$$u_p = -\frac{\left(\dfrac{1 - \nu^2}{E}\right)\rho\omega^2 r^3}{(8 - (3 + \nu)s)}$$

so that the general solution is

$$u = C_1 r^{\lambda_1} + C_2 r^{\lambda_2} - n\Phi r^3$$

6-7 VARIABLE-THICKNESS ROTATING DISKS

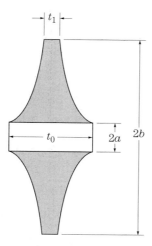

FIGURE 6-30 Variable-thickness disk.

where $\Phi = (1-\nu^2)\rho\omega^2/8E$ and $n = 8/(8-(3+\nu)s)$. The stresses can be expressed as

$$\sigma_r = \frac{E}{1-\nu^2}(C_1(\lambda_1+\nu)r^{\lambda_1-1} + C_2(\lambda_2+\nu)r^{\lambda_2-1}) - \frac{E}{1-\nu^2}(3+\nu)\Phi n r^2$$

$$\sigma_\theta = \frac{E}{1-\nu^2}(C_1(1+\lambda_1\nu)r^{\lambda_1-1} + C_2(1+\lambda_2\nu)r^{\lambda_2-1}) - \frac{E}{1-\nu^2}(1+3\nu)\Phi n r^2$$

These results can be used in the solution of various problems of interest involving variable-thickness rotating disks.

◆**EXAMPLE 6-9** Consider the problem of a hollow variable-thickness disk, as shown in Figure 6-30. The form of the solution is taken as

$$u = C_1\left(\frac{r}{a}\right)^{\lambda_1} + C_2\left(\frac{r}{a}\right)^{\lambda_2} - \Phi n r^3$$

$$\sigma_r = \frac{E}{1-\nu^2}\left(\frac{C_1}{a}(\lambda_1+\nu)\left(\frac{r}{a}\right)^{\lambda_1-1} + \frac{C_2}{a}(\lambda_2+\nu)\left(\frac{r}{a}\right)^{\lambda_2-1}\right) - \frac{E}{1-\nu^2}(3+\nu)\Phi n r^2$$

The boundary conditions are

$$\sigma_r\bigg|_{r=a} = 0$$

and

$$\sigma_r\bigg|_{r=b} = 0$$

resulting in

$$d_1 + d_2 = Ba^2$$

$$d_1\left(\frac{b}{a}\right)^{\lambda_1-1} + d_2\left(\frac{b}{a}\right)^{\lambda_2-1} = Bb^2$$

where $d_1 = C_1(\lambda_1 + \nu)/a$, $d_2 = C_2(\lambda_2 + \nu)/a$ and $B = (3 + \nu)\Phi n$. The solutions are

$$d_1 = B\frac{a^2\left(\frac{b}{a}\right)^{\lambda_2-1} - b^2}{\left(\frac{b}{a}\right)^{\lambda_2-1} - \left(\frac{b}{a}\right)^{\lambda_1-1}} \quad \text{and} \quad d_2 = B\frac{b^2 - a^2\left(\frac{b}{a}\right)^{\lambda_1-1}}{\left(\frac{b}{a}\right)^{\lambda_2-1} - \left(\frac{b}{a}\right)^{\lambda_1-1}}$$

Substitution yields

$$\frac{\sigma_r}{\Sigma} = n\left[\frac{\left(\frac{b}{a}\right)^{\lambda_2+\lambda_1-4} - \left(\frac{b}{a}\right)^{\lambda_1-1}}{\left(\frac{b}{a}\right)^{\lambda_2-1} - \left(\frac{b}{a}\right)^{\lambda_1-1}}\left(\frac{r}{b}\right)^{\lambda_1-1}\right.$$

$$\left. + \frac{\left(\frac{b}{a}\right)^{\lambda_2-1} - \left(\frac{b}{a}\right)^{\lambda_2+\lambda_1-4}}{\left(\frac{b}{a}\right)^{\lambda_2-1} - \left(\frac{b}{a}\right)^{\lambda_1-1}}\left(\frac{r}{b}\right)^{\lambda_2-1} - \left(\frac{r^2}{b^2}\right)\right]$$

$$\frac{\sigma_\theta}{\Sigma} = n\left[\left(\frac{1+\lambda_1\nu}{\lambda_1+\nu}\right) + \frac{\left(\frac{b}{a}\right)^{\lambda_2+\lambda_1-4} - \left(\frac{b}{a}\right)^{\lambda_1-1}}{\left(\frac{b}{a}\right)^{\lambda_2-1} - \left(\frac{b}{a}\right)^{\lambda_1-1}}\left(\frac{r}{b}\right)^{\lambda_1-1}\right.$$

$$\left. + \left(\frac{1+\lambda_2\nu}{\lambda_2+\nu}\right)\frac{\left(\frac{b}{a}\right)^{\lambda_2-1} - \left(\frac{b}{a}\right)^{\lambda_2+\lambda_1-4}}{\left(\frac{b}{a}\right)^{\lambda_2-1} - \left(\frac{b}{a}\right)^{\lambda_1-1}}\left(\frac{r}{b}\right)^{\lambda_2-1} - \frac{1+3\nu}{1+\nu}\left(\frac{r^2}{b^2}\right)\right]$$

where $\Sigma = (3+\nu)\rho\omega^2 b^2/8$.

Consider the specific case where $b/a = 5$ and $t_1/t_0 = 0.25$. It follows that $s = \log(0.25)/\log(0.2) = 0.8614$, and subsequently that $\lambda_1 = -0.7709$ and $\lambda_2 = 1.6323$. The radial and hoop stresses can then be computed and are shown in Figure 6-31. For comparison, the radial and hoop stresses in a corresponding constant-thickness disk are also shown. The maximum hoop stress of approximately $1.2\,\Sigma$ in the variable-thickness disk compared to the maximum of $2.0\,\Sigma$ in the constant-thickness disk shows clearly the advantage of using the variable thickness geometry.

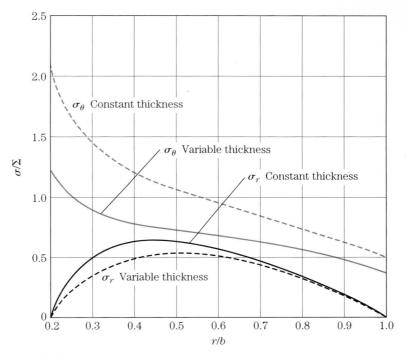

FIGURE 6-31 Stresses in a variable-thickness rotating disk.

6-8 Finite-Element Solutions

When the actual dependence of the thickness is not such that it can be represented with sufficient accuracy by the power-law assumption presented in section 6-7, the analysis can be less accurate than desired. Additionally, the algebra associated with the analytical solution is very tedious. For these reasons there is merit in seeking direct accurate numerical approximations. The finite-element model used in this section for this purpose is essentially a Ritz approximation of the sort discussed in Chapter 2. The basic underlying principle is stationary potential energy.

The total potential energy for the radially symmetrical problem shown in Figure 6-32 can be written in terms of the displacements as

$$\frac{V}{2\pi} = \int_a^b \left[\frac{E}{2(1-\nu^2)} \left(u'^2 + \left(\frac{u}{r}\right)^2 + 2\nu u' \frac{u}{r} \right) \right.$$
$$\left. - \frac{E\alpha T}{1-\nu} \left(u' + \frac{u}{r} \right) - \rho r \omega^2 u \right] tr\, dr - p_i at(a) u(a) + p_o bt(b) u(b)$$

The interval $a \leq r \leq b$ is discretized as shown in Figure 6-33a, that is, by taking N equal-length elements on the interval. As indicated in Figure 6-33b, it is then assumed that u

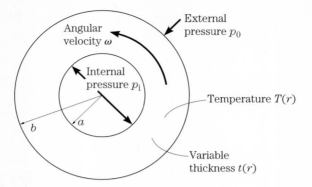

FIGURE 6-32 Geometry and inputs for the symmetrical problem.

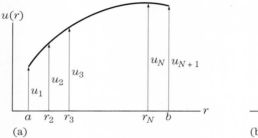

FIGURE 6-33 Discretization and interpolation.

can be approximated on a typical interval or element $r_i \leq r \leq r_{i+1}$ according to

$$u_e(x) = u_i N_i + u_{i+1} N_{i+1} = \mathbf{u_e^T N}$$

where $\mathbf{N}$ is the interpolation vector discussed in Chapter 2. Expressing the integral appearing in the representation of the potential energy as

$$\int_a^b F\, dr = \int_a^{r_2} F\, dr + \int_{r_2}^{r_3} F\, dr + \cdots + \int_{r_N}^b F\, dr$$

and inserting the representation for $u_e(x)$ on each interval leads after some algebra to

$$\frac{V}{2\pi} = \sum_e \frac{1}{2} \mathbf{u_e^T k_e u_e} - \sum_e \mathbf{u_e^T f_e} - p_i a t(a) u_1 + p_o b t(b) u_{N+1}$$

where

$$\mathbf{k_e} = \int_{r_i}^{r_{i+1}} \left[\mathbf{N}' rt \mathbf{N}'^{\mathbf{T}} + \mathbf{N} \frac{t}{r} \mathbf{N}^{\mathbf{T}} + \nu(\mathbf{N}' t \mathbf{N}^{\mathbf{T}} + \mathbf{N} t \mathbf{N}'^{\mathbf{T}}) \right] dr$$

and

$$\mathbf{f_e} = \int_{r_i}^{r_{i+1}} \left[\frac{E\alpha T}{1-\nu}(rt\mathbf{N}'^{\mathbf{T}} + t\mathbf{N}) + \rho\omega^2 r^2 t \mathbf{N} \right] dr$$

Carrying out the assembly indicated by the sums over the elements gives rise to an $N + 1$ by $N + 1$ set of equilibrium equations

$$\mathbf{K_G u_G = F_G}$$

Any constraints are enforced, after which the displacements are determined by solving the constrained set of approximate equilibrium equations. The stresses at the midpoints of the elements can then be approximated according to

$$\sigma_r = \frac{E}{1 - \nu^2}\left(\frac{u_{i+1} - u_i}{\ell_e} + \nu\frac{u_{i+1} + u_i}{r_{i+1} + r_i}\right) - \frac{E\alpha(T_{i+1} + T_i)}{2(1 - \nu)}$$

$$\sigma_\theta = \frac{E}{1 - \nu^2}\left(\frac{u_{i+1} + u_i}{r_{i+1} + r_i} + \nu\frac{u_{i+1} - u_i}{\ell_e}\right) - \frac{E\alpha(T_{i+1} + T_i)}{2(1 - \nu)}$$

where $\ell_e = r_{i+1} - r_i$.

◆**EXAMPLE 6-10** Consider the variable-thickness rotating disk problem of Example 6-9, shown again here in Figure 6-34. We will again take $b/a = 5$ and $t_1/t_0 = 0.25$. For this specific problem, the stiffness and load matrices can be written as

$$\mathbf{k_e} = \int_{r_i}^{r_{i+1}}\left[\mathbf{N'}rt\mathbf{N'^T} + \mathbf{N}\frac{t}{r}\mathbf{N^T} + \nu(\mathbf{N'}t\mathbf{N^T} + \mathbf{N}t\mathbf{N'^T})\right]dr$$

and

$$\mathbf{f_e} = \int_{r_i}^{r_{i+1}} \Phi r^2 t \mathbf{N}\, dr$$

where $\Phi = (1 - \nu^2)\rho\omega^2/E$. The calculations were carried out for the cases of four, eight and sixteen elements. The results are displayed in Figures 6-35 and 6-36. It is clearly seen that the finite-element solutions for each of the meshes investigated, that is, four, eight and sixteen elements, essentially mirror the analytical solution for both the hoop and radial stresses.

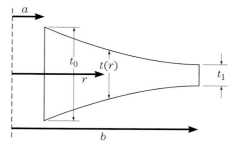

FIGURE 6-34 Geometry of a variable-thickness rotating disk.

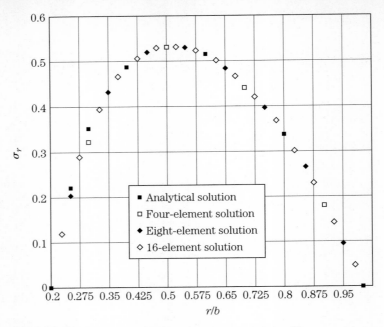

FIGURE 6-35 Radial stresses in a variable-thickness rotating disk.

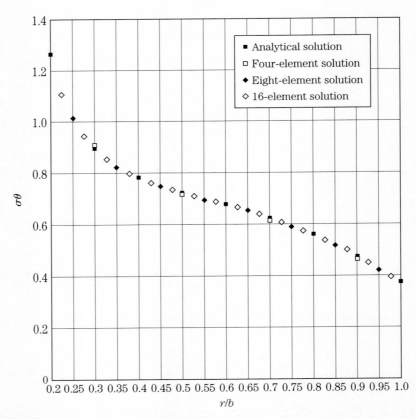

FIGURE 6-36 Hoop stresses in a variable-thickness rotating disk.

6-9 Summary

Finally, the finite-element formulation can be used irrespective of the dependence of the thickness on the radial coordinate, offering a means for obtaining accurate approximate solutions for a rotating disk of completely general shape. It is important to point out, however, that the finite-element method should *not* be applied to problems for which the basic underlying theory is not applicable.

6-9 Summary

The assumptions upon which many of the results obtained in this chapter are based are the following:

1. The region is axisymmetric bounded by two radii $r = a$ and $r = b$ and by two planes parallel to a z axis. See Figure 6-37.
2. The external loadings do not depend on θ and z.
3. The displacements and strains are small with the strain-displacement relations given by

$$\varepsilon_r = \frac{du}{dr} = u' \quad \text{and} \quad \varepsilon_\theta = \frac{u}{r}$$

   where $u(r)$ is the radial displacement.
4. The material is isotropic and linearly elastic with

$$\epsilon_r = \frac{\sigma_r - \nu(\sigma_\theta + \sigma_z)}{E} + \alpha T$$

$$\epsilon_\theta = \frac{\sigma_\theta - \nu(\sigma_z + \sigma_r)}{E} + \alpha T$$

$$\epsilon_z = \frac{\sigma_z - \nu(\sigma_r + \sigma_\theta)}{E} + \alpha T$$

5. Based on small displacement assumptions, the equilibrium equation is

$$\frac{d\sigma_r}{dr} + \frac{\sigma_r - \sigma_\theta}{r} + R = 0$$

   where R is the body force in the radial direction.

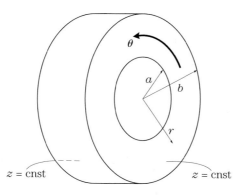

FIGURE 6-37 Axisymmetric region.

Plane stress formulation: $\sigma_z = 0$

Stress-strain:
$$\sigma_r = \frac{E}{1-\nu^2}\left(u' + \nu\frac{u}{r}\right) - \frac{E\alpha T}{1-\nu}$$

and
$$\sigma_\theta = \frac{E}{1-\nu^2}\left(\frac{u}{r} + \nu u'\right) - \frac{E\alpha T}{1-\nu}$$

Equilibrium equation:
$$u'' + \frac{u'}{r} - \frac{u}{r^2} = (1+\nu)\alpha T' - \frac{1-\nu^2}{E}R$$

Plane stress formulation: $\epsilon_z = 0$

Stress-strain:
$$\sigma_r = \frac{E}{(1+\nu)(1-2\nu)}\left((1-\nu)u' + \nu\frac{u}{r}\right) - \frac{E\alpha T}{1-2\nu}$$

and
$$\sigma_\theta = \frac{E}{(1+\nu)(1-2\nu)}\left((1-\nu)\frac{u}{r} + \nu u'\right) - \frac{E\alpha T}{1-2\nu}$$

Equilibrium equation:
$$u'' + \frac{u'}{r} - \frac{u}{r^2} = \frac{1+\nu}{1-\nu}\alpha T' - \frac{(1+\nu)(1-2\nu)}{E(1-\nu)}R$$

In any case, the solution can be written as

$$u(r) = C_1 r + \frac{C_2}{r} + u_p(r)$$

where $u_p(r)$ is a particular solution arising from the thermal and/or body force inputs. Boundary conditions on displacement or stress are then used to determine the two constants C_1 and C_2, after which calculations can be carried out.

REFERENCES

1. Timoshenko, S. P., and Goodier, J. N. *Theory of Elasticity*, 2nd ed. McGraw-Hill, New York, 1951.
2. Boresi, A. P., and Lynn, P. P. *Elasticity in Engineering Mechanics*. Prentice-Hall, Englewood Cliffs, N.J., 1974.

EXERCISES

Sections 6-2–6-3

1. A thin disk with inside radius a and outside radius $b = ka$ is subjected to an internal pressure $p_i = p$. Determine and plot the ratio of the maximum to the minimum circumferential stress as a function of k.
2. Repeat problem 1 with an external pressure $p_o = p$.
3. A thin disk with inside radius a and outside radius $b = ka$ is subjected to an internal pressure $p_i = p$. Determine and plot the ratio of the maximum circumferential stress to the absolute value of the maximum radial stress as a function of k.
4. Repeat problem 3 with an external pressure $p_o = p$.
5. A thin disk with inside radius a and outside radius $b = 1.5a$ is subjected to an internal pressure $p_i = p$, and is composed of a ductile material. If the yield stress determined from a uniaxial tension test is Y, determine $p_{\max}$ (a) on the basis of the maximum shear-stress theory of failure and (b) on the basis of the von Mises theory of failure.

EXERCISES

6. A thin disk with inside radius a and outside radius $b = 1.5a$ is subjected to an internal pressure $p_i = p$, and is composed of a brittle material. If the ultimate stress determined from a uniaxial tension test is F, determine p_{max} on the basis of the Coulomb-Mohr theory of failure.

7. A thin disk with inside radius a and outside radius $b = 2a$ is subjected to an external pressure $p_0 = p$ is composed of a ductile material. If the yield stress determined from a uniaxial tension test is Y determine p_{max} (a) on the basis of the maximum shear stress theory of failure, and (b) on the basis of the von Mises theory of failure.

8. A thin disk with inside radius a and outside radius $b = 2a$ is subjected to an external pressure $p_0 = p$, and is composed of a brittle material. If the ultimate stress determined from a uniaxial tension test is F, determine p_{max} on the basis of the Coulomb-Mohr theory of failure.

9. A thin, hollow disk with outside radius 6 in and inside radius 3 in is to be shrunk onto a solid disk of the same material with outside radius 3.002 in. Determine the circumferential and radial stresses in both components. Take $E = 3 \times 10^7$ psi and $\nu = 0.3$.

10. Repeat problem 9 for a thin, hollow disk with outside radius 0.150 m and inside radius 0.1 m shrunk onto a solid disk of the same material with outside radius 0.1001 m. Take $E = 200$ GPa and $\nu = 0.3$.

11. A thin, hollow disk with outside radius 6 in and inside radius 3 in is to be shrunk onto a hollow disk of the same material with outside radius 3.002 in and inside radius 2 in. Determine the circumferential and radial stresses in both components. Take $E = 3 \times 10^7$ psi and $\nu = 0.3$.

12. Repeat problem 11 for a thin, hollow disk with outside radius 0.150 m and inside radius 0.1 m shrunk onto a hollow disk of the same material with outside radius 0.1001 m and inside radius 0.05 m. Take $E = 200$ GPa and $\nu = 0.3$.

13. A thin, hollow aluminum disk with outside radius 6 in and inside radius 3 in is to be shrunk onto a solid steel disk with outside radius 3.002 in. Determine the circumferential and radial stresses in both components. Take $E_{al} = 1 \times 10^7$ psi, $\nu_{al} = 0.3$, $E_{st} = 3 \times 10^7$ psi and $\nu_{st} = 0.3$.

14. Repeat problem 13 for a thin, hollow steel disk with outside radius 0.150 m and inside radius 0.1 m shrunk onto a solid aluminum disk with outside radius 0.1001 m. Take $E_{st} = 200$ GPa, $\nu_{st} = 0.3$, $E_{al} = 70$ GPa and $\nu_{al} = 0.3$.

15. A thin, hollow disk is to be shrunk onto a solid disk of the same material. For $b = 2c$, determine the permissible difference in the interface diameter δ if the maximum tensile stress in the hollow disk is not to exceed Σ.

16. A thin, hollow disk is to be shrunk onto a solid disk of the same material. For $b = 2c$, determine the permissible difference in the interface diameter δ if failure is not to occur on the basis of (a) the maximum shear-stress theory and (b) the von Mises theory. Take the yield stress as determined from a uniaxial tension test to be Y.

17. A thin, hollow disk is to be shrunk onto another hollow disk of the same material. Determine the permissible difference in the interface diameter δ if failure is not to occur on the basis of (a) the maximum shear-stress theory and (b) the von Mises theory. Take the yield stress as determined from a uniaxial tension test to be Y. Also, in connection with Figure 6-9, take $b = 1.5c = 2a$.

18. A thin, hollow disk with outside radius 0.150 m and inside radius 0.1 m is to be shrunk onto a hollow disk of the same material with outside radius 0.1001 m and inside radius 0.05 m. If a pressure of 70 MPa is then applied to the internal disk, determine the circumferential and radial stresses in both components. Take $E = 200$ GPa and $\nu = 0.3$.

19. Repeat problem 18 for a thin, hollow disk with outside radius 6 in and inside radius 4 in shrunk onto a hollow disk of the same material with outside radius 4.002 in and inside radius 3.0 in. Take the internal pressure to be 10 ksi with $E = 10^7$ psi and $\nu = 0.3$.

20. A thin, hollow disk with outside radius 6 in, inside radius 3 in and thickness 1 in is to be shrunk onto a solid shaft of the same material with outside radius 3.002 in. If the coefficient of friction at the interface is 0.25, estimate the torque required to loosen the outer disk. Take $E = 3 \times 10^7$ psi and $\nu = 0.3$.

21. Repeat problem 20 for a thin, hollow disk with outside radius 0.15 m, inside radius 0.08 m and thickness 0.03 m to be shrunk onto a solid shaft of the same material with outside radius 0.0305 m. Take the coefficient of friction at the interface to be 0.30, $E = 200$ GPa and $\nu = 0.3$.

22. A thin, hollow disk with outside radius 6 in, inside radius 3 in and thickness 1 in is to be shrunk onto a solid shaft of the same material with outside radius 3.002 in. If the shaft is then subjected to a tensile force of 270 kips, estimate the pressure at the interface. Take $E = 3 \times 10^7$ psi and $\nu = 0.3$.

23. Repeat problem 22 for a thin, hollow disk with outside radius 0.15 m, inside radius 0.08 m and thickness 0.03 m shrunk onto a solid shaft of the same material with outside radius 0.0805 m. Take the tensile force to be 1400 kN, $E = 200$ GPa and $\nu = 0.3$.

Section 6-4

24. A long open cylinder with $a \leq r \leq b$ is subjected to an internal pressure $p_i = p$. Determine the final values of the inner and outer radii and the change in the length L of the cylinder.

25. Repeat problem 24 with an external pressure $p_0 = p$.

26. A long open cylinder with $a \leq r \leq b$ is subjected to an internal pressure $p_i = p$ and an axial stress σ_z. Determine the relation between p and σ_z if the length L is to remain constant. Also determine the final inner and outer radii.

27. Repeat problem 26 with an external pressure $p_i = p$ and an axial stress σ_z.

28. A long closed cylinder with outside radius 10 in and inside radius 4 in is subjected to an internal pressure of 10 ksi. Determine the maximum radial, circumferential and axial stresses near the middle of the cylinder. Take $E = 3 \times 10^7$ psi and $\nu = 0.3$.

29. Repeat problem 28 for a long closed cylinder with outside radius 300 mm and inside radius 175 mm that is subjected to an internal pressure of 60 MPa. Take $E = 100$ GPa and $\nu = 0.3$.

30. For the cylinder described in problem 28, determine the final dimensions of the inner and outer radii near the middle of the cylinder and the change in length.

31. For the cylinder described in problem 29, determine the final dimensions of the inner and outer radii near the middle of the cylinder and the change in length.

Section 6-5

32. Derive an expression for the change in radius of a thin ring subjected to a uniform temperature increase T_0.

33. Show that when the temperature is increased an amount ΔT in a solid disk with zero stress at the outer boundary, no stresses result. Also determine the radial displacement $u(r)$ and the change in the axial dimension of the disk.

34. Show that when the temperature is increased an amount ΔT in a hollow disk with zero stress at the boundaries, no stresses result. Also determine the radial displacement $u(r)$ and the change in the axial dimension of the disk.

EXERCISES

35. A solid disk of radius b is constrained against radial displacement at its outer boundary and is subjected to a uniform temperature increase T_0. Determine the stresses and displacements in the disk.

36. A hollow disk with $a \leq r \leq b$ is constrained against radial displacement at its outer boundary and is subjected to a uniform temperature increase T_0. Determine the stresses and displacements in the disk.

37. A solid disk of length L is placed between two rigid walls, as shown below. If the walls prevent any motion in the axial direction, determine the stresses in the disk as a result of a uniform temperature increase T_0. Also determine the change in the radius b.

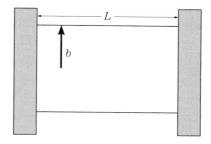

38. Repeat problem 37 if the walls can act only in compression.

39. A thin ring of aluminum is shrunk onto a solid steel disk. The initial diameter of the steel disk is 14 in with the inner and outer diameters of the aluminum ring 13.997 in and 15 in, respectively.

   a. By what minimum uniform temperature must the ring be increased in order to slip the ring over the disk?
   b. What are the stresses in the ring and disk after the assembly returns to room temperature?
   c. To what uniform temperature must the assembly be raised in order to loosen the ring on the disk?
   d. What are the maximum circumferential stresses in the ring and in the disk at the temperature determined in question (c)?

   Take $E_{st} = 30 \times 10^6$ psi, $E_{al} = 10 \times 10^6$ psi, $\alpha_{st} = 6.5 \times 10^{-6}$ in/in/°F, $\alpha_{al} = 12.8 \times 10^{-6}$ in/in/°F, $\nu_{st} = \nu_{al} = 0.3$.

40. Repeat problem 39 if the steel disk is hollow with an inner diameter of 11 in.

41. A long, hollow cylinder with closed ends is subjected to an internal pressure $p_i = p$ and a uniform temperature change T_0. What must be the relation between the pressure p and the temperature T_0 in order that the cylinder not change length? What is the corresponding change in the outer radius near the center of the cylinder?

42. A long, hollow cylinder with closed ends is subjected to an internal pressure $p_i = p$, an axial load P and a uniform temperature change T_0. What must be the relation between the pressure p, the axial load P and the temperature T_0 in order that the cylinder not change length? What is the corresponding change in the outer radius near the center of the cylinder?

43. A long, hollow pipe with $a \leq r \leq b$ carries a liquid at temperature T_0 under a pressure p with no axial load. Take the temperature distribution in the pipe to be in steady state, that is, $T(r) = T_0 \ln(r/b)/\ln(a/b)$, and determine the following:

a. The maximum circumferential stress in the pipe.
b. The radial displacement at $r = b$.
c. The change in length of the pipe

Take $a = 0.1$ m, $b = 0.3$ m, $L = 10$ m, $p = 20$ MPa, $T_0 = 200°C$, $E = 200$ GPa, $\alpha = 11.7 \times 10^{-6}$ m/m/°C and $\nu = 0.3$.

44. Repeat problem 43 for $a = 4$ in, $b = 6$ in, $L = 6$ ft, $p = 3$ ksi, $T_i = 350°F$, $T_0 = 70°F$, $E = 30 \times 10^6$ psi, $\alpha = 6.5 \times 10^{-6}$ m/m/°F and $\nu = 0.3$.

Section 6-6

45. Derive an expression for the hoop stress in a thin ring of radius R due to rotation. (Hint: Draw an FBD of a typical segment $\Delta\theta \ll 1$, as shown below, and use Newton's second law in the form $\Sigma F_r = ma_r$.)

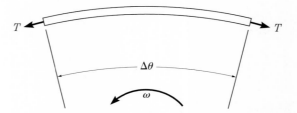

46. Derive an expression for the increase in the radius R of a thin ring due to rotation.

47. Show by taking $b = a + t$, $t \ll a$, that the equation for σ_θ in section 6-6 for the hollow disk reduces to those developed in problem 45.

48. Show by taking $b = a + t$, $t \ll a$, that the equation for the radial displacement $u(r)$ in section 6-6 for the hollow disk reduces to those developed in problem 46.

49. Determine the maximum permissible angular velocity that a thin, hollow disk with outer and inner radii of 6 in and 4 in, respectively, can withstand if the maximum circumferential stress is not to exceed 18 ksi.

50. For a thin, hollow disk with fixed outer radius b, determine the relationship between the angular velocity and the value of the inside radius a in order that the maximum circumferential stress not exceed σ_0.

51. For a thin, hollow disk with fixed inner radius a, determine the relationship between the angular velocity and the value of the outside radius b in order that the maximum circumferential stress not exceed σ_0.

52. A solid cast-iron disk with an outside radius of 10 in has a steel rim with an outside diameter of 15 in shrunk onto it. If at 8000 rpm the pressure between the rim and disk is zero, calculate the interference. Take $\nu = 0.3$, $\gamma = 0.283$ lbf/in³ for both materials, $E_{st} = 29 \times 10^6$ psi and $E_{ci} = 15 \times 10^6$ psi.

53. A solid cast-iron disk with an outside radius of 250 mm has a steel rim with an outside diameter of 375 mm shrunk onto it. If at 9000 rpm the pressure between the rim and disk is zero, calculate the interference. Take $\nu = 0.3$ and $\rho = 7.71$ Mg/m³ for both materials, $E_{st} = 210$ GPa and $E_{ci} = 110$ GPa.

54. A disk with blades, as shown in the figure below, can be approximated as a disk with an appropriate tensile stress at the radius $r = b$.

EXERCISES

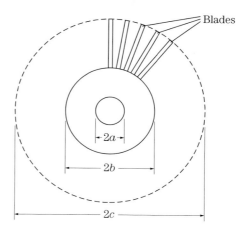

a. Show that the average stress at the root of a single blade is given by $\sigma = \rho\omega^2(c^2 - b^2)/2$.
b. With A as the area of a single blade and n as the number of blades, show that an equivalent radial stress σ_r at the radius $r = b$ can be taken to be $\sigma_r = n\rho A\omega^2(c^2 - b^2)/(2(2\pi bt))$ where t is the axial thickness of the disk.
c. Determine the circumferential stress at the inner radius.
d. Determine the circumferential stress at the outer radius.

In each of problems 55 through 58, apply the results of problem 54.

55. For a steel disk and blades, take $a = 3$ in, $b = 6$ in, $c = 9$ in, and $n = 40$ and determine the maximum allowable angular velocity ω in order that the maximum circumferential stress not exceed 10 ksi. Take $\rho = 7.33 \times 10^{-4}$ lbf-sec^2/in^4, $t = 1$ in and $A = 0.5$ in^2.

56. Repeat problem 55 with $a = 0.08$ m, $b = 0.16$ m, $c = 0.24$ m, $n = 36$, $\sigma_{max} = 70$ MPa, $\rho = 7.87$ Mg/m^3, $t = 0.025$ m and $A = 3.2 \times 10^{-4}$ m^2.

57. For a steel disk and blades, take $a = 3$ in, $b = 6$ in, $n = 40$ and $\omega = 5000$ rpm and determine the maximum allowable value of c in order that the maximum circumferential stress not exceed 10 ksi. Take $\rho = 7.33 \times 10^{-4}$ lbf-sec^2/in^4, $t = 1$ in and $A = 0.5$ in^2.

58. Repeat problem 57 with $a = 0.08$ m, $b = 0.16$ m, $n = 36$, $\omega = 5000$ rpm, $\sigma_{max} = 70$ MPa, $\rho = 7.87$ Mg/m^3, $t = 0.025$ m and $A = 3.2 \times 10^{-4}$ m^2.

Section 6-7

59. Verify that the equilibrium equation for a variable-thickness rotating disk is

$$\frac{d(tr\sigma_r)}{dr} - t\sigma_\theta + t\rho\omega^2 r^2 = 0$$

60. Verify that the governing equation for the displacements in a variable-thickness rotating disk is

$$u'' + \frac{u'}{r} - \frac{u}{r^2} + \frac{1}{t}\frac{dt}{dr}\left(u' + \nu\frac{u}{r}\right) = -(1 - \nu^2)\rho\omega^2 r/E$$

61. For the disk shown below, $a = 2$ in, $b = 10$ in, $t_0 = 4$ in and $t_1 = 1$ in. Assuming a power-law dependence of the thickness on the radius of the form $(t_1/t_0) = (r_0/r_1)^s$, determine the allowable value of ω in order that the maximum circumferential stress not exceed 10 ksi. Assume $\sigma_r(a) = 0 = \sigma_r(b)$. Take $\rho = 7.33 \times 10^{-4}$ lbf-sec^2/in^4, $\nu = 0.3$ and $E = 3 \times 10^7$ psi.

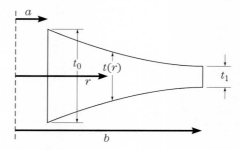

62. Repeat problem 61 if the boundary conditions are replaced by $u(a) = 0 = \sigma_r(b)$, that is, by a rigid core at $r = a$.

63. Repeat problem 61 with $a = 0.05$ m, $b = 0.25$ m, $t_0 = 0.1$ m and $t_1 = 0.025$ m, $\sigma_{max} = 70$ MPa, $\rho = 7.87$ Mg/m^3, $\nu = 0.3$ and $E = 200$ GPa.

64. Repeat problem 63 if the boundary conditions are replaced by $u(a) = 0 = \sigma_r(b)$, that is, by a rigid core at $r = a$.

65. For a hollow disk with $a = 2$ in, $b = 10$ in, $t_0 = 4$ in and $t_1 = 1$ in, the actual thickness is given by the linear function $t(r) = t_0 + (t_1 - t_0)[(r - a)/(b - a)]$. Decide on a means for approximating this by a suitable power-law replacement function of the form $(t_1/t_0) = (a/b)^s$ and hence use the results of section 6-7 to approximate an allowable value of ω in order that the maximum circumferential stress not exceed 10 ksi. Assume $\sigma_r(a) = 0 = \sigma_r(b)$. Take $\rho = 7.33 \times 10^{-4}$ lbf-sec^2/in^4, $\nu = 0.3$ and $E = 3 \times 10^7$ psi.

66. For a hollow disk with $a = 2$ in, $b = 10$ in, $t_0 = 4$ in and $t_1 = 1$ in, the actual thickness is given by the parabolic function $t(r) = t_0 + (t_1 - t_0)[(r - a)/(b - a)]^2$. Decide on a means for approximating this by a suitable power-law replacement function of the form $(t_1/t_0) = (a/b)^s$ and hence use the results of section 6-7 to approximate an allowable value of ω in order that the maximum circumferential stress not exceed 10 ksi. Assume $\sigma_r(a) = 0 = \sigma_r(b)$. Take $\rho = 7.33 \times 10^{-4}$ lbf-sec^2/in^4, $\nu = 0.3$ and $E = 3 \times 10^7$ psi.

Variational Formulation Exercises

67. Show that the total potential energy for the radially symmetrical disk shown below can be written as

$$\frac{V}{2\pi t} = \int_a^b \left[\frac{E}{2(1-\nu^2)} (\epsilon_r^2 + \epsilon_\theta^2 + 2\nu\epsilon_r\epsilon_\theta) \right.$$
$$\left. - \frac{E\alpha T}{1-\nu}(\epsilon_r + \epsilon_\theta) - \rho r \omega^2 u \right] r\, dr - p_i a u(a) + p_o b u(b)$$

or, in terms of the displacements, as

$$\frac{V}{2\pi t} = \int_a^b \left[\frac{E}{2(1-\nu^2)} \left(u'^2 + \left(\frac{u}{r}\right)^2 + 2\nu u' \frac{u}{r} \right) \right.$$
$$\left. - \frac{E\alpha T}{1-\nu}\left(u' + \frac{u}{r}\right) - \rho r \omega^2 u \right] r\, dr - p_i a u(a) + p_o b u(b)$$

Use the calculus of variations to determine the equilibrium equation and the boundary conditions.

EXERCISES

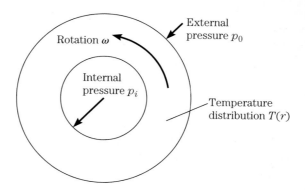

68. For the problem of a hollow disk with $a \le r \le b$ subjected to an internal pressure p, take $u = A + Br$ and use the Ritz method to determine A and B. Compare your results to the exact solution.

69. For the problem of a hollow disk with $a \le r \le b$ subjected to an internal pressure p, take $u = Ar + Br^2$ and use the Ritz method to determine A and B. Compare your results to the exact solution.

70. For the problem of a hollow rotating disk with $a \le r \le b$, take $u = A + Br$ and use the Ritz method to determine A and B. Compare your results to the exact solution.

71. For the problem of a hollow rotating disk with $a \le r \le b$, take $u = Ar + Br^2$ and use the Ritz method to determine A and B. Compare your results to the exact solution.

72. For the problem of a solid rotating disk with $0 \le r \le b$, take $u = Ar + Br^2$ and use the Ritz method to determine A and B. Compare your results to the exact solution.

73. For the problem of a hollow disk with $a \le r \le b$ subjected to a temperature increase given by $T(r) = T_0 \ln(r/b)/\ln(a/b)$, take $u = A + Br$ and use the Ritz method to determine A and B. Compare your results to the exact solution.

74. For the problem of a hollow disk with $a \le r \le b$ subjected to a temperature increase given by $T(r) = T_0 \ln(r/b)/\ln(a/b)$, take $u = Ar + Br^2$ and use the Ritz method to determine A and B. Compare your results to the exact solution.

75. For the problem of the variable-thickness rotating disk shown in the figure below, take $u = A + Br$ and use the Ritz method to determine A and B. Compare your results to the exact solution given in section 6-7. Take $r_1/r_0 = 5$ and $t_1/t_0 = 0.25$.

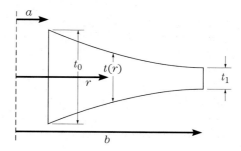

76. For the problem of the variable-thickness rotating disk shown in problem 75, take $u = Ar + Br^2$ and use the Ritz method to determine A and B. Compare your results to the exact solution given in section 6-7. Take $b/a = 5$ and $t_1/t_0 = 0.25$.

7 Plates

Part I   Rectangular Plates
7-1   Introduction
7-2   Force Resultants and Equilibrium
7-3   Deformations: Strain-Displacement Relations
7-4   Stress-Strain and Combination
7-5   Solutions to Rectangular-Plate Problems
Part II   Circular Plates
7-6   Introduction
7-7   Force Resultants and Equilibrium
7-8   Deformations: Strain-Displacement Relations
7-9   Stress-Strain and Combination
7-10   Solutions to Circular-Plate Problems
7-11   Summary
         References
         Exercises

Part I   Rectangular Plates

7-1   Introduction

There are many practical applications where it is necessary to support loads that are applied perpendicular to a flat surface. The floor of a multistory office building is an example. Another example would be a flat area of the surface of an airplane wing. Structures such as these are commonly referred to as *plates* or as *flat plates*. We define the geometry of a flat plate to be such that a single plane, usually the $z = 0$ plane, lies completely within the region occupied by the plate. We further restrict the geometry to be such that the material of the plate is distributed symmetrically about the $z = 0$ plane. The $z = 0$ plane is then referred to as the *middle surface* of the plate. The dimension of the plate perpendicular to the z axis is called the *thickness* of the plate, usually denoted by h. Very frequently the thickness is a constant. These ideas are illustrated in Figure 7-1.

    Also shown in the figure are typical transverse loadings, consisting of distributed and concentrated loads, as well as the idea that the plate must be supported in some fashion. This chapter will be devoted to developing the equations necessary to investigate the relations between the external loadings, the internal force resultants, the stresses and the displacements for rectangular and circular plates.

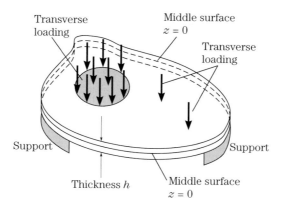

FIGURE 7-1 Geometry of a flat plate.

7-2 Force Resultants and Equilibrium

Recall that the internal forces by which a beam resists transverse loading, as shown in Figure 7-2a, are generally shear forces and bending moments. These are indicated in Figure 7-2b. To begin to see the type of force resultants that might be transmitted by a flat plate, consider a flat rectangular panel loaded by a concentrated force near the middle of the plate, as shown in Figure 7-3a. The dotted lines represent straight lines before deformation, and the solid lines represent the corresponding deformed geometry. The extent of the deformations is greatly exaggerated for clarity. It is seen that bending deformations take place in two perpendicular directions, the x and y directions in Figure 7-3a, resulting in substantially increased load carry capacity by the plate. Shown in Figure 7-3b are typical shears and bending moments that would logically accompany the bending about the x and y axes. It turns out that there are other force resultants, referred to as twisting moments, that arise due to the two-dimensional aspects of the problem.

The force resultants that are transmitted throughout the plate are logically related to the shear and normal stresses in the plate. The relations between these variables are defined using a typical segment of the plate bounded by nearby parallel y and z planes, as shown in Figures 7-4a and 7-4b. The relations between the stresses and the force resultants are as follows:

$$M_x \, dy = -\int_{-h/2}^{h/2} z\sigma_x \, dy \, dz \quad \text{or} \quad M_x = -\int_{-h/2}^{h/2} z\sigma_x \, dz \tag{7-1a}$$

FIGURE 7-2 Force resultants in a beam.

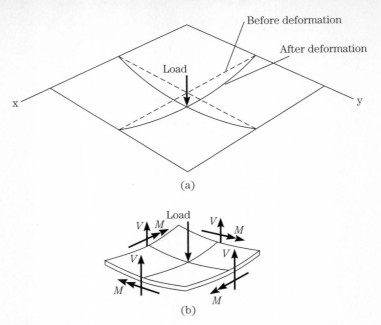

FIGURE 7-3 Deformations and force resultants for a flat panel.

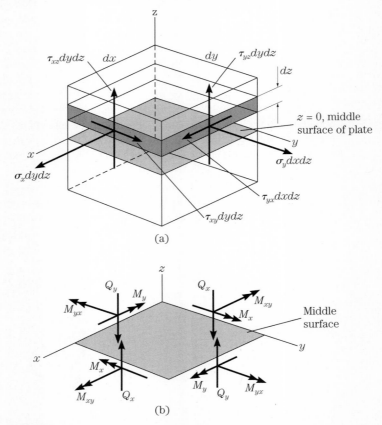

FIGURE 7-4 Force resultant–stress relations.

where M_x is a moment per unit length, that is, the dimensions are $FL/L = F$. Similarly,

$$M_y = -\int_{-h/2}^{h/2} z\sigma_y \, dz \qquad (7\text{-}1\text{b})$$

$$M_{xy} = -\int_{-h/2}^{h/2} z\tau_{xy} \, dz \qquad (7\text{-}1\text{c})$$

$$M_{yx} = \int_{-h/2}^{h/2} z\tau_{xy} \, dz \qquad (7\text{-}1\text{d})$$

$$Q_x = \int_{-h/2}^{h/2} \tau_{xz} \, dz \qquad (7\text{-}1\text{e})$$

$$Q_y = \int_{-h/2}^{h/2} \tau_{yz} \, dz \qquad (7\text{-}1\text{f})$$

M_x and M_y are referred to as *bending* moments, whereas M_{xy} and M_{yx} are generally called *twisting* moments. All the bending and twisting moments have dimensions FL/L, usually indicated as in-lbf/in in the US system of units and as N-m/m in the SI system. The shear forces Q_x and Q_y have dimensions F/L, that is, of force per unit length.

The equations of equilibrium satisfied by these force resultants are developed using the free-body diagram shown in Figure 7-5. Summing forces in the z direction yields

$$-Q_x \, dy - Q_y \, dx + \left(Q_x + \frac{\partial Q_x}{\partial x} dx\right) dy + \left(Q_y + \frac{\partial Q_y}{\partial y} dy\right) dx + q \, dx \, dy = 0$$

or

$$\frac{\partial Q_x}{\partial x} + \frac{\partial Q_y}{\partial y} + q = 0$$

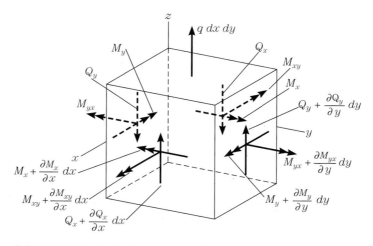

FIGURE 7-5 Free-body diagram showing force resultants and external loading.

Summing moments about the x axis yields

$$-M_{xy}\,dy - M_y\,dx + \left(M_{xy} + \frac{\partial M_{xy}}{\partial x}dx\right)dy + \left(M_y + \frac{\partial M_y}{\partial y}dy\right)dx + (Q_y\,dx)\,dy = 0$$

or

$$\frac{\partial M_y}{\partial y} + \frac{\partial M_{xy}}{\partial x} + Q_y = 0$$

Last, summing moments about the y axis yields

$$-M_{yx}\,dx + M_x\,dy + \left(M_{yx} + \frac{\partial M_{yx}}{\partial y}dy\right)dx - \left(M_x + \frac{\partial M_x}{\partial x}dx\right)dy - (Q_x\,dy)\,dx = 0$$

or

$$\frac{\partial M_x}{\partial x} - \frac{\partial M_{yx}}{\partial y} + Q_x = 0$$

Collectively, the three equations

$$\frac{\partial Q_x}{\partial x} + \frac{\partial Q_y}{\partial y} + q = 0 \tag{7-2a}$$

$$\frac{\partial M_y}{\partial y} + \frac{\partial M_{xy}}{\partial x} + Q_y = 0 \tag{7-2b}$$

and

$$\frac{\partial M_x}{\partial x} - \frac{\partial M_{yx}}{\partial y} + Q_x = 0 \tag{7-2c}$$

are the equilibrium equations. Another form of the equilibrium equations that we will find useful is obtained by eliminating the shear forces Q_x and Q_y. This is accomplished by taking the derivative of the second equation with respect to y and the derivative of the third equation with respect to x and substituting into the first equation to obtain

$$\frac{\partial^2 M_x}{\partial x^2} + \frac{\partial^2 M_x}{\partial x^2} + \frac{\partial^2 M_{xy}}{\partial y\,\partial x} - \frac{\partial^2 M_{yx}}{\partial x\,\partial y} = q \tag{7-3}$$

7-3   Deformations: Strain-Displacement Relations

As a result of the application of transverse loads, the plate displaces perpendicular to the plane of the plate. These transverse displacements are generally accompanied by horizontal displacements in the x and y directions because of the rotations of an element about the x and y axes. Consider these deformations, as shown in Figure 7-6, from a view along the y axis. Shown are u, the x component of displacement, and w, the z component of displacement, of a point P located a distance z from the middle surface of

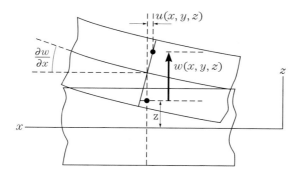

FIGURE 7-6 Deformations in a plate.

the plate. A similar figure can be drawn from a view along the x axis, indicating w and v, the y component of displacement. In a manner analogous to the treatment in Chapter 4 of displacements in a beam, assume that

$$w(x, y, z) = w_0(x, y) + zw_1(x, y)$$
$$u(x, y, z) = u_0(x, y) + zu_1(x, y)$$
$$v(x, y, z) = v_0(x, y) + zv_1(x, y)$$

where w_0, u_0 and v_0 represent the displacements of the middle surface ($z = 0$) and w_1, u_1 and v_1 represent the displacement relative to the middle surface. Essentially, the representations for each of the displacements can be thought of as the first two terms in a Taylor series about the middle surface of the plate. The assumption that u and v depend linearly on the z coordinate is essentially the classical assumption that "plane sections remain plane."

Assume small displacements and rotations so that the linear portions of the strain-displacement relations are adequate for analyzing the deformations. Thus,

$$\epsilon_x = \frac{\partial u}{\partial x} \qquad \epsilon_y = \frac{\partial v}{\partial y} \qquad \epsilon_z = \frac{\partial w}{\partial z}$$

and

$$\gamma_{xy} = \frac{\partial u}{\partial y} + \frac{\partial v}{\partial x} \qquad \gamma_{yz} = \frac{\partial w}{\partial y} + \frac{\partial v}{\partial z} \qquad \gamma_{xz} = \frac{\partial u}{\partial z} + \frac{\partial w}{\partial x}$$

Substitution yields

$$\epsilon_x = \frac{\partial u_0}{\partial x} + z\frac{\partial u_1}{\partial x} \qquad \epsilon_y = \frac{\partial v_0}{\partial y} + z\frac{\partial v_1}{\partial y} \qquad \epsilon_z = w_1(x, y)$$

Assume that $\epsilon_z \approx 0$, that is, that there is negligible change in the thickness of the plate, leading to $w_1(x, y) = 0$. Also assume that the in-plane displacements of the middle surface of the plate, represented by $u_0(x, y)$ and $v_0(x, y)$, are zero. Computing the shear strains then leads to

$$\gamma_{xy} = z\left[\frac{\partial v_1(x,y)}{\partial x} + \frac{\partial u_1(x,y)}{\partial y}\right]$$

$$\gamma_{yz} = \frac{\partial w_0(x,y)}{\partial y} + v_1(x,y)$$

$$\gamma_{xz} = \frac{\partial w_0(x,y)}{\partial x} + u_1(x,y)$$

Further assume that transverse shear deformation is negligible by taking $\gamma_{yz} = 0$ and $\gamma_{xz} = 0$, from which

$$v_1 = -\frac{\partial w_0(x,y)}{\partial y} \quad \text{and} \quad u_1 = -\frac{\partial w_0(x,y)}{\partial x}$$

Neglecting shear deformation forces the tangents and normals to the middle surface to rotate through the same angles and is classically known as the Kirchoff assumption.

Finally, then, the displacements and strain-displacement relations can be expressed as

$$u = -z\frac{\partial w}{\partial x} \quad v = -z\frac{\partial w}{\partial y} \quad w = w(x,y)$$

and

$$\epsilon_x = -z\frac{\partial^2 w}{\partial x^2} \quad \epsilon_y = -z\frac{\partial^2 w}{\partial y^2} \quad \epsilon_z = 0$$

$$\gamma_{xy} = -2z\frac{\partial^2 w}{\partial x\, \partial y} \quad \gamma_{yz} = 0 \quad \gamma_{xz} = 0 \quad \text{(7-4a-f)}$$

where the zero subscript on w has been dropped for brevity. It should be remembered that $w = w(x,y)$ now represents the transverse displacement of the middle surface.

7-4 Stress-Strain and Combination

The stress-strain relations appropriate for thin-plate theory are the plane stress expressions given by

$$\epsilon_x = \frac{\sigma_x - \nu\sigma_y}{E} \quad \epsilon_y = \frac{\sigma_y - \nu\sigma_x}{E} \quad \gamma_{xy} = \frac{\tau_{xy}}{G} = \frac{2(1+\nu)\tau_{xy}}{E}$$

or, upon solving for the stresses,

$$\sigma_x = \frac{E}{1-\nu^2}(\epsilon_x + \nu\epsilon_y) \tag{7-5a}$$

$$\sigma_y = \frac{E}{1-\nu^2}(\epsilon_y + \nu\epsilon_x) \tag{7-5b}$$

$$\tau_{xy} = \frac{E}{2(1+\nu)}\gamma_{xy} \tag{7-5c}$$

7-4 STRESS-STRAIN AND COMBINATION

Combination. First eliminate the strains between Equations (7-4) and (7-5) to obtain

$$\sigma_x = -\frac{Ez}{1-\nu^2}\left(\frac{\partial^2 w}{\partial x^2} + \nu \frac{\partial^2 w}{\partial y^2}\right) \quad (7\text{-}6a)$$

$$\sigma_y = -\frac{Ez}{1-\nu^2}\left(\frac{\partial^2 w}{\partial y^2} + \nu \frac{\partial^2 w}{\partial x^2}\right) \quad (7\text{-}6b)$$

$$\tau_{xy} = -\frac{Ez}{1-\nu^2}(1-\nu)\frac{\partial^2 w}{\partial x\, \partial y} \quad (7\text{-}6c)$$

These can then be substituted into Equations (7-1) to yield

$$M_x = -\int_{-h/2}^{h/2} z\sigma_x\, dz = D\left(\frac{\partial^2 w}{\partial x^2} + \nu \frac{\partial^2 w}{\partial y^2}\right) \quad (7\text{-}7a)$$

$$M_y = -\int_{-h/2}^{h/2} z\sigma_y\, dz = D\left(\frac{\partial^2 w}{\partial y^2} + \nu \frac{\partial^2 w}{\partial x^2}\right) \quad (7\text{-}7b)$$

$$M_{xy} = -\int_{-h/2}^{h/2} z\tau_{xy}\, dz = D(1-\nu)\frac{\partial^2 w}{\partial x\, \partial y} \quad (7\text{-}7c)$$

$$M_{yx} = -\int_{-h/2}^{h/2} z\tau_{xy}\, dz = -D(1-\nu)\frac{\partial^2 w}{\partial x\, \partial y} = -M_{xy} \quad (7\text{-}7d)$$

where $D = Eh^3/12(1-\nu^2)$ is referred to as the *flexural rigidity*. It is the counterpart of the EI term in beam theory. Note that by eliminating the displacement terms between Equations (7-6) and (7-7) there results

$$\sigma_x = -\frac{12z}{h^3}M_x$$

so that the maximum bending stress associated with M_x is

$$(\sigma_x)_{\max} = \pm \frac{6M_x}{h^2} \quad (7\text{-}8a)$$

Similarly,

$$(\sigma_y)_{\max} = \pm \frac{6M_y}{h^2} \quad (7\text{-}8b)$$

and

$$(\tau_{xy})_{\max} = \pm \frac{6M_{xy}}{h^2} \quad (7\text{-}8c)$$

These bending stress equations are the counterparts of the classic $\sigma = My/I$ equation for the beam equation.

The last step is to substitute Equations (7-7) into Equation (7-3), namely,

$$\frac{\partial^2 M_x}{\partial x^2} + \frac{\partial^2 M_x}{\partial x^2} + \frac{\partial^2 M_{xy}}{\partial y\, \partial x} - \frac{\partial^2 M_{yx}}{\partial x\, \partial y} = q$$

to obtain

$$\frac{\partial^2}{\partial x^2}\left(D\left(\frac{\partial^2 w}{\partial x^2} + \nu \frac{\partial^2 w}{\partial y^2}\right)\right) + \frac{\partial^2}{\partial y^2}\left(D\left(\frac{\partial^2 w}{\partial y^2} + \nu \frac{\partial^2 w}{\partial y^2}\right)\right)$$
$$+ \frac{\partial^2}{\partial y \, \partial x}\left(D(1-\nu)\frac{\partial^2 w}{\partial x \, \partial y}\right) - \frac{\partial^2}{\partial x \, \partial y}\left(-D(1-\nu)\frac{\partial^2 w}{\partial x \, \partial y}\right) = q$$

or, after simplification,

$$D\left(\frac{\partial^4 w}{\partial x^4} + 2\frac{\partial^4 w}{\partial x^2 \, \partial y^2} + \frac{\partial^4 w}{\partial y^4}\right) = q \tag{7-9}$$

as the equation of equilibrium, expressed in terms of the displacements, for the plate. Equation (7-9), which can also be expressed as,

$$D\nabla^2(\nabla^2 w) = q$$

is frequently referred to as the *biharmonic equation*.

Boundary Conditions. In this section the boundary conditions appropriate for a rectangular or square plate will be discussed. For plates having other boundaries, the reader is referred to the references.

For rectangular or square plates, boundary conditions on lines $x = $ const and on lines $y = $ const must be specified. The boundary conditions for the three cases of simply supported edges, clamped edges, and free edges for a rectangular plate will be discussed. For boundary conditions that are appropriate when an edge is not parallel to either of the x or the y axes, the reader is referred to the references.

Simply supported edge: For a simply supported edge, the two conditions that must be satisfied are that the displacement w must be zero and the moment whose direction coincides with the direction of the edge must be zero. Consulting Figure (7-7), it is seen

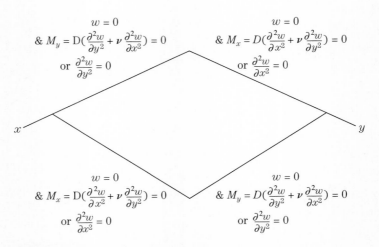

FIGURE 7-7 Boundary conditions for simply supported edges.

that on an edge $y = $ const

$$w = 0$$

$$M_y = D\left(\frac{\partial^2 w}{\partial y^2} + \nu \frac{\partial^2 w}{\partial x^2}\right) = 0$$

Note that since $w = 0$ along the edge $y = $ const, it follows that $\partial^2 w/\partial x^2 \equiv 0$ so that it is possible to replace the above boundary conditions with

$$w = 0 \quad \text{and} \quad \frac{\partial^2 w}{\partial y^2} = 0$$

On an edge $y = $ const, the slope $\partial w/\partial y$ and the shear Q_y will not be zero in general.

On an edge $x = $ const

$$w = 0$$

$$M_x = D\left(\frac{\partial^2 w}{\partial x^2} + \nu \frac{\partial^2 w}{\partial y^2}\right) = 0$$

Here, since $\partial^2 w/\partial y^2 \equiv 0$, it is possible to replace the above boundary conditions with

$$w = 0 \quad \text{and} \quad \frac{\partial^2 w}{\partial x^2} = 0$$

On an edge $x = $ const, the slope $\partial w/\partial x$ and the shear Q_x will not be zero in general.

Clamped edge: For a clamped or fixed edge, the two conditions that must be satisfied are that the displacement w must be zero and the slope of the line perpendicular to the edge must be zero. Consulting Figure (7-8), it is seen that on an edge $y = $ const

$$w = 0$$

$$\frac{\partial w}{\partial y} = 0$$

On an edge $y = $ const, the moment M_y and the shear Q_y will not be zero in general.

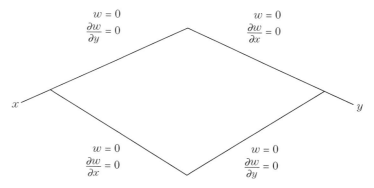

FIGURE 7-8 Boundary conditions for clamped or fixed edges.

On an edge $x = \text{const}$

$$w = 0$$

$$\frac{\partial w}{\partial x} = 0$$

On an edge $x = \text{const}$, the moment M_x and the shear Q_x will not be zero in general.

Free edge: For a free edge $x = \text{const}$, the two conditions that must be satisfied are that the moment M_x must be zero and that the quantity $V_x = Q_x - \partial M_{xy}/\partial y$ on that edge must be zero. From Equation (7-2c), Q_x can be expressed as

$$Q_x = \frac{\partial M_{yx}}{\partial y} - \frac{\partial M_x}{\partial x}$$

so that with $M_{xy} = -M_{yx}$

$$V_x = -\frac{\partial M_x}{\partial x} - 2\frac{\partial M_{xy}}{\partial y}$$

$$= -D\left(\frac{\partial^3 w}{\partial x^3} + (2-\nu)\frac{\partial^3 w}{\partial x\, \partial y^2}\right)$$

Thus the two conditions on a free edge $x = \text{const}$ are

$$M_x = D\left(\frac{\partial^2 w}{\partial x^2} + \nu\frac{\partial^2 w}{\partial y^2}\right) = 0$$

and

$$V_x = -D\left(\frac{\partial^3 w}{\partial x^3} + (2-\nu)\frac{\partial^3 w}{\partial x\, \partial y^2}\right) = 0$$

For a free edge $y = \text{const}$, the two boundary conditions that must be satisfied are

$$M_y = D\left(\frac{\partial^2 w}{\partial y^2} + \nu\frac{\partial^2 w}{\partial x^2}\right) = 0$$

and

$$V_y = Q_y + \frac{\partial M_{yx}}{\partial x} = 0$$

Using

$$Q_y = -\frac{\partial M_y}{\partial y} - \frac{\partial M_{xy}}{\partial x}$$

and $M_{xy} = -M_{yx}$

$$V_y = -\frac{\partial M_y}{\partial y} - 2\frac{\partial M_{xy}}{\partial x}$$

$$= -D\left(\frac{\partial^3 w}{\partial y^3} + (2-\nu)\frac{\partial^3 w}{\partial x^2\, \partial y}\right)$$

In summary, then, in order to analyze the displacements and stresses in a rectangular plate the differential equation

$$D\left(\frac{\partial^4 w}{\partial x^4} + 2\frac{\partial^4 w}{\partial x^2 \partial y^2} + \frac{\partial^4 w}{\partial y^4}\right) = q$$

must be solved in the region occupied by the plate subject to boundary conditions on the four sides. With $w(x, y)$ known, the moments can then be determined using

$$M_x = D\left(\frac{\partial^2 w}{\partial x^2} + \nu \frac{\partial^2 w}{\partial y^2}\right)$$

$$M_y = D\left(\frac{\partial^2 w}{\partial y^2} + \nu \frac{\partial^2 w}{\partial x^2}\right)$$

$$M_{xy} = D(1 - \nu)\frac{\partial^2 w}{\partial x \, \partial y}$$

with the stresses then computed using

$$(\sigma_x)_{\max} = \pm \frac{6M_x}{h^2}$$

$$(\sigma_y)_{\max} = \pm \frac{6M_y}{h^2}$$

and

$$(\tau_{xy})_{\max} = \pm \frac{6M_{xy}}{h^2}$$

The theory developed in the preceding sections is known as the *small displacement theory* of thin, flat plates. This means that the displacements and stresses predicted by the theory are valid only as long as the transverse displacements $w(x, y)$ are "small" with respect to the thickness of the plate. Specifically for a rectangular plate, significant in-plane forces are developed in the plate when the transverse deflection exceeds approximately half the thickness of the plate. When the displacement exceeds this limit, the actual deflections and bending stresses as predicted by the large displacement theory and as measured by experiments are less than those given by the small displacement theory. The reader is referred to the references [1] for the details.

EXAMPLE 7-1 Consider the case of the square simply supported, uniformly loaded rectangular plate for which (see the next section)

$$w_{\max} = 0.0443 \frac{q_0 a^4}{E h^3}$$

If we set $w_{\max} = h/2$, there results

$$\left(\frac{h}{a}\right)^4 = 0.0868 \frac{q_0}{E}$$

Further assuming that $h/a = 1/100$ and that the plate is made from steel with $E = 3 \times 10^7$ psi, it follows that $q_0 = 3.46$ psi is the maximum allowable uniform load. From the next section the corresponding maximum bending stress is

$$\sigma_{max} = \frac{6}{h^2} 0.0479 q_0 a^2 = 9.94 \text{ ksi}$$

For an aluminum plate with the same h/a ratio and $E = 10^7$ psi, the corresponding results are $q_0 = 1.15$ psi and $\sigma_{max} = 3.31$ ksi. Both these results show clearly that it is possible for stresses to have permissible values compared to allowables for the material in question, but for the corresponding displacements to exceed those for which the theory is valid. ◆

7-5 Solutions to Rectangular-Plate Problems

In this section the problem of a rectangular plate simply supported on two opposite edges with several combinations of boundary conditions on the other two edges will be considered. The two types of solutions associated with Navier and Levy are discussed in some detail.

Navier Solution. If we take the solution for the transverse displacement $w(x, y)$ to be of the form

$$w(x, y) = w_{mn} \sin \frac{m\pi x}{a} \sin \frac{n\pi y}{b}$$

it is easily seen that all the boundary conditions for a simply supported plate, namely,

$$w(0, y) = \frac{\partial^2 w(0, y)}{\partial x^2} = 0 \quad \text{on } x = 0 \text{ and on } x = a$$

and

$$w(x, 0) = \frac{\partial^2 w(x, 0)}{\partial y^2} = 0 \quad \text{on } y = 0 \text{ and on } y = b$$

are satisfied. Superposition can then be used to conclude that

$$w(x, y) = \sum_{m=1}^{\infty} \sum_{n=1}^{\infty} w_{mn} \sin \frac{m\pi x}{a} \sin \frac{n\pi y}{b}$$

also satisfies all the boundary conditions for a rectangular plate simply supported on all edges. Represent the loading $q(x, y)$ on the plate as

$$q(x, y) = \sum_{m=1}^{\infty} \sum_{n=1}^{\infty} q_{mn} \sin \frac{m\pi x}{a} \sin \frac{n\pi y}{b} \qquad (7\text{-}10)$$

so that substitution of the solution w into the differential equation of equilibrium $D\nabla^2(\nabla^2 w) = q$ yields

$$D \sum_{m=1}^{\infty} \sum_{n=1}^{\infty} w_{mn} \left(\left(\frac{m\pi}{a}\right)^2 + \left(\frac{n\pi}{b}\right)^2\right)^2 \sin \frac{m\pi x}{a} \sin \frac{n\pi y}{b} =$$

$$\sum_{m=1}^{\infty} \sum_{n=1}^{\infty} q_{mn} \sin \frac{m\pi x}{a} \sin \frac{n\pi y}{b}$$

7-5 SOLUTIONS TO RECTANGULAR-PLATE PROBLEMS

or, upon comparing coefficients,

$$w_{mn} = \frac{q_{mn}}{D\left(\left(\frac{m\pi}{a}\right)^2 + \left(\frac{n\pi}{b}\right)^2\right)^2}$$

so that

$$w = \sum_{m=1}^{\infty}\sum_{n=1}^{\infty} \frac{q_{mn}}{D\left(\left(\frac{m\pi}{a}\right)^2 + \left(\frac{n\pi}{b}\right)^2\right)^2} \sin\frac{m\pi x}{a} \sin\frac{n\pi y}{b}$$

Using orthogonality of the two sets of functions $\sin(m\pi x/a)$ and $\sin(n\pi y/b)$ and the loading function $q(x, y)$, the coefficients q_{mn} are determined in the usual way from Equation (7-10) as

$$\int_0^a \int_0^b q(x, y) \sin\frac{M\pi x}{a} \sin\frac{M\pi y}{b}\, dy\, dx = q_{MN}\frac{a}{2}\frac{b}{2}$$

EXAMPLE 7-2 Consider the specific case $q = q_0$, that is, the case of a uniformly loaded plate. Evaluating the integrals gives

$$q_{MN} = \frac{16q_0}{MN\pi^2} \quad \text{if } M \text{ odd and } N \text{ odd}$$

or

$$q_{MN} = 0 \quad \text{otherwise}$$

so that

$$w = \frac{16q_0}{\pi^6 D} \sum_{m \text{ odd}} \sum_{n \text{ odd}} \frac{\sin\frac{m\pi x}{a}\sin\frac{n\pi y}{b}}{mn\left(\left(\frac{m}{a}\right)^2 + \left(\frac{n}{b}\right)^2\right)^2}$$

The maximum occurs at the center $x = a/2$, $y = b/2$ with

$$w\left(\frac{a}{2}, \frac{b}{2}\right) = \frac{16q_0}{\pi^6 D} \sum_{m \text{ odd}} \sum_{n \text{ odd}} \frac{(-1)^{(m+n)/2-1}}{mn\left(\left(\frac{m}{a}\right)^2 + \left(\frac{n}{b}\right)^2\right)^2}$$

Taking only the first term of the series and assuming a square plate with $\nu = 0.3$, there results

$$w\left(\frac{a}{2}, \frac{b}{2}\right) \approx \frac{0.0454 q_0 a^4}{E h^3}$$

Taking the first four terms of the series is sufficient to give essentially the exact value

$$w_{max} = \frac{0.0443 q_0 a^4}{E h^3}$$

The bending and twisting moments are easily computed to be

$$M_x = D\left(\frac{\partial^2 w}{\partial x^2} + \nu \frac{\partial^2 w}{\partial y^2}\right) = -\frac{16q_0}{\pi^4} \sum_{m \text{ odd}} \sum_{n \text{ odd}} \frac{\left(\frac{m^2}{a^2} + \nu\left(\frac{n^2}{b^2}\right)\right)\sin\frac{m\pi x}{a}\sin\frac{n\pi y}{b}}{mn\left(\left(\frac{m}{a}\right)^2 + \left(\frac{n}{b}\right)^2\right)^2}$$

$$M_y = D\left(\frac{\partial^2 w}{\partial y^2} + \nu \frac{\partial^2 w}{\partial x^2}\right) = -\frac{16q_0}{\pi^4} \sum_{m \text{ odd}} \sum_{n \text{ odd}} \frac{\left(\frac{n^2}{b^2} + \nu\left(\frac{m^2}{a^2}\right)\right)\sin\frac{m\pi x}{a}\sin\frac{n\pi y}{b}}{mn\left(\left(\frac{m}{a}\right)^2 + \left(\frac{n}{b}\right)^2\right)^2}$$

$$M_{xy} = D(1-\nu)\frac{\partial^2 w}{\partial x\, \partial y} = \frac{16(1-\nu)q_0}{\pi^4 ab} \sum_{m \text{ odd}} \sum_{n \text{ odd}} \frac{\cos\frac{m\pi x}{a}\cos\frac{n\pi y}{b}}{\left(\left(\frac{m}{a}\right)^2 + \left(\frac{n}{b}\right)^2\right)^2}$$

The maximum value of the bending moment M_x occurs at the center $x = a/2$, $y = b/2$ with

$$(M_x)_{\max} = -\frac{16q_0}{\pi^4} \sum_{m \text{ odd}} \sum_{n \text{ odd}} \frac{\left(\frac{m^2}{a^2} + \nu\left(\frac{n^2}{b^2}\right)\right)(-1)^{(m+n)/2-1}}{mn\left(\left(\frac{m}{a}\right)^2 + \left(\frac{n}{b}\right)^2\right)^2}$$

For a square plate with $\nu = 0.3$, the first term of the series for M_x gives $(M_x)_{\max} \approx -0.0534 q_0 a^2$ with the corresponding maximum bending stress at the top of the plate given by

$$(\sigma_x)_{\max} = -\frac{6(M_x)_{\max}}{h^2} \approx 0.320 q_0 \left(\frac{a}{h}\right)^2$$

Taking the first four terms of the series give $(M_x)_{\max} \approx -0.0469 q_0 a^2$ with

$$(\sigma_x)_{\max} = -\frac{6(M_x)_{\max}}{h^2} \approx 0.281 q_0 \left(\frac{a}{h}\right)^2$$

which is within 2% of the exact value $(\sigma_x)_{\max} = 0.287 q_0 (a/h)^2$. The quantities to remember are that for a uniformly loaded square plate $w_{\max} \approx q_0 a^4/22.5\, h^3$ and $\sigma_{\max} \approx q_0 a^2/3.5 h^2$. ◆

Levy Solution. For a plate that is simply supported on two opposite edges—say, $x = 0$ and $x = a$—the boundary conditions can be satisfied by taking

$$w(x, y) = \sum_{1}^{\infty} w_n(y)\sin\left(\frac{n\pi x}{a}\right)$$

Substituting into the equation of equilibrium gives

$$\sum_{1}^{\infty}\left\{\frac{d^4 w_n}{dy^4} - \alpha_n^2 \frac{d^2 w_n}{dy^2} + \alpha_n^4 w_n\right\}\sin\left(\frac{n\pi x}{a}\right) = q/D$$

7-5 SOLUTIONS TO RECTANGULAR-PLATE PROBLEMS

where $\alpha_n = n\pi/a$. It is also necessary to represent q as

$$q = \sum_{1}^{\infty} q_n(y) \sin\left(\frac{n\pi x}{a}\right)$$

where

$$q_n(y) = \frac{2}{a} \int_0^a q(x, y) \sin\left(\frac{n\pi x}{a}\right) dx$$

The equation satisfied by w_n is then

$$\frac{d^4 w_n}{dy^4} - \alpha_n^2 \frac{d^2 w_n}{dy^2} + \alpha_n^4 w_n = \frac{q_n(y)}{D}$$

The homogeneous solution is

$$w_{nh} = A_n \cosh(\alpha_n y) + B_n \sinh(\alpha_n y) + C_n \alpha_n y \cosh(\alpha_n y) + D_n \alpha_n y \sinh(\alpha_n y)$$

EXAMPLE 7-3 Consider the specific case of a uniformly loaded plate with $q(x, y) = q_0$ so that

$$q_n = \frac{2}{a} \int_0^a q_0 \sin\left(\frac{n\pi x}{a}\right) dx = \frac{4q_0}{n\pi} \quad \text{if } n \text{ odd}$$

or

$$q_n = 0 \quad \text{if } n \text{ even}$$

The particular solution is then $w_{np} = q_n/D\alpha_n^4$ with $w_n = w_{nh} + w_{np}$ or

$$w_n = (A_n + \alpha_n y C_n)\cosh(\alpha_n y) + (B_n + \alpha_n y D_n)\sinh(\alpha_n y) + q_n/D\alpha_n^4$$

For the case where the plate is simply supported on the edges $y = 0$ and $y = b$, the displacement will be symmetrical with respect to the center of the plate. It is then possible to locate $y = 0$ at the center of the plate and use only the symmetric portion of the solution, namely,

$$w_n = a_n \cosh(\alpha_n y) + \alpha_n y b_n \sinh(\alpha_n y) + \frac{q_n}{D\alpha_n^4}$$

involving only two constants. The boundary conditions that must be satisfied are then

$$w\left(\frac{b}{2}\right) = \frac{\partial^2 w\left(\frac{b}{2}\right)}{\partial y^2} = 0$$

leading to

$$a_n \cosh\left(\frac{\alpha_n b}{2}\right) + \alpha_n\left(\frac{b}{2}\right) b_n \sinh\left(\frac{\alpha_n b}{2}\right) + \frac{q_n}{\alpha_n^4} = 0$$

and

$$\alpha_n^2 \left[a_n \cosh\left(\frac{\alpha_n b}{2}\right) + b_n \left\{ \alpha_n\left(\frac{b}{2}\right) \sinh\left(\frac{\alpha_n b}{2}\right) + 2 \cosh\left(\frac{\alpha_n b}{2}\right) \right\} \right] = 0$$

from which

$$b_n = \frac{q_n}{2\cosh(u_n)D\alpha_n^4}$$

and

$$a_n = -\frac{q_n(2\cosh(u_n) + u_n\sinh(u_n))}{2\cosh^2(u_n)D\alpha_n^4}$$

where $u_n = \alpha_n(b/2)$. It then follows that

$$w_n(y) = \frac{q_n}{D\alpha_n^4}\left[\frac{\alpha_n y\,\sinh(\alpha_n y)}{2\cosh(u_n)} - \cosh(\alpha_n y)\frac{u_n\tanh(u_n) + 2}{2\cosh(u_n)} + 1\right]$$

and

$$w(x, y) = \sum_{\text{odd}}^{\infty} w_n(y)\sin\left(\frac{n\pi x}{a}\right)$$

The maximum displacement at $x = a/2$, $y = 0$ is

$$w\left(\frac{a}{2}, 0\right) = \frac{4q_0 a^4}{\pi^5 D}\sum_{\text{odd}}^{\infty}\frac{(-1)^{(n-1)/2}}{n^5}\left(1 - \frac{u_n\tanh(u_n) + 2}{\cosh(u_n)}\right)$$

Taking only the first term of the series yields

$$w_{\max} \approx \frac{4q_0 a^4}{\pi^5 D}\left(1 - \frac{\left(\frac{\pi}{2}\right)\tanh\left(\frac{\pi}{2}\right) + 2}{\cosh\left(\frac{\pi}{2}\right)}\right)$$

$$= 0.00411\frac{q_0 a^4}{D} = 0.0449\frac{q_0 a^4}{Eh^3}$$

approximately 1.4% in error. It is left for the student to show that the bending moment at the center of the plate is

$$M_x\left(\frac{a}{2}, 0\right) = D\left(\frac{\partial^2 w}{\partial x^2} + \nu\frac{\partial^2 w}{\partial y^2}\right)\left(\frac{a}{2}, 0\right)$$

$$= \frac{4q_0 a^2}{\pi^3}\sum_{\text{odd}}^{\infty}\frac{(-1)^{(n-1)/2}}{n^3}\left[-1 + (1-\nu)\frac{u_n\tanh(u_n) + 2}{2\cosh(u_n)} + \frac{\nu}{2\cosh(u_n)}\right]$$

Taking only the first terms results in

$$M_x\left(\frac{a}{2}, 0\right) \approx -0.0517 q_0 a^2$$

which is approximately 8% in error compared to the exact value of $-0.0479 q_0 a^2$. Taking the first three terms gives $M_x(a/2, 0) = -0.0481 q_0 a^2$ with an error of 0.4%, showing the more rapid convergence of the Levy solution compared to that of the Navier solution.

7-6 INTRODUCTION

TABLE 7-1 Constants for Determining Displacements and Stresses in a Uniformly Loaded Rectangular Plate Simply Supported on All Sides

b/a	C_1	C_2	C_3
1.0	0.0443	0.0479	0.0479
1.25	0.0658	0.0661	0.0503
1.50	0.0843	0.0812	0.0498
2.0	0.1106	0.1017	0.0463
4.0	0.1400	0.1235	0.0384

Values for the maximum displacement and maximum bending stresses in a uniformly loaded rectangular plate can be represented as follows:

$$w_{max} = C_1 \frac{q_0 a^4}{E h^3}$$

$$(M_x)_{max} = C_2 \, q_0 a^2$$

$$(M_y)_{max} = C_3 \, q_0 a^2$$

Table 7-1 contains values for the constants C_1 and C_2 for a range of values of the aspect ratio a/b of the plate.

In the exercises the student is asked to investigate displacements and stresses in rectangular plates with other combinations of boundary conditions.

Part II Circular Plates

7-6 Introduction

Circular plates frequently serve as a component of a structure. As indicated in Figure 7-9a, one of the most common occurrences involves a circular plate that forms the end of a cylindrical container. The cylindrical container is often pressurized, resulting

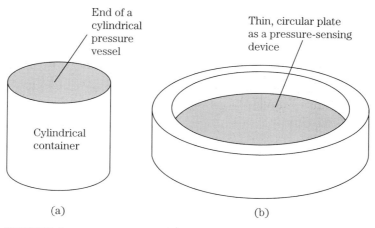

FIGURE 7-9 Examples of circular plates.

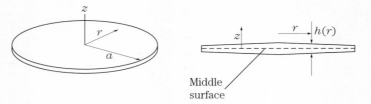

FIGURE 7-10 Geometry of a circular plate.

in stresses and deformations in the circular plate. Another example involves the use of a circular plate as a pressure-sensing device, as shown in Figure 7-9b. Circular plate theory can be used to relate deflections of the center of the plate to the pressure difference between the two surfaces of the plate.

For the theory to be developed and applied in this chapter, the following assumptions are made in connection with Figure 7-10 regarding the geometry and loading:

1. The thickness h, which can vary gradually with the radius r, is small compared to the radius a of the plate. The thickness extends equally in both directions from the so-called middle surface of the plate, defined by $z = 0$.
2. External loads are applied in the z direction and are dependent only on the radial coordinate. These include normal or transverse pressures $p(r)$, concentrated loads at the center of the plate and line loads at some radius $r = r_0$.
3. Supports occur on $r = $ constant circles and have the same character, that is, clamped, simply supported, or whatever, at all points around the periphery of the circle.

If any of these assumptions is not satisfied for the application in question, the theory is in general invalid.

7-7 Force Resultants and Equilibrium

In response to axisymmetric tranverse loads that are applied to the circular plate, the stresses and internal force resultants that result are as shown on a typical differential element in Figure 7-11. The primary nonzero stress components are σ_r, σ_θ and τ_{rz}. The relations between the stresses shown in Figure 7-11a and the internal force resultants shown in Figure 7-11b are easily seen to be

$$M_r = \int_{-h/2}^{h/2} \sigma_r z \, dz \tag{7-11a}$$

$$M_\theta = -\int_{-h/2}^{h/2} \sigma_\theta z \, dz \tag{7-11b}$$

$$Q = \int_{-h/2}^{h/2} \tau_{rz} \, dz \tag{7-11c}$$

All of these force resultants are measured per unit length around the periphery, that is, the dimensions are $[M_r] = FL/L = F$, $[M_\theta] = FL/L = F$, $[Q] = F/L$. In what follows, M_r, M_θ and Q will have the positive directions shown in Figure 7-11b.

7-7 FORCE RESULTANTS AND EQUILIBRIUM

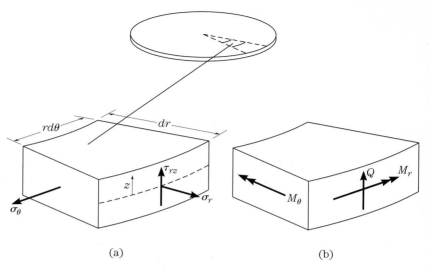

FIGURE 7-11 Stresses and internal force resultants.

The equilibrium equations are expressed in terms of the force resultants M_r, M_θ and Q and the external loading $q(r)$ by considering the free-body diagram of a differential element $dVol = tr\,dr\,d\theta$, as shown in Figure 7-12. Summing forces in the vertical direction yields

$$-rQ\,d\theta + \left[rQ\,d\theta + \frac{d(rQ)}{dr}dr\,d\theta\right] + q(r)r\,dr\,d\theta = 0$$

or

$$(rQ)' + rq(r) = 0$$

A sum of moments about a circumferential direction yields

$$-rM_r\,d\theta + \left[rM_r\,d\theta + \frac{d(rM_r)}{dr}dr\,d\theta\right] - dr\,(rQ\,d\theta) + 2\left[M_\theta\,dr\,\sin\frac{d\theta}{2}\right] = 0$$

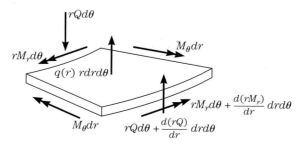

FIGURE 7-12 Free-body diagram of a differential element.

Passing to the limit yields

$$\frac{d(rM_r)}{dr} + M_\theta - rQ = 0$$

A sum of moments about a radial line is satisfied identically so that collectively the equations of equilibrium are

$$(rQ)' + rq(r) = 0 \tag{7-12a}$$

and

$$\frac{d(rM_r)}{dr} + M_\theta - rQ = 0 \tag{7-12b}$$

constituting two equations for the three unknowns M_r, M_θ and Q.

7-8 Deformations: Strain-Displacement Relations

The observable deformations that result from the symmetric loading of a circular plate are in the r and z directions, that is, the displacement $u(r, z)$ in the radial direction and the displacement $w(r, z)$ normal to the middle surface of the plate. As shown in Figure 7-13, where the deformation of a diametral section is pictured, the displacements in the radial direction are due in part to a translation and in part to a rotation of the plate about a circumferential direction.

The basic assumption that is made is analogous to the developments in Chapter 4 for the beam and in Part I of this chapter for the rectangular plate. Assume that

$$u(r, z) = u_0(r) + zu_1(r)$$

and

$$w(r, z) = w_0(r) + zw_1(r) \tag{7-13}$$

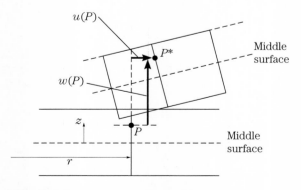

FIGURE 7-13 Deformations of a circular plate.

where $w_0(r)$ and $u_0(r)$ represent the displacements of the middle surface with u_1 and w_1 accounting for the deformation relative to the middle surface. The circumferential component of displacement is absent for the symmetric problem.

With no dependence of the displacements on θ, the strain-displacement relations in cylindrical coordinates reduce to

$$\epsilon_r = \frac{\partial u}{\partial r} \qquad \epsilon_\theta = \frac{u}{r} \qquad \epsilon_z = \frac{\partial w}{\partial z}$$

$$\gamma_{r\theta} = 0 \qquad \gamma_{\theta z} = 0 \qquad \gamma_{rz} = \frac{\partial u}{\partial z} + \frac{\partial w}{\partial r}$$

Substitution of Equations (7-13) into the strain-displacement relations yields

$$\epsilon_r = u_0'(r) + z u_1'(r) \qquad \epsilon_\theta = \frac{u_0(r) + z u_1(r)}{r} \qquad \epsilon_z = w_1$$

$$\gamma_{r\theta} = 0 \qquad \gamma_{\theta z} = 0 \qquad \gamma_{rz} = u_1 + w_0'(r) + z w_1'(r)$$

Assume that the extensional strain in the direction normal to the plate is negligible, leading to $w_1 = 0$. This essentially states that the thickness of the plate is unchanged as a result of the deformation. It is then assumed that shear deformations can be ignored, a reasonable assumption for a thin plate. With $w_1 = 0$ this leads to

$$\gamma_{rz} = u_1 + w_0'(r) = 0$$

or

$$u_1 = -w_0'(r)$$

This amounts to stating that a normal to the middle surface remains normal to the middle surface (no shear strain), frequently referred to as the Kirchoff assumption. Thus, finally, the displacements can be represented as

$$u(r, z) = u_0(r) - z w_0'(r)$$
$$w(r, z) = w_0(r)$$

and the strains as

$$\epsilon_r = u_0'(r) - z w_0''(r) \qquad (7\text{-}14)$$
$$\epsilon_\theta = \frac{u_0(r) - z w_0'(r)}{r}$$

showing that, on the basis of the assumptions made, both strains are linear functions of the z coordinate.

Last, it is important to note that unless the transverse displacement is everywhere small compared to the thickness of the plate, significant in-plane forces may be developed. The appropriate equations, which show that there is coupling between the in-plane and out-of-plane variables, are nonlinear. The interested reader is referred to reference [1] for details. In any event, the important point is that if the calculations show that the transverse displacement exceeds the thickness of the plate, the calculations are in fact suspect, so that in order to use the theory developed in this chapter, the additional assumption that the displacements not exceed the thickness must be observed.

7-9 Stress-Strain and Combination

The additional equations necessary for relating the kinematic variables ϵ_r, ϵ_θ, u_0 and w_0 to the force variables σ_r, σ_θ, τ_{r2}, M_r, M_θ and Q come from making assumptions regarding the material behavior. Assume a linearly elastic isotropic solid under plane stress conditions for which

$$\epsilon_r = \frac{\sigma_r - \nu\sigma_\theta}{E}, \quad \epsilon_\theta = \frac{\sigma_\theta - \nu\sigma_r}{E} \quad \text{and} \quad \gamma_{r2} = \frac{\tau_{r2}}{G} \tag{7-15}$$

which are additional equations relating the kinematic and kinetic variables. Solving Equations (7-15) for the stresses in terms of the strains yields

$$\sigma_r = \frac{E}{1-\nu^2}(\epsilon_r + \nu\epsilon_\theta)$$

$$\sigma_\theta = \frac{E}{1-\nu^2}(\epsilon_\theta + \nu\epsilon_r)$$

and

$$\tau_{r2} = G\gamma_{r2} \tag{7-16}$$

Substituting Equations (7-14) into Equations (7-16) yields

$$\sigma_r = \frac{E}{1-\nu^2}\left[u_0' + \nu\frac{u_0}{r} - z\left(w_0'' + \nu\frac{w_0'}{r}\right)\right]$$

$$\sigma_\theta = \frac{E}{1-\nu^2}\left[\frac{u_0}{r} + \nu u_0' - z\left(\frac{w_0'}{r} + \nu w_0''\right)\right]$$

Substituting these equations into Equations (7-11a and b) and performing the integrations then yields

$$M_r = -D\left(w_0'' + \nu\frac{w_0'}{r}\right) \tag{7-17a}$$

and

$$M_\theta = D\left(\frac{w_0'}{r} + \nu w_0''\right) \tag{7-17b}$$

where $D = Eh^3/(12(1-\nu^2))$ is termed the *flexural rigidity* or *bending stiffness* of the plate. It is analogous to the quantity EI for the beam problem. The terms in the above expressions for σ_r and σ_θ that involve the in-plane displacement u_0 have been covered in Chapter 6 for the axisymmetric in-plane problems. They are uncoupled from the transverse terms as long as the transverse deflections of the plate are small, that is, roughly less than the thickness of the plate. Assuming a small deflection problem, discard the in-plane terms in what follows with the result that

$$\sigma_r = -\frac{Ez}{1-\nu^2}\left(w_0'' + \nu\frac{w_0'}{r}\right) \tag{7-18a}$$

$$\sigma_\theta = -\frac{Ez}{1-\nu^2}\left(\frac{w_0'}{r} + \nu w_0''\right) \tag{7-18b}$$

Eliminating w_0 between Equations (7-17) and (7-18) yields

$$\sigma_r = \frac{Ez}{1-\nu^2}\frac{M_r}{D} = 12\,M_r\frac{z}{h^3} \tag{7-19a}$$

and

$$\sigma_\theta = -\frac{Ez}{1-\nu^2}\frac{M_\theta}{D} = -12\,M_\theta\frac{z}{h^3} \tag{7-19b}$$

having maximum values at the top and bottom surfaces of the plate, respectively, of

$$(\sigma_r)_{\max} = \pm\frac{6M_r}{h^2} \quad\text{and}\quad (\sigma_\theta)_{\max} = \mp\frac{6M_\theta}{h^2}$$

By virtue of the fact that one of the kinematic assumptions was that $\gamma_{r2} = 0$, it follows that τ_{r2} and hence Q are zero. However, on the basis of substituting Equations (7-17a and b) into Equation (7-16e), it is seen that

$$rQ = \frac{d(rM_r)}{dr} + M_\theta$$

$$= -D\left(rw_0''' + w_0'' - \frac{w_0'}{r}\right)$$

or

$$Q = -D\frac{d}{dr}\left(\frac{1}{r}\frac{d}{dr}(rw_0')\right) \tag{7-20}$$

where h has been assumed to be a constant. Thus, there is in general an obvious contradiction contained within the governing equations in that for a general loading it is clear that a shear force Q must be transmitted for equilibrium. This contradiction is a result of the Kirchoff assumption regarding the shear deformation.

Temporarily sidestep this issue as follows. Eliminate the shear resultant Q from Equations (7-16d and e) to obtain

$$\frac{d^2}{dr^2}(rM_r) + \frac{d}{dr}(M_\theta) + rq(r) = 0$$

which, upon using Equations (7-17a and b) and assuming that $h = $ constant, yields

$$D\left(\frac{d^4w_0}{dr^4} + \frac{2}{r}\frac{d^3w_0}{dr^3} - \frac{1}{r^2}\frac{d^2w_0}{dr^2} + \frac{1}{r^3}\frac{dw_0}{dr}\right) - q = 0$$

as the governing equilibrium equation in terms of the midsurface transverse displacement w_0. For the purpose of integrating this equation, it is convenient to express it as

$$D\left[\frac{1}{r}\frac{d}{dr}\left(r\frac{d}{dr}\left(\frac{1}{r}\frac{d}{dr}\left(r\frac{dw_0}{dr}\right)\right)\right)\right] - q = 0$$

By recognizing that the Laplacian for cylindrical coordinates can be written as

$$\nabla^2 = \frac{d^2}{dr^2} + \frac{1}{r}\frac{d}{dr} = \frac{1}{r}\frac{d}{dr}\left(r\frac{d}{dr}\right)$$

it follows that the governing equilibrium equation can also be expressed as

$$D\nabla^2\nabla^2 w_0 - q = D\nabla^4 w_0 - q = 0 \tag{7-21}$$

which is precisely the same equation as for the rectangular plate. This equation must be solved for w_0 with appropriate boundary conditions then enforced.

The governing equation of equilibrium is a fourth-order linear ordinary differential equation requiring four boundary conditions to determine the four constants of integration. For an annular plate with $b \leq r \leq a$, two boundary conditions must be enforced at each boundary. For a solid plate there is no inner boundary and $r = 0$ is a point in the plate at which stresses and displacements must have finite values. Thus, for a solid plate the two boundary conditions at $r = 0$ are replaced by two *boundedness conditions*, with four equations again available for determining the four constants of integration.

At a boundary $r = a$ or $r = b$, appropriate boundary conditions are as described below.

Simply Supported. A simply supported edge, which is shown symbolically in Figure 7-14, is accomplished by restraining the displacement w_0 to be zero while at the same time allowing complete freedom of the rotation w_0'. This rotational freedom is accomplished by setting the corresponding moment M_r equal to zero. In equation form, the two boundary conditions at $r = a$ are

$$w_0 \bigg|_{r=a} = 0$$

$$M_r \bigg|_{r=a} = -D\left(w_0'' + \nu\frac{w_0'}{r}\right)\bigg|_{r=a} = 0$$

with the boundary condition at b completely similar in form.

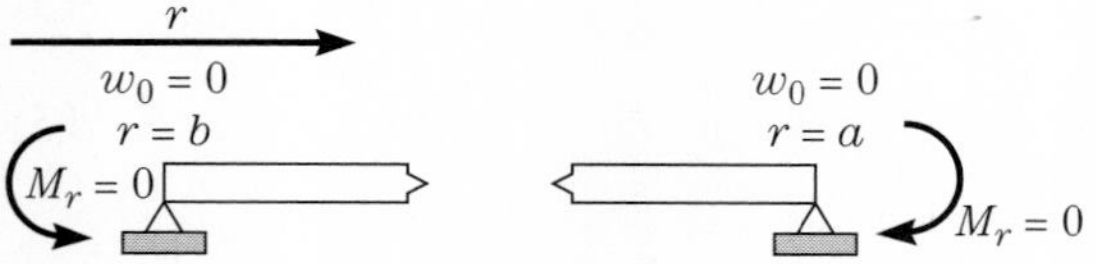

FIGURE 7-14 Simply supported edge.

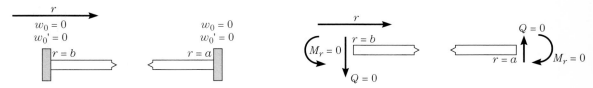

FIGURE 7-15 Clamped edge.

FIGURE 7-16 Free edge.

Clamped. A clamped edge, which is shown symbolically in Figure 7-15, is accomplished by restraining the displacement w_0 and the rotation w_0' to be zero. The corresponding equations are

$$w_0 \Big|_{r=a} = 0$$

$$w_0' \Big|_{r=a} = 0$$

with the boundary condition at b completely similar in form.

Free. A free edge, which is shown symbolically in Figure 7-16, is accomplished by completely unrestraining the displacement w_0 and the rotation w_0'. This leads to setting the corresponding force resultants equal to zero, namely,

$$Q \Big|_{r=a} = -D \frac{d}{dr}\left(\frac{1}{r}\frac{d}{dr}(rw_0')\right)\Big|_{r=a} = 0$$

and

$$M_r \Big|_{r=a} = -D\left(w_0'' + \nu \frac{w_0'}{r}\right)\Big|_{r=a} = 0$$

with the form of the boundary condition at b completely similar.

Once w_0 is known, the shear force Q can be determined from Equation (7-20). It is left for the student to show in the exercises that the shear stresses can be computed in terms of the shear resultant Q on the basis of

$$\tau_{rz} = \frac{3Q}{2h}\left(1 - \left(\frac{2z}{h}\right)^2\right) \tag{7-22}$$

that is, a parabolic distribution vanishing at the surfaces $z = \pm h/2$.

7-10 Solutions to Circular-Plate Problems

The governing differential equation for the transverse displacement of a uniform-thickness circular plate can be written as

$$\frac{1}{r}\frac{d}{dr}\left(r\frac{d}{dr}\left(\frac{1}{r}\frac{d}{dr}(rw')\right)\right) - \frac{q}{D} = 0 \tag{7-23}$$

where w_0 has been replaced by w for notational simplicity. The solution of this linear differential equation can be decomposed into homogeneous and particular parts according to $w = w_h + w_p$. The student should verify that with $q = 0$, Equation (7-23) can be integrated to yield

$$w_h = C_1 + C_2 \ln\left(\frac{r}{a}\right) + C_3 r^2 + C_4 r^2 \ln\left(\frac{r}{a}\right) \tag{7-24}$$

so that the general solution to Equation (7-23) can be expressed as

$$w = C_1 + C_2 \ln\left(\frac{r}{a}\right) + C_3 r^2 + C_4 r^2 \ln\left(\frac{r}{a}\right) + w_p$$

where w_p is any solution to Equation (7-23) corresponding to the homogeneous term $q \neq 0$. We compute successively

$$w' = \frac{C_2}{r} + 2C_3 r + C_4\left(2r \ln\left(\frac{r}{a}\right) + r\right) + w_p'$$

$$\frac{M_r}{D} = C_2 \frac{(1-\nu)}{r^2} - 2C_3(1+\nu) - C_4\left(3 + \nu + 2(1+\nu)\ln\left(\frac{r}{a}\right)\right) + \frac{M_{rp}}{D}$$

$$\frac{M_\theta}{D} = C_2 \frac{(1-\nu)}{r^2} + 2C_3(1+\nu) + C_4\left(1 + 3\nu + 2(1+\nu)\ln\left(\frac{r}{a}\right)\right) + \frac{M_{\theta p}}{D}$$

$$\frac{Q}{D} = -\frac{4C_4}{r} + \frac{Q_p}{D}$$

where $M_{rp}, M_{\theta p}$ and Q_p correspond to the particular solution w_p.

Solid Plates. For a solid plate $r = 0$ is a valid value of the independent variable at which the displacements and force resultants are expected to be bounded. Thus, by inspection of w, w', M_r, M_θ, and Q, it is seen that it is necessary to take $C_2 = C_4 = 0$ with the result that

$$w = C_1 + C_3 r^2 + w_p$$
$$w' = 2C_3 r + w_p'$$
$$\frac{M_r}{D} = -2C_3(1+\nu) + \frac{M_{rp}}{D}$$
$$\frac{M_\theta}{D} = 2C_3(1+\nu) + \frac{M_{\theta p}}{D}$$
$$\frac{Q}{D} = \frac{Q_p}{D}$$

The solution for a particular loading is then generated by determining a particular solution and enforcing two appropriate boundary conditions at the edge $r = a$ to determine C_1 and C_3.

EXAMPLE 7-4 A solid plate is loaded by edge moments M_0 distributed around the periphery $r = a$, as shown in Figure 7-17. Determine the displacements and stresses.

7-10 SOLUTIONS TO CIRCULAR-PLATE PROBLEMS

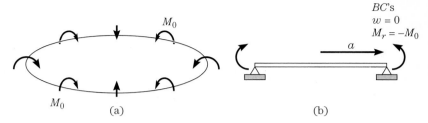

FIGURE 7-17 Edge moments on a circular plate.

Solution: With the transverse loading $q = 0$, $w_p = 0$ and the boundary conditions indicated in Figure 7-17b can be expressed in equation form as

$$w\bigg|_{r=a} = C_1 + C_3 a^2 = 0$$

and

$$M_r\bigg|_{r=a} = -2DC_3(1 + \nu) = -M_0$$

from which $C_3 = M_0/2D(1 + \nu)$ and $C_1 = -M_0 a^2/2D(1 + \nu)$ with

$$w = \frac{M_0(r^2 - a^2)}{2D(1 + \nu)}$$

$$M_r = -M_0$$
$$M_\theta = M_0$$
$$Q = 0$$

with the parabolic shape of the deformed plate as shown in Figure 7-18. The maximum displacement at the center of the plate is

$$w_{max} = w\bigg|_{r=0} = -\frac{M_0 a^2}{2D(1 + \nu)}$$

with the minus indicating a downward motion, as would be expected. With the bending moments $M_r = M_\theta$ everywhere constant in the plate, the maximum bending stress oc-

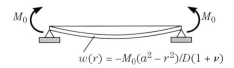

$w(r) = -M_0(a^2 - r^2)/D(1 + \nu)$

FIGURE 7-18 Diametral section showing transverse displacements.

curs at the top and bottom surfaces with

$$\sigma_r\Big|_{z=h/2} = \sigma_\theta\Big|_{z=h/2} = -\frac{6M_0}{h^2} \qquad 0 \le r \le a$$

$$\sigma_r\Big|_{z=-h/2} = \sigma_\theta\Big|_{z=-h/2} = \frac{6M_0}{h^2} \qquad 0 \le r \le a$$

◆

◆EXAMPLE 7-5 A clamped solid plate is uniformly loaded, as shown in Figure 7-19. Determine the stresses and deflections.

Solution: The particular solution corresponding to $q(r) = -q_0$ must satisfy

$$\frac{1}{r}\frac{d}{dr}\left(r\frac{d}{dr}\left(\frac{1}{r}\frac{d}{dr}\left(r\frac{dw_p}{dr}\right)\right)\right) = -\frac{q_0}{D}$$

The student should show that by successive integrations $w_p = -q_0 r^4/64D$ so that

$$w = C_1 + C_3 r^2 - \frac{q_0 r^4}{64D}$$

and

$$w' = 2C_3 r - \frac{q_0 r^3}{16D}$$

so that satisfying the boundary conditions indicated in Figure 7-19 leads to

$$w\Big|_{r=a} = C_1 + C_3 a^2 - \frac{q_0 a^4}{64D} = 0$$

and

$$w'\Big|_{r=a} = 2C_3 a - \frac{q_0 a^3}{16D} = 0$$

from which $C_1 = -q_0 a^4/64D$ and $C_3 = q_0 a^2/32D$. The transverse displacement is then

$$w(r) = -\frac{q_0}{64D}(r^2 - a^2)^2$$

with $w_{max} = w(0) = -q_0 a^4/64D$. The bending moments can be computed as

$$M_r = -\frac{q_0}{16}[(1+\nu)a^2 - (3+\nu)r^2]$$

and

$$M_\theta = \frac{q_0}{16}[(1+\nu)a^2 - (1+3\nu)r^2]$$

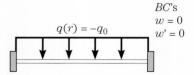

FIGURE 7-19 Uniformly loaded clamped plate.

7-10 SOLUTIONS TO CIRCULAR-PLATE PROBLEMS

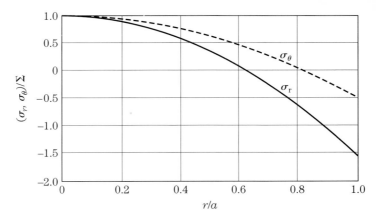

FIGURE 7-20 Bending stresses σ_r and σ_θ in a clamped plate.

The corresponding bending stresses at the lower surface are given by $\sigma_r = 6M_r/h^2$ and $\sigma_\theta = -6M_\theta/h^2$. In particular,

$$\sigma_r \bigg|_{\substack{r=0 \\ z=-h/2}} = -\frac{6M_r(0)}{h^2} = \frac{3q_0 a^2(1+\nu)}{8h^2}$$

and

$$\sigma_r \bigg|_{\substack{r=a \\ z=-h/2}} = -\frac{6M_r(a)}{h^2} = -\frac{3q_0 a^2}{4h^2}$$

With $\Sigma = 3q_0 a^2(1+\nu)/(8h^2)$, a plot of σ_r and σ_θ at $z = -h/2$ is shown in Figure 7-20. ◆

Annular Plates. Annular plates have an inside radius b and an outside radius a so that it is necessary to enforce two boundary conditions at each edge. These four boundary conditions are sufficient to determine the four constants of integration in Equation (7-24). Several examples are presented in what follows.

EXAMPLE 7-6 An annular plate is simply supported at the outer edge and loaded by a central force applied to a rigid member, as shown in Figure 7-21. There is no external loading for $b \leq r \leq a$ so that the solution for the transverse displacement can be expressed as

$$w = C_1 + C_2 \ln\left(\frac{r}{a}\right) + C_3 r^2 + C_4 r^2 \ln\left(\frac{r}{a}\right) \tag{7-25}$$

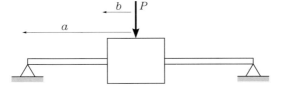

FIGURE 7-21 Centrally loaded annular plate.

where the natural log terms have been written in the form shown for convenience in solving for the constants. The outer edge is simply supported with the boundary conditions as

$$w\bigg|_{r=a} = 0$$

$$M_r\bigg|_{r=a} = -D\left(w'' + \frac{\nu}{r}w'\right)\bigg|_{r=a} = 0$$

At the inner edge the plate is assumed to be fixed against rotation by being attached to the rigid member. Equilibrium of the inner rigid member shows that the intensity of the shear resultant applied by the plate is $P/2\pi b$ so that the boundary conditions at the inner edge are

$$w'\bigg|_{r=b} = 0$$

$$Q\bigg|_{r=b} = \frac{P}{2\pi b}$$

The four equations for determining the constants can then be written as

$$C_1 + C_3 a^2 = 0$$

$$\frac{C_2(1-\nu)}{a^2} - 2C_3(1+\nu) - C_4(3+\nu) = 0$$

$$\frac{C_2}{b} + 2C_3 b + C_4\left(2b\ln\left(\frac{b}{a}\right) + b\right) = 0$$

$$-\frac{4DC_4}{b} = \frac{P}{2\pi b}$$

with solution

$$C_1 = -\frac{3+\nu+(1-\nu)\phi^2\left(1+2\ln\left(\frac{b}{a}\right)\right)}{2(1+\nu+(1-\nu)\phi^2)}\frac{Pa^2}{8\pi D} = f_1\frac{Pa^2}{8\pi D}$$

$$C_2 = -\frac{3+\nu-(1+\nu)\left(1+2\ln\left(\frac{b}{a}\right)\right)}{1+\nu+(1-\nu)\phi^2}\frac{Pb^2}{8\pi D} = f_2\frac{Pb^2}{8\pi D}$$

$$C_3 = \frac{3+\nu+(1-\nu)\phi^2\left(1+2\ln\left(\frac{b}{a}\right)\right)}{2(1+\nu+(1-\nu)\phi^2)}\frac{P}{8\pi D} = -f_1\frac{P}{8\pi D}$$

$$C_4 = -\frac{P}{8\pi D}$$

where f_1 and f_2 are defined in the obvious ways and where $\phi = b/a$. The displacement can be reconstructed by substituting for C_1, C_2, C_3 and C_4 into Equation (7-25), leading to

$$w(b) = \frac{Pa^2}{8\pi D}[f_1 + \phi^2(f_2 - f_1 + \ln(\phi))]$$

$$= \frac{Pa^2}{8\pi D}F(\nu, \phi)$$

which, after substituting for $D = Eh^3/12(1 - \nu^2)$, can be represented as

$$w(b) = k_1(\nu, \phi)\frac{Pa^2}{Eh^3} \tag{7-26}$$

Values of $k_1(\nu, \phi)$ for use in Equation (7-26) are given in Table 7-2 later in this section for $\nu = 0.3$ and for a range of a/b ratios.

The bending moments can also be constructed analytically if desired. For the specific case of $a/b = 2$ and $\nu = 0.3$, M_r and M_θ are shown as functions of position in Figure 7-22. The maximum positive bending stress occurs at $z = -h/2$ for $r = b$ and is given by

$$\sigma_{max} = \sigma_r \bigg|_{\substack{r=b \\ z=-h/2}} = \frac{6}{h^2}3.156\frac{P}{8\pi} = 0.753\frac{P}{h^2}$$

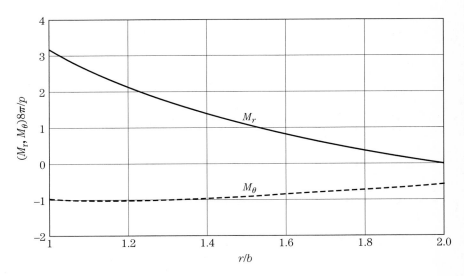

FIGURE 7-22 Bending moments throughout the plate for $a/b = 2$.

For any value of the ratio a/b, the maximum bending stress occurs at $r = b$ and can be represented as

$$\sigma_{max} = k_2(\nu, \phi)\frac{P}{h^2} \tag{7-27}$$

For $\nu = 0.3$, Table 7-2 (later in this section) can be consulted for the appropriate value of $k_2(\nu, \phi)$ given as a function of the ratio a/b to be used in Equation (7-27) for determining the maximum stress.

EXAMPLE 7-7 An annular plate is simply supported at the outer edge and loaded by a shear load P around the inner edge, as shown in Figure 7-23. Determine the displacements and stresses.

Solution: With the transverse loading $q = 0$, the differential equation and boundary conditions can be stated as

$$\nabla^4 w = 0 \qquad b \leq r \leq a$$

$$M_r(b) = 0, \qquad Q(b) = \frac{P}{2\pi b}$$

$$w(a) = 0, \qquad M_r(a) = 0$$

Taking the solution of the form

$$w = C_1 + C_2 \ln\left(\frac{r}{a}\right) + C_3 r^2 + C_4 r^2 \ln\left(\frac{r}{a}\right)$$

and satisfying the boundary conditions gives the equations

$$\frac{C_2(1-\nu)}{b^2} - 2C_3(1+\nu) - C_4\left(3 + \nu + 2(1+\nu)\ln\left(\frac{b}{a}\right)\right) = 0$$

$$-\frac{4C_4 D}{b} = \frac{P}{2\pi b}$$

$$C_1 + C_3 a^2 = 0$$

$$\frac{C_2(1-\nu)}{a^2} - 2C_3(1+\nu) - C_4(3+\nu) = 0$$

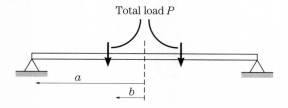

FIGURE 7-23 Edge-loaded annular plate.

7-10 SOLUTIONS TO CIRCULAR-PLATE PROBLEMS

with solution

$$C_4 = -\frac{P}{8\pi D}$$

$$C_2 = \frac{2(1+\nu)La^2b^2}{(1-\nu)(a^2-b^2)}C_4$$

$$C_3 = -\frac{C_4}{2(1+\nu)}\left[3+\nu - \frac{2(1+\nu)Lb^2}{a^2-b^2}\right]$$

$$C_1 = -C_3 a^2$$

where $L = \ln(b/a)$. The displacement w, the bending moments M_r and M_θ and the stresses σ_r and σ_θ can then be determined by substituting for the constants C_1, C_2, C_3 and C_4. The results are shown for a range of a/b ratios in Table 7-2 (later in this section).

Alternative solution: It is not always necessary or desirable to work with the fourth order equation $D\nabla^4 w - q = 0$. In the present example, for instance, summing forces on the free-body diagram of a portion of the plate between $r = b$ and an intermediate point $a \leq r \leq b$, as shown in Figure 7-24, yields the result $Q = P/(2\pi r)$. Using Equation (7-20) there results

$$-D\frac{d}{dr}\left(\frac{1}{r}\frac{d}{dr}(rw')\right) = \frac{P}{2\pi r}$$

Successive integrations yield

$$w = \frac{C_1 r^2}{4} + C_2 \ln\left(\frac{r}{a}\right) + C_3 - \frac{Pr^2}{8\pi D}\left(\ln\left(\frac{r}{a}\right) - 1\right)$$

The boundary conditions that must be applied are $M_r(b) = 0$, $M_r(a) = 0$ and $w(a) = 0$, leading to

$$\frac{1-\nu}{b^2}C_2 - \frac{1+\nu}{2}C_1 + \frac{P}{8\pi D}[1 - \nu + 2(1+\nu)L] = 0$$

$$\frac{1-\nu}{a^2}C_2 - \frac{1+\nu}{2}C_1 + \frac{P}{8\pi D}[1 - \nu] = 0$$

and

$$C_1 \frac{a^2}{4} + C_3 + \frac{Pa^2}{8\pi D} = 0$$

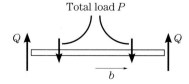

FIGURE 7-24 Free-body diagram.

with solutions

$$C_2 = -\frac{P(1+\nu)La^2b^2}{4\pi(1-\nu)D(a^2-b^2)}$$

$$C_1 = \frac{P}{4\pi(1+\nu)D(a^2-b^2)}[(1-\nu)a^2 - b^2(1-\nu+2(1+\nu)L)]$$

and

$$C_3 = -C_1\frac{a^2}{4} - \frac{Pa^2}{8\pi D}$$

This solution can be shown to coincide with the first one developed in this example. ◆

It is frequently possible to consider a particular loading and set of boundary conditions for an unknown solution to be a combination of two or more solutions for which each of the solutions is known. The process, based on the fact that the governing differential equation and boundary conditions for the circular-plate problem are linear, is commonly referred to as *superposition*. We will demonstrate the idea by means of an example.

EXAMPLE 7-8 Use superposition to develop a solution for the problem of a uniformly loaded solid plate, as shown in Figure 7-25.

Solution: It is left for the student to show in the exercises that the solution can be obtained directly by applying the two boundary conditions $w(a) = 0$ and $w''(a) + \nu w'(a)/a = 0$ to the general solution

$$w = C_1 + C_2 r^2 - \frac{q_0 r^4}{64D}$$

leading to

$$w = -\frac{q_0(a^2-r^2)}{64D}\left(\frac{5+\nu}{1+\nu}a^2 - r^2\right)$$

with

$$M_r = -\frac{q_0}{16}(3+\nu)(a^2-r^2)$$

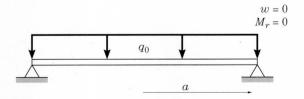

FIGURE 7-25 Uniformly loaded solid circular plate.

7-10 SOLUTIONS TO CIRCULAR-PLATE PROBLEMS

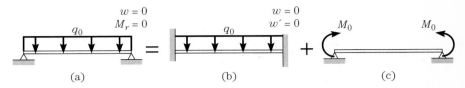

FIGURE 7-26 Superposition of known solutions.

and

$$M_\theta = \frac{q_0}{16}((3 + \nu)a^2 - (1 + 3\nu)r^2)$$

With the superposition approach, we proceed as indicated in Figure 7-26. From Example 7-5 for the clamped plate, shown as problem (b) in Figure 7-26, the edge moments are

$$M_r(a) = -\frac{q_0}{16}[(1 + \nu)a^2 - (3 + \nu)r^2]\bigg|_{r=a} = \frac{q_0 a^2}{8}$$

If we superpose the edge moments from Example 7-4 in the amount $M_0 = -q_0 a^2/8$, shown as problem (c) in Figure 7-26, the edge $r = a$ will be free of moment and the boundary conditions $w = M_r = 0$ will be satisfied at $r = a$. The corresponding equations for the three problems are indicated below.

Problem (a):
$$\nabla^4 w_1 = -\frac{q_0}{D}$$

$$w_1(a) = 0$$

$$-D\left(w_1''(a) + \frac{\nu w'(a)}{a}\right) = 0 = M_{1r}(a)$$

Problem (b):
$$\nabla^4 w_2 = -\frac{q_0}{D}$$

$$w_2(a) = 0$$

$$w_2'(a) = 0 \qquad \left[M_{2r}(a) = \frac{q_0 a^2}{8}\right]$$

Problem (c):
$$\nabla^4 w_3 = 0$$
$$w_3(a) = 0$$
$$M_{3r}(a) = -M_0 = -\frac{q_0 a^2}{8} \qquad \left[w_2'(a) = \frac{q_0 a^3}{8D(1 + \nu)}\right]$$

Superposing problems (b) and (c) yields

$$\nabla^4 w_2 + \nabla^4 w_3 = \nabla^4(w_2 + w_3) = -\frac{q_0}{D} = \nabla^4 w_1$$

$$w_2(a) + w_3(a) = w_1(a) = 0$$

$$M_{2r}(a) + M_{3r}(a) = \frac{q_0 a^2}{8} - \frac{q_0 a^2}{8} = 0 = M_{1r}(a)$$

showing that the differential equation and boundary conditions are satisfied by w_1 for the simply supported plate. From the information given it also follows that

$$w_2'(a) + w_3'(a) = w_1'(a) = \frac{q_0 a^3}{8D(1+\nu)}$$

Thus, with $M_0 = -q_0 a^2/8$, the results from Examples 7-4 and 7-5 can be used to express the desired displacement as $w_1(r) = w_2(r) + w_3(r)$

$$w_1(r) = -\frac{q_0}{64D}(r^2 - a^2)^2 + \frac{-q_0 a^2}{8}\frac{(r^2 - a^2)}{2D(1+\nu)}$$

which can be simplified to

$$w_1(r) = -\frac{q_0}{64D}(r^2 - a^2)\left(\frac{5+\nu}{1+\nu}a^2 - r^2\right)$$

In a similar manner, the moments M_r and M_θ are given by

$$M_{1r} = -\frac{q_0}{16}[(1+\nu)a^2 - (3+\nu)r^2] - \frac{q_0 a^2}{8}$$

$$= -\frac{q_0}{16}(3+\nu)(a^2 - r^2)$$

and

$$M_{1\theta} = \frac{q_0}{16}[(1+\nu)a^2 - (1+3\nu)r^2] + \frac{q_0 a^2}{8}$$

$$= \frac{q_0}{16}((3+\nu)a^2 - (1+3\nu)r^2)$$

all of which results coincide with those obtained by the direct approach above. ◆

The remainder of this section displays the results for several load cases for annular plates as shown in Figure 7-27. In all instances the results are of the form

$$w_{max} = k_1 \frac{Pa^2}{Eh^3} \quad \text{and} \quad \sigma_{max} = k_2 \frac{P}{h^2}$$

in cases where a single concentrated load P is applied or

$$w_{max} = k_1 \frac{qa^4}{Eh^3} \quad \text{and} \quad \sigma_{max} = k_2 \frac{qa^2}{h^2}$$

7-11 SUMMARY

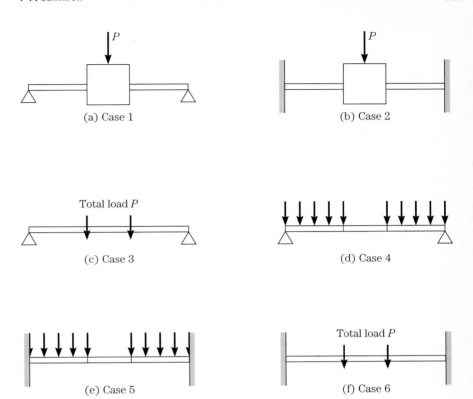

FIGURE 7-27 Load cases for annular plates.

TABLE 7-2 Coefficients for Maximum Transverse Displacement and Maximum Bending Stress*

a/b =	1.25		1.5		2.0		3.0		4.0	
Case	k_1	k_2	k_1	k_2	k_1	k_2	k_1	k_2	k_1	k_2
1	0.00510	0.227	0.0249	0.428	0.0877	0.753	0.209	1.205	0.293	1.514
2	0.00129	0.115	0.0064	0.220	0.0237	0.405	0.062	0.703	0.092	0.933
3	0.341	1.10	0.519	1.26	0.672	1.48	0.734	1.88	0.724	2.17
4	0.184	0.592	0.414	0.976	0.664	1.44	0.824	1.88	0.830	2.08
5	0.00199	0.105	0.0139	0.259	0.0575	0.48	0.130	0.657	0.162	0.710
6	0.00504	0.194	0.0242	0.320	0.0810	0.454	0.172	0.673	0.217	1.021

*For values of a/b not represented in the table, interpolation can be used to estimate the values of k_1 and k_2 for the load case in question.

in cases where a distributed loading q is acting. The numerical values for the constants k_1 and k_2 are given in Table 7-2 for a range of ratios b/a.

7-11 Summary

The displacements and stresses for a thin, flat plate undergoing transverse displacements that are small compared to the thickness of the plate are governed by the equation of equilibrium

$$D\nabla^4 w = q$$

where for a rectangular plate

$$\nabla^4 w = \left(\frac{\partial^2}{\partial x^2} + \frac{\partial^2}{\partial y^2}\right)\left(\frac{\partial^2 w}{\partial x^2} + \frac{\partial^2 w}{\partial y^2}\right)$$

and where for a circular plate

$$\nabla^4 w = \left(\frac{\partial^2}{\partial r^2} + \frac{1}{r}\frac{\partial}{\partial r}\right)\left(\frac{\partial^2 w}{\partial r^2} + \frac{1}{r}\frac{\partial w}{\partial r}\right)$$

After solving for w and enforcing the boundary conditions, the bending stresses can be calculated by using

$$\sigma_x = \frac{6M_x}{h^2} \quad \text{and} \quad \sigma_y = \frac{6M_y}{h^2}$$

with

$$M_x = D\left(\frac{\partial^2 w}{\partial x^2} + \nu \frac{\partial^2 w}{\partial y^2}\right) \quad \text{and} \quad M_y = D\left(\frac{\partial^2 w}{\partial y^2} + \nu \frac{\partial^2 w}{\partial x^2}\right)$$

for the rectangular plate and

$$\sigma_r = \frac{6M_r}{h^2} \quad \text{and} \quad \sigma_\theta = -\frac{6M_\theta}{h^2}$$

with

$$M_r = -D\left(w'' + \nu\frac{w'}{r}\right) \quad \text{and} \quad M_\theta = D\left(\frac{w'}{r} + \nu w''\right)$$

for the circular plate.

REFERENCES

1. Timoshenko, S. P., and Woinowsky-Krieger, S. *Theory of Plates and Shells*, 2nd ed., McGraw-Hill, New York, 1959.
2. Langhaar, H. L. *Energy Methods in Applied Mechanics*, Wiley, New York, 1962.
3. Szilard, R. *Theory and Analysis of Plates*, Prentice-Hall, Englewood Cliffs, N.J., 1974.
4. Ugural, A. C. *Stresses in Plates and Shells*, McGraw-Hill, New York, 1981.
5. Roark, R. J., and Young, W. C. *Formulas for Stress and Strain*, 5th ed., McGraw-Hill, New York, 1975.

EXERCISES

Part I Rectangular Plates

1. Verify that substitution of the moment expressions given by Equations (7-7) into the equilibrium Equation (7-3) gives the displacement equation of equilbrium, namely, $D\nabla^4 w = q$.

2. A rectangular plate simply supported on all edges is loaded by a pile of sand that is modeled as applying a pressure loading $q = q_0 \sin(\pi x/a)\sin(\pi y/b)$, as shown in the figure below. Take the displacement to be of the form $w = w_0 \sin(\pi x/a)\sin(\pi y/b)$ and hence determine w_0. Also determine the bending stresses at the center of the plate.

EXERCISES

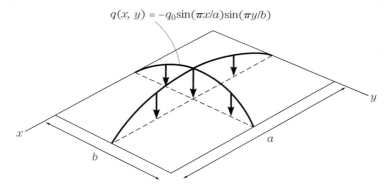

$q(x, y) = -q_0 \sin(\pi x/a)\sin(\pi y/b)$

3. A steel plate 10 ft on a side and 0.625 inches thick is loaded by a pile of sand that has the approximate shape shown in the figure accompanying problem 2. If the bending stress is not to exceed 16 ksi and the maximum displacement is limited to 0.625 in, determine the maximum height of the sand at the middle of the plate. Assume that sand has a weight of 80 lbf/ft^3.

4. For the rectangular plate described in problem 2, determine the shear resultants $V_x = -D[\partial^3 w/\partial x^3 + (2-\nu)\partial^3 w/\partial x\,\partial y^2]$ along the two $x = $ const edges and $V_y = -D[\partial^3 w/\partial y^3 + (2-\nu)\partial^3 w/\partial x^2\,\partial y]$ along the two $y = $ const edges. By integrating these along the respective edges, show that the resultant is not equal to the resultant of the distributed loading $q(x, y)$. Show also that if forces of magnitude $R = 2M_{xy}$ are applied at the corners of the plate, equilibrium is satisfied.

5. A uniformly loaded rectangular plate is simply supported on two opposite edges and clamped on the other two edges. Construct a Levy solution and determine the maximum displacement and maximum bending stress for a square plate.

6. A uniformly loaded rectangular plate is simply supported on two opposite edges, clamped on one of the other two edges and simply supported on the other. Construct a Levy solution and determine the maximum displacement and maximum bending stress for a square plate.

7. A uniformly loaded rectangular plate is simply supported on two opposite edges, free on one of the other two edges and simply supported on the other. Construct a Levy solution and determine the maximum displacement and maximum bending stress for a square plate.

8. A uniformly loaded rectangular plate is simply supported on two opposite edges, clamped on one of the other two edges and free on the other. Construct a Levy solution and determine the maximum displacement and maximum bending stress for a square plate.

9. An aluminum plate with all sides simply supported and with $a = 2.6$ m and $b = 1.8$ m is subjected to a uniform load of 10 kN/m^2. If the allowable bending stress is 70 MPa, determine the required thickness of the plate.

10. Integrate the first equation of equilibrium, namely,

$$\frac{\partial \sigma_x}{\partial x} + \frac{\partial \tau_{xy}}{\partial y} + \frac{\partial \tau_{xz}}{\partial z} = 0$$

with respect to z, require τ_{xz} to be zero at $z = \pm h/2$ and then use Equations (7-2) to show that

$$\tau_{xz} = \frac{3Q_x}{2h}\left(1 - \left(\frac{2z}{h}\right)^2\right)$$

11. Integrate the second equation of equilibrium, namely,

$$\frac{\partial \tau_{xy}}{\partial x} + \frac{\partial \sigma_y}{\partial y} + \frac{\partial \tau_{yz}}{\partial z} = 0$$

with respect to z, require τ_{yz} to be zero at $z = \pm h/2$ and then use Equations (7-2) to show that

$$\tau_{yz} = \frac{3Q_y}{2h}\left(1 - \left(\frac{2z}{h}\right)^2\right)$$

Part II Circular Plates

12. Carry out the details of showing that substitution of Equations (7-17a) and (7-17b) into the equilibrium equation $(rM_r)'' + M_\theta' + rq(r) = 0$ leads to the equilibrium equation $D\nabla^4 w_0 - q = 0$.

13. One of the equilibrium equations for cylindrical coordinates is

$$\frac{\partial \sigma_r}{\partial r} + \frac{\sigma_r - \sigma_\theta}{r} + \frac{\partial \tau_{rz}}{\partial z} = 0$$

Combine this with Equations (7-19), integrate with respect to z and satisfy the boundary conditions $\tau_{rz}(z = -h/2) = \tau_{rz}(z = +h/2) = 0$ to obtain Equation 7-22.

14. Verify that successive integration of Equation (7-23) yields the homogeneous solution given by Equations (7-24).

15. Verify that successive integration of $\nabla^4 w_p - q_0/D = 0$ yields the particular solution $w_p = q_0 r^4/64D$.

16. Use the results from Example 7-4 to determine the maximum edge moment M_0 that can be applied in order that the transverse displacement not exceed the thickness of the plate. What is the corresponding maximum bending stress?

17. Use the results from Example 7-5 to determine the maximum intensity q_0 that can be applied to a clamped circular plate in order that the transverse displacement not exceed the thickness of the plate. What is the corresponding maximum bending stress?

18. Determine the maximum allowable transverse load q_0 on a 0.1 in thick, 5 in diameter clamped aluminum plate in order that the maximum transverse displacement not exceed the thickness of the plate. What is the corresponding maximum bending stress? Take $E = 10^7$ psi and $\nu = 0.3$.

19. Determine the maximum allowable transverse load q_0 on a 3 mm thick, 0.13 m diameter clamped aluminum plate in order that the maximum bending stress not exceed 100 MPa. What is the corresponding maximum transverse displacement?

For each of the following problems, set up and solve the differential equation and boundary conditions and hence determine $w(r), M_r(r)$ and $M_\theta(r)$.

20.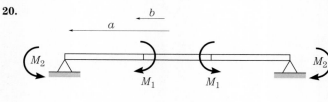

21.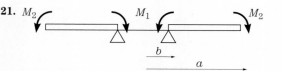

22. 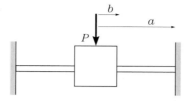 **23.**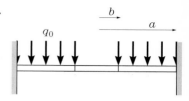

24. 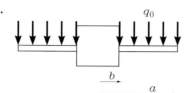 **25.**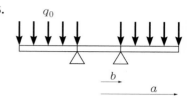

26. State specifically how the results of Example 7-6 and problem 20 should be combined to produce the results of Example 7-7.

27. Carry out the details associated with problem 26.

28. State specifically how the results of problems 21 and 24 should be combined to produce the results of problem 25.

29. Carry out the details associated with problem 28.

30. A clamped, solid plate is composed of an aluminum whose yield stress, as determined by an axial tension test, is 18 ksi. Take $h = 0.1$ in and $a = 5$ in, and determine the maximum load intensity q_0 based on the maximum shear-stress theory. Determine the corresponding maximum transverse displacement. Take $E = 10^7$ psi and $\nu = 0.3$.

31. A simply supported solid plate is composed of an aluminum whose yield stress, as determined by an axial tension test, is 15 ksi. Take $h = 0.1$ in and $a = 5$ in, and determine the maximum load intensity q_0 based on the maximum shear-stress theory. Determine the corresponding maximum transverse displacement. Take $E = 10^7$ psi and $\nu = 0.3$.

32. A steel plate is loaded and supported as in Figure 7-27a. Take $a = 24$ in, $b = 12$ in, $h = 0.375$ in and $E = 3 \times 10^7$ psi, and determine the maximum allowable value of P if the yield stress, as determined by an axial tension test, is 24 ksi. Use the maximum shear-stress theory. What is the corresponding maximum transverse displacement? Use $\nu = 0.3$.

33. A steel plate is loaded and supported as in Figure 7-27a. Take $a = 0.6$ m, $b = 0.3$ m, $h = 0.01$ m, and $E = 200$ GPa, and determine the maximum allowable value of P if the yield stress, as determined by an axial tension test, is 170 MPa. Use the maximum shear-stress theory. What is the corresponding maximum transverse displacement? Use $\nu = 0.3$.

34. For a uniformly loaded, solid, simply supported steel plate, take $a = 0.1$ m, $h = 0.002$ m, and $E = 200$ GPa, and determine the maximum allowable value of q_0 if the yield stress, as determined by an axial tension test, is 150 MPa. Use the maximum shear-stress theory. What is the corresponding maximum transverse displacement? Use $\nu = 0.3$.

35. For a uniformly loaded, solid, clamped steel plate, take $a = 6$ in, $h = 0.1$ in and $E = 30 \times 10^6$ psi, and determine the maximum allowable value of q_0 if the yield stress, as determined by an axial tension test, is 20 ksi. Use the maximum shear-stress theory. What is the corresponding maximum transverse displacement? Use $\nu = 0.3$.

Variational Formulation Exercises

36. Show that by integrating the strain energy density $U_0 = (E/(2(1-\nu^2)))(\epsilon_x^2 + \epsilon_y^2 + 2\nu\epsilon_x\epsilon_y + ((1-\nu)/2)\gamma_{xy}^2)$ over the volume, the internal potential energy for a rectangular plate can be expressed as

$$U = \frac{D}{2}\int_0^b\int_0^a\left[(\nabla^2 w)^2 - 2(1-\nu)\left\{\frac{\partial^2 w}{\partial x^2}\frac{\partial^2 w}{\partial y^2} - \left(\frac{\partial^2 w}{\partial x \partial y}\right)^2\right\}\right]dx\,dy$$

with the external potential energy as

$$\Omega = -\int_0^b\int_0^a qw\,dx\,dy$$

37. For a uniformly loaded rectangular plate simply supported on all edges, take $w = w_0 x(a-x)y(b-y)$ and use the Ritz method to determine w_0. For a square plate, compare the maximum displacement and the maximum bending moment to the corresponding exact values. Discuss the deficiencies of the assumed solution.

38. For a rectangular plate simply supported on all edges and loaded by a force P at the center, take $w = w_0 x(a-x)y(b-y)$ and use the Ritz method to determine w_0. For a square plate, compare the maximum displacement to the value $w = 0.0116 P a^2/D$.

39. For a uniformly loaded rectangular plate clamped on all edges, take $w = w_0 x^2(a-x)^2 y^2(b-y)^2$ and use the Ritz method to determine w_0. For a square plate, compare the maximum displacement to the value $w = 0.00126 q_0 a^4/D$ and the maximum bending moment at the midpoint of a side to $M = 0.0517 q_0 a^2$.

40. Show that by integrating the strain energy density $U_0 = (E/(2(1-\nu^2)))(\epsilon_r^2 + \epsilon_\theta^2 + 2\nu\epsilon_r\epsilon_\theta)$ over the volume, the internal potential energy for a circular plate can be expressed as

$$U = 2\pi\int_b^a \frac{D}{2}\left\{\left(w'' + \frac{1}{r}w'\right)^2 - 2(1-\nu)\frac{w'w''}{r}\right\}r\,dr$$

with the external potential energy as

$$\Omega = -2\pi\int_b^a qw\,r\,dr$$

41. For the problem of a uniformly loaded simply supported plate, take $w = A(a^2 - r^2)$, which satisfies the essential boundary condition. Use the Ritz method to determine A. Compare the displacement and the bending moments to the corresponding exact values. Discuss the deficiencies of the assumed solution.

42. For the problem of a uniformly loaded simply supported plate, take $w = A + Br^2 + Cr^4$ and enforce the essential boundary condition $w(a) = 0$ to eliminate one of the constants. Then use the Ritz method to determine the remaining constants. Compare the displacement and the bending moments to the corresponding exact values.

43. For the problem shown below, take $w = A(a^2 - r^2)^2$, which satisfies the essential boundary conditions, and use the Ritz method to determine A. Let $b = 0.1a$ and $q_0\pi b^2 = P$ and compare your displacement at $r = 0$ to the exact value of $Pa^2/16\pi D$.

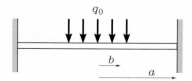

44. For the problem shown below, take $w = A(r^2 - b^2)^2$, which satisfies the essential boundary conditions, and use the Ritz method to determine A. Take $b/a = 0.5$ and compare your displacement at $r = a$ with the corresponding exact value given in Table 7-2.

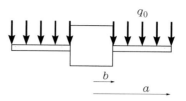

8 Membrane Shells of Revolution

8-1 Introduction
8-2 Geometry of Shells of Revolution
8-3 Governing Equations for Membrane Theory
    8-3-1 Equilibrium
    8-3-2 Deformations
    8-3-3 Material Behavior and Combination
8-4 Axisymmetrically Loaded Membrane Shells
    8-4-1 Cylindrical Shells
    8-4-2 Spherical Shells
    8-4-3 Conical Shells
8-5 Asymmetrically Loaded Membrane Shells
    8-5-1 Cylindrical Shells
    8-5-2 Conical Shells
    8-5-3 Spherical Shells
8-6 Summary
    References
    Exercises

8-1 Introduction

This chapter will be devoted to developing a theory for thin axisymmetric shells subjected to loadings that result primarily in membrane forces, that is, internal force resultants that are everywhere parallel to the tangent plane of the shell. Axisymmetric means that the shape of the shell can be considered to have been generated by revolving a continuous curve or generator around the z axis as shown in Figure 8-1a. Such shells are also referred to as *shells of revolution*. Thin refers to the fact that the thickness t is small compared to the minimum principal radius of curvature ρ, as indicated in Figure 8-1b.

In order to discuss the meaning of membrane forces, consider a segment of a shell of revolution, as shown in Figure 8-2. In a situation involving general loading of a shell, both bending resultants and shear force resultants, pictured in Figure 8-2a, and membrane forces, shown in Figure 8-2b, act simultaneously. For the theory presented in this chapter, it will be assumed that only the membrane resultants act throughout the shell.

As an example, consider a cylindrical shell acted upon by a uniform external pressure, as shown in Figure 8-3. As discussed in Chapter 6, the uniform pressure produces

8-1 INTRODUCTION

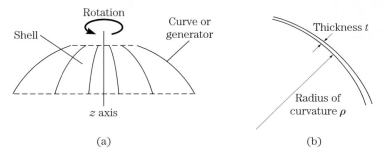

FIGURE 8-1 Thin, axisymmetric shell.

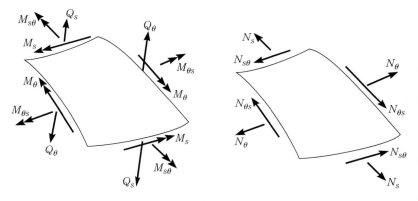

FIGURE 8-2 Bending, shear and force resultants.

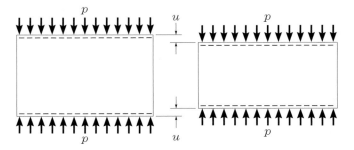

FIGURE 8-3 Uniform external pressure on a cylindrical shell.

a uniform radial displacement u and the uniform hoop stress $\sigma_\theta = pR/t$ or the hoop force resultant $N_\theta = t\sigma_\theta = pR$. No bending or shear force resultants are present. This type of loading, producing only membrane forces, is to be contrasted with the situations shown in Figure 8-4, where two types of loading on the cylindrical shell result in significant bending. The deformations shown in Figure 8-4 are greatly exaggerated for clarity. Figure 8-4a shows a situation where a line load, a load applied radially around the cir-

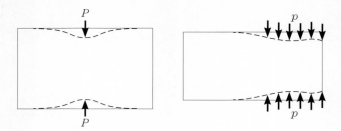

FIGURE 8-4 Loadings resulting in bending and shear force resultants.

cumference of the shell, results in bending and shear force resultants. Figure 8-4b indicates a situation where a uniform load applied over a portion of the shell results in bending type deformations and in bending and shear force resultants. In both instances it would be necessary to include these effects in order to accurately predict the stresses and displacements in the shell.

Another example of a situation where the internal forces necessary to sustain the loads are primarily membrane forces is a thin, spherical shell under an internal pressure. As indicated in Figure 8-5, this internal pressure is resisted by uniform membrane resultants given by $N = pR/2$. An inflated balloon is often approximately spherical where the only internal forces acting are the tensions in the balloon. Another example of a shell that is very effective acting as a membrane is that of an egg. It is known that an egg is quite strong with respect to loadings that produce membrane forces, such as a reasonably uniform external pressure, but significantly less resistant to concentrated loadings which produce local bending.

Whether or not a membrane theory is appropriate also depends on the type of supports or boundary conditions. As an example, consider the problem of a spherical shell segment acting under its own weight, as indicated in Figure 8-6a. If, as shown in Figure 8-6b, the shell is supported by a roller so that the forces required for equilibrium can be supplied entirely by membrane resultants, that is, reactions acting in the direction of the tangent to the shell, a membrane theory can give accurate predictions as to the displacements and stresses. If, on the other hand, the shell is clamped at the sup-

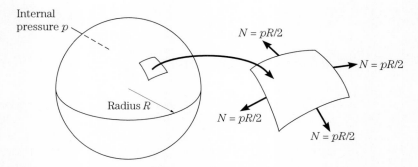

FIGURE 8-5 Thin, spherical shell under uniform internal pressure.

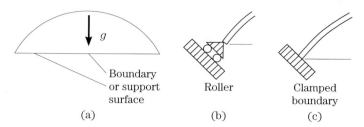

FIGURE 8-6 Spherical shell segment acting under its own weight subjected to different supports.

port, as shown in Figure 8-6c, significant bending and shear forces can result in the neighborhood of the support, negating the use of the membrane theory. In the remainder of this chapter, the equations for the membrane theory will be developed, and several typical applications will be indicated.

8-2 Geometry of Shells of Revolution

The middle surface of a shell of revolution is formed by rotating a continuous curve or generator $r = r(z)$ about the z axis, as indicated in Figure 8-7. Such a curve is also referred to as a *meridional line*. The intersection of any plane containing the z axis with the middle surface of the shell is also a generator. Relative to the middle surface, the thickness of the shell is distributed equally in both directions, as shown in Figure 8-8. At any point on the middle surface of the shell, we erect an orthogonal system of coordinates n, s and θ, as shown in Figure 8-9. The s coordinate line is tangent to a meridional line, the n coordinate is normal to the shell and the θ coordinate line is tangent to the circle that is formed by passing a plane parallel to the xy plane through

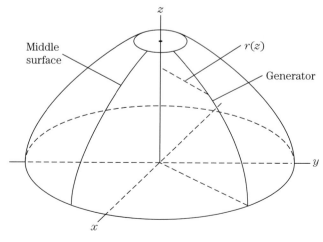

FIGURE 8-7 Generation of a shell of revolution.

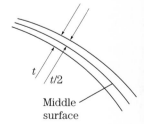

FIGURE 8-8 Middle surface and thickness of the shell.

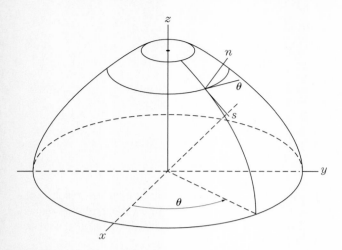

FIGURE 8-9 Shell coordinates.

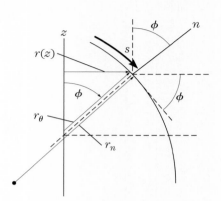

FIGURE 8-10 Shell surface geometry variables.

the point in question. Note that if the shell is spherical, the (n, s, θ) system of coordinates essentially reduces to the usual spherical coordinates (ρ, φ, θ) with $n \approx \rho$, $\varphi \approx s$ and $\theta \approx \theta$. The similarities and differences will emerge in the developments that follow.

Several other variables that are necessary in the development of the shell equations are shown in Figure 8-10. The angle ϕ is the angle between the vertical and the normal to the shell surface, that is, the usual conical angle for spherical coordinates. The distance r_θ, measured along the n direction from the midsurface of the shell to the axis, is one of the principal curvatures of the shell surface. As shown in Figure 8-10, it is related to the radius $r(z)$ according to $r(z) = r_\theta \sin \phi$, or

$$r_\theta = \frac{r(z)}{\sin \phi}$$

It is left for the student to show in the exercises that r_n, the other principal curvature of the surface, also measured along the n direction, is given by

$$\frac{1}{r_n} = \frac{d\phi}{ds} = -\frac{r''}{(1 + r'^2)^{3/2}}$$

where $r' = dr/dz$ and $r'' = d^2r/dz^2$. The meaning of r_n is best seen by writing $ds = r_n \, d\phi$, showing that r_n is the local curvature of the shell surface in the meridional plane.

It is also left for the student to show in the exercises that the derivatives of the unit vectors $\mathbf{e_n}$, $\mathbf{e_s}$ and $\mathbf{e_\theta}$, shown in Figure 8-11, are given by

$$\frac{\partial \mathbf{e_n}}{\partial n} = \mathbf{0}, \qquad \frac{\partial \mathbf{e_\theta}}{\partial n} = \mathbf{0}, \qquad \frac{\partial \mathbf{e_s}}{\partial n} = \mathbf{0},$$

$$\frac{\partial \mathbf{e_n}}{\partial \theta} = \mathbf{e_\theta} \sin \phi, \qquad \frac{\partial \mathbf{e_\theta}}{\partial \theta} = -\mathbf{e_s} \cos \phi - \mathbf{e_n} \sin \phi, \qquad \frac{\partial \mathbf{e_s}}{\partial \theta} = \mathbf{e_\theta} \cos \phi, \qquad (8\text{-}1)$$

$$\frac{\partial \mathbf{e_n}}{\partial s} = \frac{\mathbf{e_s}}{r_n}, \qquad \frac{\partial \mathbf{e_\theta}}{\partial s} = \mathbf{0}, \qquad \frac{\partial \mathbf{e_s}}{\partial s} = -\frac{\mathbf{e_n}}{r_n}$$

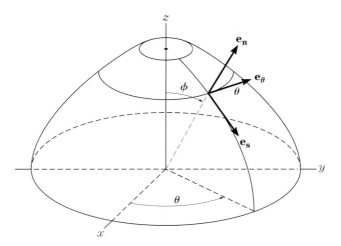

FIGURE 8-11 Unit vectors for shell surface.

These relations will be needed in the development of the equilibrium equations and the strain-displacement relations in the next sections.

8-3 Governing Equations for Membrane Theory

The governing equations for membrane theory are developed in the usual manner, that is, by careful consideration of the loading and supports, equilibrium, deformation and material behavior. Depending on the particulars of the problem to be solved and on the information desired, these equations can be combined and solved for the displacements and internal forces. Once the internal forces are known, the stresses can be determined.

8-3-1 Equilibrium The types of loadings that will be allowed are shown in Figure 8-12a. All of $p_n(\theta, s), p_s(\theta, s)$ and $p_\theta(\theta, s)$ are assumed to be continuous functions of θ and s. According to the membrane theory and as shown in Figure 8-12b, the internal forces that result are the circumferential or hoop force N_θ acting in the θ direction, the meridional force N_s acting in the s direction and the shear resultants $N_{s\theta}$ and $N_{\theta s}$. In the

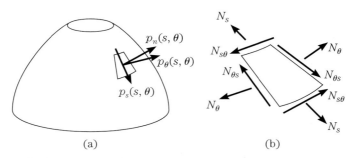

FIGURE 8-12 Surface loadings on a shell of revolution.

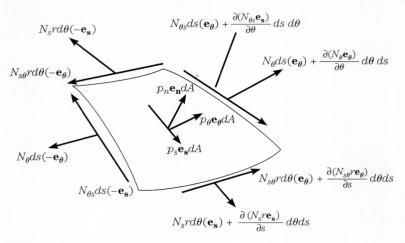

FIGURE 8-13 Membrane force resultants.

membrane theory, it is assumed that equilibrium can be satisfied with only these resultants acting.

For developing the equations of equilibrium, the membrane force resultants and their changes acting on a differential element for a general loading of the sort indicated in Figure 8-12 are shown in Figure 8-13. Requiring equilibrium, using $dA = r\, d\theta\, ds$ and canceling like terms results in

$$\frac{\partial (N_s r \mathbf{e_s})}{\partial s} + \frac{\partial (N_{s\theta} r \mathbf{e_\theta})}{\partial s} + \frac{\partial (N_\theta \mathbf{e_\theta})}{\partial \theta} + \frac{\partial (N_{\theta s} \mathbf{e_s})}{\partial \theta} + r p_s \mathbf{e_s} + r p_\theta \mathbf{e_\theta} + r p_n \mathbf{e_n} = 0$$

Expanding the first four terms and using the relations stated in the previous section for the derivatives of the unit vectors yields

$$\frac{\partial N_s}{\partial s} r \mathbf{e_s} + N_s \frac{\partial r}{\partial s} \mathbf{e_s} + N_s r \left(-\frac{\mathbf{e_n}}{r_n}\right) + \frac{\partial N_{s\theta}}{\partial s} r \mathbf{e_\theta} + N_{s\theta} \frac{\partial r}{\partial s} \mathbf{e_\theta} + N_s r(0)$$
$$+ \frac{\partial N_\theta}{\partial \theta} \mathbf{e_\theta} + N_\theta (-\mathbf{e_s} \cos \phi - \mathbf{e_n} \sin \phi) + \frac{\partial N_{\theta s}}{\partial \theta} \mathbf{e_s} + N_{\theta s}(\mathbf{e_\theta} \cos \phi)$$
$$+ r p_s \mathbf{e_s} + r p_\theta \mathbf{e_\theta} + r p_n \mathbf{e_n}$$
$$= 0$$

Using the fact that $dr/ds = \cos \phi$ and collecting coefficients results in

$$\mathbf{e_n}\left(-N_\theta \sin \phi - N_s \frac{r}{r_n} + r p_n\right)$$
$$+ \mathbf{e_\theta}\left(\frac{\partial N_\theta}{\partial \theta} + N_{s\theta} \cos \phi + r \frac{\partial N_{s\theta}}{\partial s} + N_{\theta s} \cos \phi + r p_\theta\right)$$
$$+ \mathbf{e_s}\left(r \frac{\partial N_s}{\partial s} + N_s \cos \phi + \frac{\partial N_{\theta s}}{\partial \theta} - N_\theta \cos \phi + r p_s\right) = 0$$

8-3 GOVERNING EQUATIONS FOR MEMBRANE THEORY

Thus, the equilibrium equations for N_s, N_θ, $N_{s\theta}$ and $N_{\theta s}$ can be expressed as

$$\sum F_n = -N_\theta \sin\phi - N_s \frac{r}{r_n} + r p_n = 0$$

$$\sum F_\theta = \frac{\partial N_\theta}{\partial \theta} + N_{s\theta} \cos\phi + r\frac{\partial N_{s\theta}}{\partial s} + N_{\theta s}\cos\phi + r p_\theta = 0$$

$$\sum F_s = r\frac{\partial N_s}{\partial s} + (N_s - N_\theta)\cos\phi + \frac{\partial N_{\theta s}}{\partial \theta} + r p_s = 0$$

where s and θ are considered the independent variables on the surface of the shell. It is left to the student to show that a moment equation about the n direction leads to the conclusion that $N_{s\theta} = N_{\theta s}$, so that the equations of equilibrium can be written as

$$-N_\theta \sin\phi - N_s \frac{r}{r_n} + r p_n = 0$$

$$\frac{\partial N_\theta}{\partial \theta} + N_{s\theta} \cos\phi + r\frac{\partial N_{s\theta}}{\partial s} + N_{s\theta}\cos\phi + r p_\theta = 0 \qquad (8\text{-}2)$$

$$r\frac{\partial N_s}{\partial s} + (N_s - N_\theta)\cos\phi + \frac{\partial N_{s\theta}}{\partial \theta} + r p_s = 0$$

Recognizing that $ds = r_n\, d\phi$ and that $r/\sin\phi = r_\theta$, these can also be expressed, respectively, as

$$\frac{N_\theta}{r_\theta} + \frac{N_s}{r_n} = p_n$$

$$\frac{1}{r}\frac{\partial N_\theta}{\partial \theta} + \frac{N_{s\theta}\cos\phi}{r} + \frac{1}{r_n}\frac{\partial N_{s\theta}}{\partial \phi} + \frac{N_{\theta s}\cos\phi}{r} + p_\theta = 0 \qquad (8\text{-}3)$$

$$\frac{1}{r_n}\frac{\partial N_s}{\partial \phi} + \frac{(N_s - N_\theta)\cos\phi}{r} + \frac{1}{r}\frac{\partial N_{\theta s}}{\partial \theta} + p_s = 0$$

where ϕ and θ are now taken as the independent shell variables.

Irrespective of the choice of independent variables, note that it is possible in principle to eliminate N_θ by solving the first of Equations (8-2) or (8-3) for N_θ and to then substitute into the second and third equations to obtain two equations in N_s and $N_{s\theta}$, that is, a statically determinate system if N_s is known at one of the boundaries. This approach will be demonstrated several times in the examples that follow.

8-3-2 Deformations

As shown in Figure 8-14, there are three components of displacement for the shell, namely, $w(s,\theta)$ in the normal or $\mathbf{e_n}$ direction, $u(s,\theta)$ in the meridional or $\mathbf{e_s}$ direction and $v(s,\theta)$ in the circumferential or $\mathbf{e_\theta}$ direction. Note that taking the displacements to be functions of only s and θ assumes that the displacements are constants over the thickness of the shell.

As was demonstrated in Chapter 1, the strain-displacement relations can be derived by using the definition

$$\epsilon_n(P) = \lim_{d\mathbf{r}\to 0} \frac{d\mathbf{r}^* \cdot d\mathbf{r}^* - d\mathbf{r}\cdot d\mathbf{r}}{2\, d\mathbf{r}\cdot d\mathbf{r}}$$

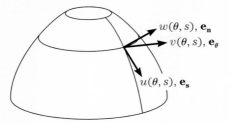

FIGURE 8-14 Shell displacements.

where

$$\mathbf{dr} = \mathbf{e_n}\, dn + \mathbf{e_\theta}\, r\, d\theta + \mathbf{e_s}\, ds$$
$$\mathbf{du} = d(w\mathbf{e_n} + u\mathbf{e_s} + v\mathbf{e_\theta})$$

and $\mathbf{dr^*} = \mathbf{dr} + \mathbf{du}$. Using the relations for the derivatives of the unit vectors, it follows that

$$\mathbf{du} = \mathbf{e_s}\left[\left(\frac{\partial u}{\partial \theta} - v\cos\phi\right)d\theta + \left(\frac{\partial u}{\partial s} + \frac{w}{r_n}\right)ds\right]$$
$$+ \mathbf{e_\theta}\left[\left(\frac{\partial v}{\partial \theta} + w\sin\phi + u\cos\phi\right)d\theta + \frac{\partial v}{\partial s}ds\right]$$
$$+ \mathbf{e_n}\left[\left(\frac{\partial w}{\partial \theta} - v\sin\phi\right)d\theta + \left(\frac{\partial w}{\partial s} - \frac{u}{r_n}\right)ds\right]$$

It is left for the student to carry out the details showing that the linear part of the strain at P is given by

$$\epsilon_n(P) = n_2 n_2 \frac{1}{r}\left(\frac{\partial v}{\partial \theta} w\sin\phi + u\cos\phi\right) + n_3 n_3\left(\frac{\partial u}{\partial s} + \frac{w}{r_n}\right)$$
$$+ n_1 n_2 \frac{1}{r}\left(\frac{\partial w}{\partial \theta} - v\sin\phi\right) + n_1 n_3\left(\frac{\partial w}{\partial s} - \frac{u}{r_n}\right)$$
$$+ n_2 n_3\left(\frac{\partial v}{\partial s} + \frac{1}{r}\left(\frac{\partial u}{\partial \theta} - v\cos\phi\right)\right)$$
$$= n_2^2 \epsilon_\theta + n_3^2 \epsilon_s + n_2 n_3 \gamma_{r\theta} + n_1 n_2 \gamma_{n\theta} + n_1 n_3 \gamma_{ns}$$

where $n_1 = dn/|\mathbf{dr}|$, $n_2 = r\, d\theta/|\mathbf{dr}|$ and $n_3 = ds/|\mathbf{dr}|$ are the direction cosines of the element $\mathbf{dr}$. The extensional and shear strains associated with membrane shell theory are then given by

$$\epsilon_\theta = \frac{1}{r}\left(\frac{\partial v}{\partial \theta} + w\sin\phi + u\cos\phi\right)$$
$$\epsilon_s = \frac{\partial u}{\partial s} + \frac{w}{r_n} \qquad (8\text{-}4)$$
$$\gamma_{s\theta} = \frac{\partial v}{\partial s} + \frac{1}{r}\left(\frac{\partial u}{\partial \theta} - v\cos\phi\right)$$

The last two shear strains, namely,

$$\gamma_{n\theta} = \frac{1}{r}\left(\frac{\partial w}{\partial \theta} - v \sin \phi\right)$$

and

$$\gamma_{ns} = \left(\frac{\partial w}{\partial s} - \frac{u}{r_n}\right)$$

correspond to force resultants associated with the bending of shells and will not be used for the membrane theory.

8-3-3 Material Behavior and Combination

The unknowns appearing in the three equations of equilibrium are the kinetic variables or membrane force resultants N_s, N_θ and $N_{s\theta}$. The development of the three strain-displacement relations gave rise to the kinematical variables $u, v, w, \epsilon_s, \epsilon_\theta$ and $\gamma_{\theta s}$, for a total of nine unknown dependent variables. The additional equations as well as the means for relating the kinetic and kinematic variables come from the stress-strain relations. Assume an isotropic material with

$$\epsilon_\theta = \frac{\sigma_\theta - \nu \sigma_s}{E}$$

$$\epsilon_s = \frac{\sigma_s - \nu \sigma_\theta}{E} \tag{8-5}$$

and

$$\gamma_{\theta s} = \frac{\tau_{\theta s}}{G} = \frac{\tau_{s\theta}}{G}$$

where E is Young's modulus, ν is Poisson's ratio and $G = E/2(1 + \nu)$ is the shear modulus. Solving for the stresses from Equation (8-5) yields

$$\sigma_\theta = \frac{E}{1 - \nu^2}(\epsilon_\theta + \nu \epsilon_s)$$

$$\sigma_s = \frac{E}{1 - \nu^2}(\epsilon_s + \nu \epsilon_\theta) \tag{8-6}$$

$$\tau_{\theta s} = G\gamma_{\theta s} = \tau_{s\theta}$$

The membrane force resultants are given by

$$N_\theta = \int_{-h/2}^{h/2} \sigma_\theta \, dn, \quad N_s = \int_{-h/2}^{h/2} \sigma_s \, dn,$$

and

$$N_{s\theta} = \int_{-h/2}^{h/2} \tau_{s\theta} \, dn \tag{8-7}$$

leading to

$$N_\theta = \frac{Eh}{1 - \nu^2}(\epsilon_\theta + \nu \epsilon_s)$$

$$N_s = \frac{Eh}{1 - \nu^2}(\epsilon_s + \nu \epsilon_\theta) \tag{8-8}$$

and

$$N_{s\theta} = \frac{Eh}{2(1+\nu)}\gamma_{\theta s}$$

which are the three additional equations relating the kinetic variables N_s, N_θ and $N_{s\theta}$ and kinematic variables ϵ_s, ϵ_θ and $\gamma_{\theta s}$. Note that by combining Equations (8-6) and (8-8) it follows that the stresses are given in terms of the force resultants as

$$\sigma_\theta = \frac{N_\theta}{h}, \quad \sigma_s = \frac{N_s}{h} \quad \text{and} \quad \tau_{s\theta} = \frac{N_{s\theta}}{h}$$

Thus, the following sets of equations are produced:

Equilibrium equations:

$$\frac{N_\theta}{r_\theta} + \frac{N_s}{r_n} = p_n$$

$$\frac{\partial N_\theta}{\partial \theta} + 2N_{s\theta}\cos\phi + r\frac{\partial N_{s\theta}}{\partial s} + rp_\theta = 0$$

$$r\frac{\partial N_s}{\partial s} + (N_s - N_\theta)\cos\phi + \frac{\partial N_{s\theta}}{\partial \theta} + rp_s = 0$$

or

$$\frac{N_\theta}{r_\theta} + \frac{N_s}{r_n} = p_n$$

$$\frac{1}{r}\frac{\partial N_\theta}{\partial \theta} + 2\frac{N_{s\theta}\cos\phi}{r} + \frac{1}{r_n}\frac{\partial N_{s\theta}}{\partial \phi} + p_\theta = 0$$

$$\frac{1}{r_n}\frac{\partial N_s}{\partial \phi} + \frac{(N_s - N_\theta)\cos\phi}{r} + \frac{1}{r}\frac{\partial N_{s\theta}}{\partial \theta} + p_s = 0$$

Strain-displacement equations:

$$\epsilon_\theta = \frac{1}{r}\left(\frac{\partial v}{\partial \theta} + w\sin\phi + u\cos\phi\right)$$

$$\epsilon_s = \frac{\partial u}{\partial s} + \frac{w}{r_n}$$

$$\gamma_{s\theta} = \frac{\partial v}{\partial s} + \frac{1}{r}\left(\frac{\partial u}{\partial \theta} - v\cos\phi\right)$$

Force-displacement relations:

$$N_\theta = \frac{Eh}{1-\nu^2}(\epsilon_\theta + \nu\epsilon_s)$$

$$N_s = \frac{Eh}{1-\nu^2}(\epsilon_s + \nu\epsilon_\theta)$$

$$N_{\theta s} = \frac{Eh}{2(1+\nu)}\gamma_{\theta s}$$

with $\sigma_\theta = N_\theta/h$, $\sigma_s = N_\theta/h$ and $\tau_{s\theta} = N_{s\theta}/h$. In principle, it would now be possible to eliminate the strains between Equations (8-4) and (8-8) so as to express the force resultants in terms of the displacements. These could then be substituted into the equilibrium Equations (8-2) or (8-3) to yield a set of three coupled equations in terms of the three displacement components $u(\theta, s)$, $v(\theta, s)$ and $w(\theta, s)$. Except for shells with special geometries such as cylindrical, spherical and conical, these equations are not particularly useful, and it turns out that it is preferable to proceed in a different manner. Several examples will be presented in the following sections.

8-4 Axisymmetrically Loaded Membrane Shells

There are many important applications where a shell of revolution is loaded and supported in such a manner as to result in displacements and internal forces that depend only on the meridional coordinate, that is, all derivatives with respect to the circumferential coordinate θ vanish. In particular, if $p_\theta = 0$ and both p_n and p_s depend only on the meridional coordinate s, the membrane shell equations reduce to the following:

Equilibrium equations:

$$\frac{N_\theta}{r_\theta} + \frac{N_s}{r_n} = p_n \tag{8-9a}$$

$$r\frac{\partial N_s}{\partial s} + (N_s - N_\theta)\cos\phi + rp_s = 0 \tag{8-9b}$$

$$2N_{s\theta}\cos\phi + r\frac{\partial N_{s\theta}}{\partial s} = 0 \tag{8-9c}$$

Strain-displacement equations:

$$\epsilon_\theta = \frac{1}{r}(w\sin\phi + u\cos\phi) \tag{8-10a}$$

$$\epsilon_s = \frac{du}{ds} + \frac{w}{r_n} \tag{8-10b}$$

$$\gamma_{s\theta} = \frac{\partial v}{\partial s} - \frac{v\cos\phi}{r} \tag{8-10c}$$

Force-displacement relations:

$$N_\theta = \frac{Eh}{1-\nu^2}(\epsilon_\theta + \nu\epsilon_s) \tag{8-11a}$$

$$N_s = \frac{Eh}{1-\nu^2}(\epsilon_s + \nu\epsilon_\theta) \tag{8-11b}$$

$$N_{s\theta} = \frac{Eh}{2(1+\nu)}\gamma_{\theta s} \tag{8-11c}$$

Note that the assumption of axisymmetry uncouples the equilibrium equation into a set containing $N_s, N_\theta, \epsilon_s, \epsilon_\theta, u$ and w, and a set containing $N_{s\theta}, \gamma_{s\theta}$ and v. The equation set (8-9c), (8-10c) and (8-11c) containing $N_{s\theta}, \gamma_{s\theta}$ and v corresponds to the problem of the torsion of a shell of revolution, which is left for the student to investigate in the exercises. In what follows problems governed by the set of equations (8-9a and b), (8-10a and b) and (8-11a and b) will be considered. These are sometimes referred to as the *torsionless axisymmetric membrane shell equations*.

For a statically determinate problem, the solution of these equations consists of two steps.

1. Solve for the force resultants from the equilibrium equations. There are two approaches that can be used. The first consists of integrating the equilibrium equations. This approach can be carried out in the following manner. First, eliminate the hoop force N_θ in the equilibrium equations to yield

$$\frac{dN_s}{ds} + \frac{\cos\phi}{r}\left(1 + \frac{r_\theta}{r_n}\right)N_s = -p_s(s) + \frac{r_\theta \cos\phi}{r}p_n(s)$$

which can in principle be integrated to determine N_s. N_θ would then be determined from

$$N_\theta = r_\theta\left(p_n(s) - \frac{N_s}{r_n}\right)$$

The other approach involves consideration of the free-body diagram shown in Figure 8-15. Shown are the meridional membrane forces $N_{\phi o}$ assumed to be known at the bottom of the shell and N_ϕ at an arbitrary location, and the loads p_n and p_s acting on the shell. A sum of forces in the axial direction yields

$$-N_{\phi o}\cdot 2\pi r_0 \cdot \sin\phi_o + N_\phi \cdot 2\pi r \cdot \sin\phi + 2\pi \int_0^s (p_n \cos\phi + p_s \sin\phi)r\,ds = 0$$

which can be solved for N_ϕ. Note that it would also be possible to develop a similar equation if the known meridional forces occurred at the top of the shell. This approach is frequently simpler than integrating the equation of equilibrium, although both should yield the same results.

For a statically indeterminate problem, the values of the meridional membrane forces at the ends of the shell would be unknown with an arbitrary constant occurring in the solutions for N_ϕ and N_θ. This unknown would ultimately be determined from the boundary conditions on the displacements.

2. The second step consists of integrating the strain-displacement equations to determine w and u. This can be carried out in the following manner. By eliminating the displacement w from Equation (8-10b) according to $w = r_n(\epsilon_s - du/ds)$, the remain-

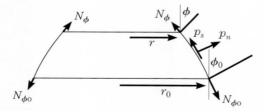

FIGURE 8-15 Free-body diagram of a portion of the shell.

ing strain-displacement equation can be expressed as

$$\frac{du}{d\phi} - u \cot \phi = r_n \epsilon_s - r_\theta \epsilon_\theta$$

where $ds = r_n \, ds$ has been used. Solving the force displacement relations for ϵ_s and ϵ_θ in terms of N_s and N_θ yields

$$\epsilon_s = \frac{1}{Eh}(N_s - \nu N_\theta) \quad \text{and} \quad \epsilon_\theta = \frac{1}{Eh}(N_\theta - \nu N_s)$$

from which

$$\frac{du}{d\phi} - u \cot \phi = \frac{1}{Eh}[N_s(r_n + \nu r_\theta) - N_\theta(r_\theta + \nu r_n)]$$

With N_s and N_θ having been determined in step 1, the right-hand side is known, and this equation can be integrated to yield the meridional displacement u. The normal displacement w can then be determined from either of the two strain-displacement relations. The problem of determining force resultants and displacements for several special geometries will be considered in what follows.

8-4-1 Cylindrical Shells For a cylindrical shell $r_\theta = R$, the radius of the cylinder, $r_n = \infty$ and $\phi = \pi/2$, as shown in Figure 8-16. All loadings, displacements and membrane forces can be considered to depend only on the coordinate s, which will be taken as the independent variable. The equations reduce to

$$N_\theta = R p_n(s)$$

$$\frac{dN_s}{ds} + p_s(s) = 0$$

$$\epsilon_\theta = w/R$$

$$\epsilon_s = \frac{du}{ds}$$

$$N_\theta = \frac{Eh}{1-\nu^2}(\epsilon_\theta + \nu \epsilon_s)$$

$$N_s = \frac{Eh}{1-\nu^2}(\epsilon_s + \nu \epsilon_\theta)$$

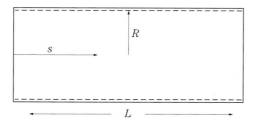

FIGURE 8-16 Cylindrical shell geometry.

The hoop membrane force N_θ can be determined immediately from the first equation. N_s can be determined by integrating the second equation. Knowing N_θ and N_s, the last two equations can be used to solve for ϵ_θ and ϵ_s, after which the middle two equations can be integrated to determine the displacements.

EXAMPLE 8-1 A cylindrical shell hangs under its own weight and is supported by rollers that allow the shell to move freely in the radial direction, as shown in Figure 8-17a. Determine the internal forces and displacements.

Solution: With γ as the weight density of the shell, the loadings are $p_n = 0$ and $p_s = -\gamma h$ where h is the uniform thickness of the shell. It follows that $N_\theta = 0$ and that

$$\frac{dN_s}{ds} = \gamma h$$

Integration yields $N_s = \gamma h s + C_1$. The constant of integration is evaluated on the basis that $N_s(0) = 0$, yielding $C_1 = 0$ and $N_s = \gamma h s$. Using the force resultant displacement equations, it follows from $N_\theta = 0$ that

$$0 = \epsilon_\theta + \nu \epsilon_s$$

and from $N_s = \gamma h s$ that

$$\gamma h s = \frac{Eh}{1 - \nu^2}(\epsilon_s + \nu \epsilon_\theta)$$

from which by eliminating $\epsilon_\theta = -\nu \epsilon_s$ gives

$$\epsilon_s = \frac{\gamma s}{E} = \frac{du}{ds}$$

Integration yields

$$u = \frac{\gamma s^2}{2E} + C_2$$

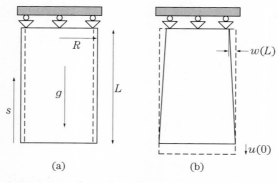

FIGURE 8-17 Hanging cylindrical shell.

which with $u(L) = 0$ gives $C_2 = -\gamma L^2/2E$ and

$$u = \frac{\gamma(s^2 - L^2)}{2E}$$

The radial displacement is then determined by returning to

$$\epsilon_\theta = -\nu\epsilon_s = -\nu\frac{\gamma s}{E} = \frac{w}{R}$$

from which $w = -\nu\gamma Rs/E$.

The maximum axial displacement is at the bottom of the cylinder with the value $u_{\max} = u(0) = -\gamma L^2/2E$, the minus indicating a downward displacement. The maximum radial displacement occurs at the top of the cylinder with $w_{\max} = w(L) = -\nu\gamma RL/E$, with the minus sign indicating a decrease in the radius of the cylinder. Greatly exaggerated, the deformed configuration appears in Figure 8-17b. ◆

8-4-2 Spherical Shells The geometry of a spherical shell is shown in Figure 8-18, where it is seen that $r_\theta = r_n = R$, the radius of the spherical shell, and that $r = R\sin\phi$. With ϕ chosen to be the independent variable and $ds = R\,d\phi$, the general equations reduce to the following:

Equilibrium equations:

$$N_\theta + N_s = Rp_n(\phi)$$

$$\frac{dN_s}{d\phi} + (N_s - N_\theta)\cot\phi = -Rp_s(\phi)$$

Strain-displacement equations:

$$\epsilon_\theta = \frac{w + u\cot\phi}{R}$$

$$\epsilon_s = \frac{du}{ds} + \frac{w}{r_n}$$

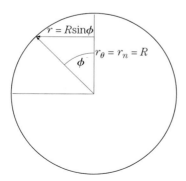

FIGURE 8-18 Spherical shell geometry.

Force-displacement relations:

$$N_\theta = \frac{Eh}{1-\nu^2}(\epsilon_\theta + \nu\epsilon_s)$$

$$N_s = \frac{Eh}{1-\nu^2}(\epsilon_s + \nu\epsilon_\theta)$$

Elimination of N_θ in the equilibrium equations results in

$$\frac{dN_s}{d\phi} + 2\cot\phi\, N_s = -R(p_s(\phi) - p_n(\phi)\cot\phi)$$

a single equation that can be integrated to determine N_s. N_θ is then given as $N_\theta = R \cdot p_n(s) - N_s$. The equation that must be integrated in order to determine the meridional displacement u becomes

$$\frac{du}{d\phi} - u\cot\phi = \frac{(1+\nu)R}{Eh}[N_s - N_\theta]$$

EXAMPLE 8-2 A spherical shell rotates about a diameter, as shown in Figure 8-19a. Determine the internal forces and displacements.

Solution: With the body force per unit area given as $f = \rho h R \omega^2$, it is easily seen that loadings p_n and p_s are given by $p_n = f\sin\phi$ and $p_s = f\cos\phi$, respectively, so that the equilibrium equation in the meridional force N_s becomes

$$\frac{dN_s}{d\phi} + 2\cot\phi\, N_s = 0$$

The student should review linear first-order ordinary differential equations to determine that this equation can be put in integrable form by multiplying by the integrating factor $\sin^2\phi$ to yield

$$\sin^2\phi\,\frac{dN_s}{d\phi} + 2\sin\phi\cos\phi\, N_s = 0$$

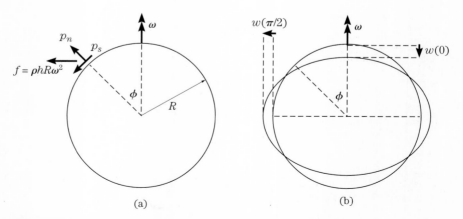

(a)   (b)

FIGURE 8-19 Rotating spherical shell.

which integrates to

$$\sin^2 \phi \, N_s = C_1$$

or

$$N_s = \frac{C_1}{\sin^2 \phi}$$

In order that N_s be finite as $\phi \to 0$, the constant of integration must be zero, leading to $N_s = 0$. The student is asked to show in the exercises that the same conclusion is reached on the basis of drawing a free-body diagram. The circumferential membrane force N_θ is then given by

$$N_\theta = R \cdot p_n = \rho h R^2 \omega^2 \sin \phi$$

which clearly has a maximum value of $\rho h R^2 \omega^2$ at the equator $\phi = \pi/2$. For the displacements, the equation

$$\frac{du}{d\phi} - u \cot \phi = \frac{(1+\nu)R}{Eh}[N_s - N_\theta] = -\frac{(1+\nu)\rho R^3 \omega^2 \sin \phi}{E}$$

must be integrated. The integrating factor μ for this equation is determined by integrating the equation

$$\frac{d\mu}{\mu} = -\frac{\cos \phi}{\sin \phi}$$

from which $\ln(\mu) + \ln(\sin \phi) = \ln C$, or $\mu = C/\sin \phi$. The equation in u then becomes

$$\frac{1}{\sin \phi}\frac{du}{d\phi} - u\frac{\cos \phi}{\sin^2 \phi} = -\frac{(1+\nu)\rho R^3 \omega^2}{E}$$

or

$$\frac{d}{d\phi}\left(\frac{u}{\sin \phi}\right) = -\frac{(1+\nu)\rho R^3 \omega^2}{E}$$

from which

$$u = -\frac{(1+\nu)\rho R^3 \omega^2 \phi \sin \phi}{E} + C_2 \sin \phi$$

Requiring that the meridional displacement vanish at $\phi = \pi/2$ leads to $C_2 = \pi(1+\nu)\rho R^3 \omega^2 \phi \sin \phi / 2E$ and

$$u = \frac{(1+\nu)\rho R^3 \omega^2}{E}\left(\frac{\pi}{2} - \phi\right)\sin \phi$$

The radial component of displacement can then be determined from the equations

$$\epsilon_\theta = \frac{1}{R \cos \phi}(w \cos \phi - u \sin \phi) \quad \text{and} \quad \epsilon_\theta = \frac{1}{Eh}(N_\theta - \nu N_s)$$

which can be combined to yield

$$w = \frac{\rho R^3 \omega^2}{E}\left(\sin\phi - (1+\nu)\left(\frac{\pi}{2} - \phi\right)\cos\phi\right)$$

It follows that $w(\pi/2) = \rho R^3\omega^2/E$ and that $w(0) = -(1+\nu)\pi\rho R^3\omega^2/2E = -((1+\nu)\pi/2)w(\pi/2) \approx 2w(\pi/2)$. The deformed configuration of the shell is shown, greatly exaggerated, in Figure 8-18b.

8-4-3 Conical Shells

The geometry of a conical shell is shown in Figure 8-20. With $r_n = \infty$, $r(s) = r_0 + s\cdot\cos\phi = r_0 + (r_1 - r_0)s/L$ and $r_\theta = r(s)/\sin\phi$, the governing equations can be written as follows:

Equilibrium equations:

$$N_\theta = r_\theta p_n(s)$$

$$\frac{dN_s}{ds} + \frac{N_s \cos\phi}{r(s)} = -p_s(s) + \cot\phi\, p_n(s)$$

Strain-displacement equations:

$$\epsilon_\theta = \frac{1}{r(s)}(w\sin\phi + u\cos\phi)$$

$$\epsilon_s = \frac{du}{ds}$$

Force-displacement relations:

$$N_\theta = \frac{Eh}{1-\nu^2}(\epsilon_\theta + \nu\epsilon_s)$$

$$N_s = \frac{Eh}{1-\nu^2}(\epsilon_s + \nu\epsilon_\theta)$$

Generally, the strategy is to solve for N_s and N_θ using the equilibrium equations. The force-displacement equations can then be used to write

$$\epsilon_s = \frac{N_s - \nu N_\theta}{En} = \frac{du}{ds}$$

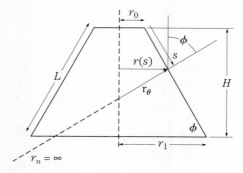

FIGURE 8-20 Conical shell geometry.

8-4 AXISYMMETRICALLY LOADED MEMBRANE SHELLS

which can be integrated to determine $u(s)$, after which the remaining equation

$$\epsilon_\theta = \frac{N_\theta - \nu N_s}{En} = \frac{1}{r(s)}(w \sin \phi + u \cos \phi)$$

can be solved for w. These steps will be demonstrated in the example that follows.

EXAMPLE 8-3 A conical shell is filled with a liquid of weight density γ, as shown in Figure 8-21. Determine the membrane forces and displacements.

Solution: For the coordinate s shown in Figure 8-21, $r(s) = r_1 s/L$ and $r_\theta = r_1 s/L \sin \phi$. With the pressures given by $p_n(s) = \gamma(H - s \cdot \sin \phi) = \gamma \sin \phi(L - s)$ and $p_s(s) = 0$, the equilibrium equations become

$$N_\theta = r_\theta p_n(s) = (r_1 s/L \sin \phi) \gamma \sin \phi (L - s) = \frac{\gamma r_1}{L} s(L - s)$$

and

$$r \frac{dN_s}{ds} + N_s \cos \phi = \frac{\gamma r_1 \cos \phi}{L} s(L - s)$$

Dividing by $\cos \phi$ and using $L \cos \phi = r_1$, this equation can be written as

$$s \frac{dN_s}{ds} + N_s = \frac{\gamma r_1}{L} s(L - s)$$

which is immediatedly integrable to

$$sN_s = \frac{\gamma r_1}{L}\left(\frac{Ls^2}{2} - \frac{s^3}{3}\right) + C_1$$

or

$$N_s = \frac{\gamma r_1}{L}\left(\frac{Ls}{2} - \frac{s^2}{3}\right) + \frac{C_1}{s}$$

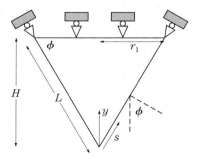

FIGURE 8-21 Conical shell containing liquid.

In order that N_s be bounded at the apex $s = 0$, it is necessary to take $C_1 = 0$ so that finally there results

$$N_s = \frac{\gamma r_1}{L}\left(\frac{Ls}{2} - \frac{s^2}{3}\right) \quad \text{and} \quad N_\theta = \frac{\gamma r_1}{L}s(L - s)$$

for the membrane forces. In terms of the coordinate $y = s \sin \phi$ shown in Figure 8-21, these can also be represented as

$$N_s = \frac{\gamma \cos \phi}{2 \sin^2 \phi} y\left(H - \frac{2y}{3}\right) \quad \text{and} \quad N_\theta = \frac{\gamma \cos \phi}{\sin^2 \phi} y(H - y)$$

where H, shown in Figure 8-21, is the height of the shell. The displacement u is obtained by integrating

$$\frac{du}{ds} = \frac{N_s - \nu N_\theta}{Et} = \frac{\gamma r_1}{EtL}\left[\frac{Ls}{2} - \frac{s^2}{3} - \nu s(L - s)\right]$$

from which

$$u = \frac{\gamma r_1}{EtL}\left[\frac{Ls^2}{4} - \frac{s^3}{9} - \nu\left(\frac{Ls^2}{2} - \frac{s^3}{3}\right)\right] + C_2$$

Satisfying the boundary condition $u(0) = 0$ gives $C_2 = -(5 - 6\nu)\gamma r_1 L^2/36Et$ and

$$u = \frac{\gamma r_1}{EtL}\left[\frac{Ls^2}{4} - \frac{s^3}{9} - \nu\left(\frac{Ls^2}{2} - \frac{s^3}{3}\right) - \frac{L^3(5 - 6\nu)}{36}\right]$$

The task of determining and investigating the character of the radial displacement w is left for the student to consider in the exercises. ◆

8-5 Asymmetrically Loaded Membrane Shells

The general problem of asymmetric loadings where $p_n = p_n(s, \theta)$, $p_s = p_s(s, \theta)$ and $p_\theta = p_\theta(s, \theta)$ are arbitrary functions of s and θ, is beyond the scope of this text. There are, however, several important types of loadings that result in relatively simple solutions of the membrane shell equations. Several of these will be investigated in this section, while somewhat more detailed cases are left to the exercises at the end of the chapter.

8-5-1 Cylindrical Shells As indicated in Figure 8-22, the two independent variables are the distance s along the cylinder and the circumferential coordinate θ. With $r_n = \infty$, $r = r_n = R$ and $\phi = 0$, the governing equations are as follows:

Equilibrium equations:

$$N_\theta = Rp_n$$

$$\frac{\partial N_\theta}{\partial \theta} + R\frac{\partial N_{\theta s}}{\partial s} + Rp_\theta = 0$$

$$R\frac{\partial N_s}{\partial s} + \frac{\partial N_{\theta s}}{\partial \theta} + Rp_s = 0$$

8-5 ASYMMETRICALLY LOADED MEMBRANE SHELLS

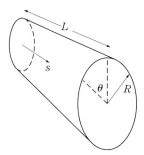

FIGURE 8-22 Cylindrical shell coordinates.

Strain-displacement and force-resultant equations:

$$\epsilon_\theta = \frac{1}{R}\left(\frac{\partial v}{\partial \theta} + w\right) = \frac{1}{Eh}(N_\theta - \nu N_s)$$

$$\epsilon_s = \frac{\partial u}{\partial s} = \frac{1}{Eh}(N_s - \nu N_\theta)$$

$$\gamma_{s\theta} = \frac{\partial v}{\partial s} + \frac{1}{R}\frac{\partial u}{\partial \theta} = \frac{2(1+\nu)N_{\theta s}}{Eh}$$

Again, the basic strategy is to first solve the equilibrium equations for N_θ, N_s and $N_{\theta s}$ and then to integrate the strain-displacement relations to determine the displacements.

◆EXAMPLE 8-4 A cylindrical shell is supported at one end and loaded by its own weight, as shown in Figure 8-23. The exact nature of the roller supports will be discussed presently. Determine the internal forces and the displacements.

Internal forces: As seen in Figure 8-23b, the loadings are $p_n = -\gamma h \cos\theta$ and $p_\theta = \gamma h \sin\theta$ with $p_s = 0$. Thus, the equilibrium equations become

$$N_\theta = -\gamma R h \cos\theta \tag{8-12a}$$

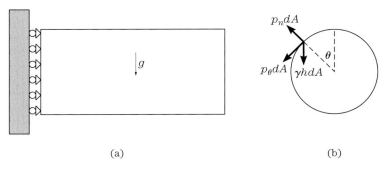

FIGURE 8-23 Cylindrical shell under its own weight.

CHAPTER 8 MEMBRANE SHELLS OF REVOLUTION

$$\frac{\partial N_\theta}{\partial \theta} + R\frac{\partial N_{\theta s}}{\partial s} + \gamma Rh\ \sin\theta = 0 \tag{8-12b}$$

$$R\frac{\partial N_s}{\partial s} + \frac{\partial N_{\theta s}}{\partial \theta} = 0 \tag{8-12c}$$

Combining the first two equations yields

$$\frac{\partial N_{\theta s}}{\partial s} + 2\gamma h\ \sin\theta = 0$$

Integration yields

$$N_{\theta s} + 2\gamma hs\ \sin\theta = f_1(\theta) \tag{8-13}$$

where $f_1(\theta)$ is an arbitrary function of θ. Requiring $N_{\theta s}(L) = 0$ in Equation (8-13) gives $f_1(\theta) = 2\gamma hL\ \sin\theta$ so that

$$N_{\theta s} = 2\gamma h(L - s)\sin\theta$$

Equation (8-12c) can then be expressed as

$$\frac{\partial N_s}{\partial s} + \frac{2\gamma h(L - s)\cos\theta}{R} = 0$$

Another integration yields

$$N_s - \frac{\gamma h(L - s)^2 \cos\theta}{R} = f_2(\theta) \tag{8-14}$$

Requiring $N_s(L) = 0$ in Equation (8-14) gives $f_2(\theta) = 0$ so that

$$N_s = \frac{\gamma h(L - s)^2 \cos\theta}{R}$$

It is left for the student to show in the homework exercises that the expressions for N_s and $N_{s\theta}$ coincide with the results from the elementary theory.

Displacements: Substituting for the known membrane forces in

$$\epsilon_s = \frac{\partial u}{\partial s} = \frac{1}{Eh}(N_s - \nu N_\theta)$$

there results

$$\frac{\partial u}{\partial s} = \frac{\gamma\ \cos\theta}{E}\left(\frac{(L - s)^2}{R} + \nu R\right)$$

Integration yields

$$u(s, \theta) = \frac{\gamma\ \cos\theta}{E}\left(-\frac{(L - s)^3}{3R} + \nu Rs\right) + f_3(\theta)$$

With $u(0, \theta) = 0$, it follows that $f_3(\theta) = \gamma L^3 \cos\theta/3ER$ and, after simplification, that

$$u(s, \theta) = \frac{\gamma\ \cos\theta}{3ER}(3L^2 s - 3Ls^2 + s^3) + \frac{\gamma\ \cos\theta}{E}\nu Rs$$

8-5 ASYMMETRICALLY LOADED MEMBRANE SHELLS

Then from

$$\gamma_{s\theta} = \frac{\partial v}{\partial s} + \frac{1}{R}\frac{\partial u}{\partial \theta} = \frac{2(1+\nu)N_{\theta s}}{Eh} = \frac{N_{\theta s}}{Gh}$$

it follows that

$$\frac{\partial v}{\partial s} + \frac{1}{R}\frac{\partial u}{\partial \theta} = \frac{2\gamma(L-s)\sin\theta}{G}$$

and upon substitution for u,

$$\frac{\partial v}{\partial s} = \frac{2\gamma(L-s)\sin\theta}{G} + \frac{1}{R}\frac{\gamma\sin\theta}{ER}\left(L^2 s - Ls^2 + \frac{s^3}{3} + \nu R^2 s\right)$$

An integration yields

$$v(s,\theta) = -\frac{\gamma(L-s)^2\sin\theta}{G} + \frac{\gamma\sin\theta}{ER^2}\left(\frac{L^2 s^2}{2} - \frac{Ls^3}{3} + \frac{s^4}{12} + \frac{\nu R^2 s^2}{2}\right) + f_4(\theta) \tag{8-15}$$

At this point it is necessary to discuss the type of roller support indicated in Figure 8-23b. Viewed from the rear, as shown in Figure 8-24, the support must prevent displacements in the axial direction and in the direction tangent to the cylinder at all points around the circumference. This can be accomplished in principle by allowing the rollers to move only in the radial direction, thus preventing any shear forces from developing. The constraint against motion in the circumferential direction will support the membrane force $N_{\theta s} = 2\gamma h(L-s)\sin\theta$. If this condition is not realized, there will be local bending at the support.

In particular, the circumferential displacement v at $s = 0$ and $\theta = \pi/2$ is zero, leading to $f_4(\theta) = \gamma L^2/G$ and then to

$$v(s,\theta) = \frac{\gamma\sin\theta}{12ER^2}(6L^2 s^2 - 4Ls^3 + s^4) + \frac{\gamma\sin\theta}{2E}\nu s^2 + \frac{\gamma(2Ls - s^2)\sin\theta}{G}$$

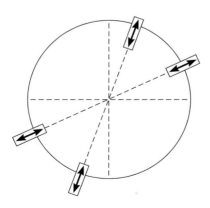

FIGURE 8-24 Radial roller supports at $s = 0$.

Last, from

$$\frac{1}{R}\left(\frac{\partial v}{\partial \theta} + w\right) = \frac{1}{Eh}(N_\theta - \nu N_s)$$

it is possible to solve for w as

$$w(s,\theta) = -\frac{\gamma \cos\theta}{12ER^2}(6L^2s^2 - 4Ls^3 + s^4)$$

$$- \frac{\gamma \cos\theta}{E}\left[R^2 + \nu(L-s)^2 + \frac{\nu s^2}{2}\right] - \frac{\gamma}{G}(2Ls - s^2)$$

Further investigation and discussion of the displacements is left to the homework exercises. ◆

8-5-2 Conical Shells

As shown in Figure 8-25, the two independent variables are the distance s along a generator of the cylinder and the circumferential coordinate θ. With $r_n = \infty$, $r_\theta = r(s)/\sin\phi$ and $r(s) = r_0 + (r_1 - r_0)(s/L) = r_0 + s\cos\phi$, the governing equations can be expressed as follows:

Equilibrium equations:

$$N_\theta = r_\theta p_n$$

$$\frac{\partial}{\partial s}(r(s)N_{s\theta}) + \frac{\partial N_\theta}{\partial \theta} + N_{\theta s}\cos\phi + r(s)p_\theta = 0$$

$$\frac{\partial}{\partial s}(r(s)N_s) + \frac{\partial N_{s\theta}}{\partial \theta} - N_\theta \cos\phi + r(s)p_s = 0$$

Strain-displacement and force-resultant equations:

$$\epsilon_\theta = \frac{1}{r(s)}\left(\frac{\partial v}{\partial \theta} + w\sin\phi + u\cos\phi\right) = \frac{1}{Eh}(N_\theta - \nu N_s)$$

$$\epsilon_s = \frac{\partial u}{\partial s} = \frac{1}{Eh}(N_s - \nu N_\theta)$$

$$\gamma_{s\theta} = \frac{\partial v}{\partial s} + \frac{1}{r(s)}\left(\frac{\partial u}{\partial \theta} - v\cos\phi\right) = \frac{2(1+\nu)N_{\theta s}}{Eh}$$

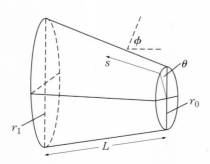

FIGURE 8-25 Conical shell coordinates.

8-5 ASYMMETRICALLY LOADED MEMBRANE SHELLS

Again, the basic strategy is to first solve the equilibrium equations for N_θ, N_s and $N_{\theta s}$ and then to integrate the strain-displacement relations to determine the displacements.

EXAMPLE 8-5 A conical shell is supported at one end and loaded by its own weight, as shown in Figure 8-26. Determine the internal forces and the displacements.

Internal forces: The loadings are given by $p_n = -\gamma h \cos\theta \sin\phi$, $p_\theta = \gamma h \sin\theta$ and $p_s = -\gamma h \cos\theta \cos\phi$, leading to

$$N_\theta = -\gamma h r \cos\theta$$

$$\frac{\partial}{\partial s}(rN_{s\theta}) + \frac{\partial N_\theta}{\partial \theta} + N_{\theta s}\cos\phi + \gamma h r \sin\theta = 0$$

$$\frac{\partial}{\partial s}(rN_s) + \frac{\partial N_{s\theta}}{\partial \theta} - N_\theta \cos\phi - \gamma h r \cos\theta \cos\phi = 0$$

Eliminating N_θ yields

$$\frac{\partial}{\partial s}(rN_{s\theta}) + N_{s\theta}\cos\phi + 2\gamma h r \sin\theta = 0 \qquad (8\text{-}16\text{a})$$

$$\frac{\partial}{\partial s}(rN_s) + \frac{\partial N_{s\theta}}{\partial \theta} = 0 \qquad (8\text{-}16\text{b})$$

The first equation will be used to determine $N_{s\theta}$, after which the second equation can be used to solve for N_s. Recalling that $r = r_0 + s\cos\phi$, it follows that $\partial r/\partial s = \cos\phi$ and that after multiplying by r, Equation (8-16a) can be rewritten as

$$r^2 \frac{\partial N_{s\theta}}{\partial s} + 2r \frac{\partial r}{\partial s} N_{s\theta} + 2\gamma h r^2 \sin\theta = 0$$

or

$$\frac{\partial}{\partial s}(r^2 N_{s\theta}) + 2\gamma h r^2 \sin\theta = 0$$

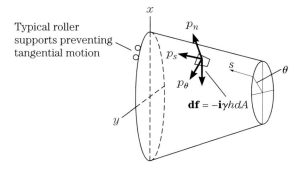

FIGURE 8-26 Hanging conical shell.

which can be integrated to yield

$$r^2 N_{s\theta} + \frac{2\gamma h r^3 \sin\theta}{3 \cos\phi} = F_1(\theta)$$

Requiring that $N_{\theta s}(0) = 0$ gives $F_1 = 2\gamma h r_0^3 \sin\theta / 3 \cos\phi$ and

$$N_{s\theta} = \frac{2\gamma h \sin\theta}{3 \cos\phi} \left(\frac{r_0^3 - r^3}{r^2} \right)$$

Equation (8-16b) can then be written as

$$\frac{\partial}{\partial s}(rN_s) + \frac{2\gamma h \cos\theta}{3 \cos\phi} \left(\frac{r_0^3 - r^3}{r^2} \right) = 0$$

and is integrated to

$$rN_s + \frac{2\gamma h \cos\theta}{3 \cos\phi} \left(-\frac{r_0^3}{r \cos\phi} - \frac{r^2}{2 \cos\phi} \right) = F_2(\theta)$$

Requiring that the N_s vanish at the free end yields $F_2 = -\gamma h r_0^2 \cos\theta / \cos^2\phi$ and then

$$N_s = \frac{\gamma h \cos\theta}{3 \cos^2\phi} \left(\frac{2r_0^3}{r^2} + r - \frac{3r_0^2}{r} \right)$$

The maximum values of $N_{s\theta}$ and N_s occur at the support with the values

$$(N_{s\theta})_{\max} = N_{s\theta}\left(L, \frac{\pi}{2}\right) = -\frac{2\gamma h}{3 \cos\phi} \left(\frac{r_1^3 - r_0^3}{r_1^2} \right)$$

$$(N_s)_{\max} = N_s(L, 0) = \frac{\gamma h}{3 \cos^2\phi} \left(\frac{2r_0^3}{r_1^2} + r_1 - \frac{3r_0^2}{r_1} \right)$$

These values as well as the distributions are shown in Figure 8-27. The membrane forces shown in this figure are the counterparts of the corresponding shear and bending membrane forces of the previous example.

It is left for the student to investigate the displacements in the homework exercises.

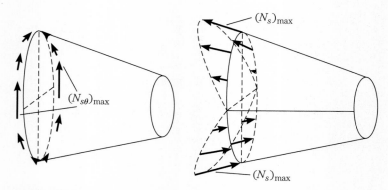

FIGURE 8-27 Maximum membrane forces.

8-5-3 Spherical Shells

For spherical shells the geometrical parameters are $r = R \sin \phi$ and $r_\theta = r_n = R$ so that the equilibrium equations reduce to

$$N_\theta + N_s = Rp_n$$

$$\frac{1}{\sin \phi} \frac{\partial N_\theta}{\partial \theta} + 2N_{s\theta} \cot \phi + \frac{\partial N_{s\theta}}{\partial \phi} + Rp_\theta = 0$$

$$\frac{\partial N_s}{\partial \phi} + (N_s - N_\theta) \cot \phi + \frac{1}{\sin \phi} \frac{\partial N_{\theta s}}{\partial \theta} + Rp_\phi = 0$$

with the strain-displacement relations as

$$\epsilon_\theta = \frac{1}{R \sin \phi} \left(\frac{\partial v}{\partial \theta} + w \sin \phi + u \cos \phi \right)$$

$$\epsilon_\phi = \frac{1}{R} \left(\frac{\partial u}{\partial \phi} + w \right)$$

$$\gamma_{\phi\theta} = \frac{1}{R} \left(\frac{\partial v}{\partial \phi} + \frac{1}{\sin \phi} \left(\frac{\partial u}{\partial \theta} - v \cos \phi \right) \right)$$

and the stress-strain relations as

$$\epsilon_\theta = \frac{\sigma_\theta - \nu \sigma_\phi}{E}, \quad \epsilon_\phi = \frac{\sigma_\phi - \nu \sigma_\theta}{E} \quad \text{and} \quad \gamma_{\phi\theta} = \frac{\tau_{\phi\theta}}{G}$$

The internal force resultants are determined from the equilibrium equations, after which the strain-displacement and stress-strain relations are used to determine the displacements.

EXAMPLE 8-6 A hemispherical shell is supported along the equator and loaded by its own weight, as shown in Figure 8-28. Determine the membrane forces. Take the specific weight to be γ.

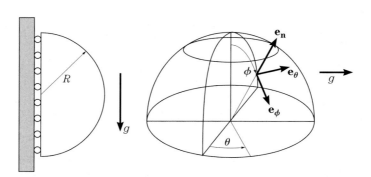

FIGURE 8-28 Hemispherical shell under gravity loading.

Solution: From the geometry the pressures are $p_n = \gamma h \sin\theta \sin\phi$, $p_\phi = \gamma h \sin\theta \cos\phi$ and $p_\theta = \gamma h \cos\theta$ so that the equilibrium equations become

$$N_\theta + N_s = \gamma R h \sin\theta \sin\phi$$

$$\frac{1}{\sin\phi}\frac{\partial N_\theta}{\partial \theta} + 2N_{s\theta}\cot\phi + \frac{\partial N_{s\theta}}{\partial \phi} + \gamma R h \cos\theta = 0$$

$$\frac{\partial N_s}{\partial \phi} + (N_s - N_\theta)\cot\phi + \frac{1}{\sin\phi}\frac{\partial N_{\theta s}}{\partial \theta} + \gamma R h \sin\theta \cos\phi = 0$$

The first step is to eliminate N_θ using the first equation to obtain

$$\sin\phi \frac{\partial N_{s\theta}}{\partial \phi} + 2N_{s\theta}\cos\phi - \frac{\partial N_s}{\partial \theta} = -2\gamma R h \cos\theta \sin\phi$$

and

$$\sin\phi \frac{\partial N_s}{\partial \phi} + 2N_s \cos\phi + \frac{\partial N_{s\theta}}{\partial \theta} = 0$$

The form of these equations suggests that we separate variables by seeking solutions of the form $N_{s\theta} = A(\phi)\cos\theta$ and $N_s = B(\phi)\sin\theta$, leading to

$$\sin\phi \frac{dA}{d\phi} + 2A\cos\phi - B = -2\gamma R h \sin\phi$$

and

$$\sin\phi \frac{dB}{d\phi} + 2B\cos\phi - A = 0$$

These equations in A and B can be uncoupled by adding and subtracting and defining $U = A + B$ and $V = A - B$ to obtain

$$\sin\phi \frac{dU}{d\phi} + 2U\cos\phi - U = -2\gamma R h \sin\phi \quad (8\text{-}17a)$$

and

$$\sin\phi \frac{dV}{d\phi} + 2V\cos\phi + V = -2\gamma R h \sin\phi \quad (8\text{-}17b)$$

To solve Equation (8-17a), we first write it as

$$\frac{dU}{d\phi} + U(2\cot\phi - \csc\phi) = -2\gamma R h \quad (8\text{-}18)$$

An integrating factor μ is then determined in the standard manner from the equation

$$\mu' = \mu(2\cot\phi - \csc\phi)$$

which the student should show leads to $\mu = \sin\phi(1 + \cos\phi)$. Equation (8-18) can then be written as

$$(\sin\phi(1 + \cos\phi)U)' = -2\gamma R h \sin\phi(1 + \cos\phi)$$

8-5 ASYMMETRICALLY LOADED MEMBRANE SHELLS

which is easily integrated to yield

$$\sin\phi(1+\cos\phi)U = \gamma Rh(2\cos\phi - \sin^2\phi) + C_1$$

and then to

$$U = \frac{(1-\cos\phi)[\gamma Rh(2\cos\phi - \sin^2\phi) + C_1]}{\sin^3\phi}$$

In an entirely similar manner, the integrating factor for the V equation is $\mu = \sin\phi(1-\cos\phi)$, leading to

$$V = \frac{(1+\cos\phi)[\gamma Rh(2\cos\phi + \sin^2\phi) + C_2]}{\sin^3\phi}$$

The basic singularity at $\phi = 0$, clearly indicated by the $\sin^3\phi$ in the denominator, will be discussed presently.

The original variables A and B can then be recovered as

$$A(\phi) = \frac{U+V}{2} = \frac{\gamma Rh\cos\phi(2+\sin^2\phi) + \dfrac{C_1+C_2}{2} - \dfrac{(C_1-C_2)\cos\phi}{2}}{\sin^3\phi}$$

$$B(\phi) = \frac{U-V}{2} = \frac{-\gamma Rh(2\cos^2\phi + \sin^2\phi) + \dfrac{C_1-C_2}{2} - \dfrac{(C_1+C_2)\cos\phi}{2}}{\sin^3\phi}$$

with the membrane forces then given by $N_{s\theta} = A(\phi)\cos\theta$ and $N_s = B(\phi)\sin\theta$.

The constants C_1 and C_2 must be chosen so that the membrane forces remain bounded as $\phi \to 0$. This limit is most easily investigated by assuming that there is a small hole corresponding to $\phi = \phi_0$ and requiring that $N_{s\theta}$ and N_s vanish on the surface of the hole. This leads to

$$\gamma Rh\cos\phi_0(2+\sin^2\phi_0) + \frac{C_1+C_2}{2} - \frac{C_1-C_2}{2}\cos\phi_0 = 0$$

and

$$-\gamma Rh(2\cos^2\phi_0 + \sin^2\phi_0) + \frac{C_1-C_2}{2} - \frac{C_1+C_2}{2}\cos\phi_0 = 0$$

Solving for C_1 and C_2 yields

$$C_1 = -\frac{\gamma Rh}{1+\cos^2\phi_0}[4\cos\phi_0 + 1 - \cos^2\phi_0 - \cos^2\phi_0\sin^2\phi_0]$$

and

$$C_2 = \frac{\gamma Rh}{1+\cos^2\phi_0}[-4\cos\phi_0 + 1 - \cos^2\phi_0 - \cos^2\phi_0\sin^2\phi_0]$$

In the limit as $\phi_0 \to 0$, $C_1 \to -2\gamma Rh$ and $C_2 \to -2\gamma Rh$ so that ultimately

$$N_{s\theta} = \frac{\gamma Rh(3\cos\phi - \cos^3\phi - 2)\cos\theta}{\sin^3\phi}$$

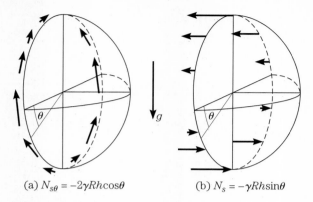

(a) $N_{s\theta} = -2\gamma Rh\cos\theta$ (b) $N_s = -\gamma Rh\sin\theta$

FIGURE 8-29 Membrane forces at the boundary.

and

$$N_s = \frac{\gamma Rh(2\cos\phi - 1 - \cos^2\phi)\sin\theta}{\sin^3\phi}$$

From the first of the equilibrium equations, the circumferential membrane force is then

$$N_\theta = Rp_n - N_s = \frac{\gamma Rh(1 - \cos\phi)(2 - \cos^2\phi - \cos^3\phi)\sin\theta}{\sin^3\phi}$$

The maximum values for all three of the membrane forces occur at the support $\phi = \pi/2$ where

$$N_{s\theta}\bigg|_{\phi=\pi/2} = -2\gamma Rh\cos\theta$$

$$N_s\bigg|_{\phi=\pi/2} = -\gamma Rh\sin\theta$$

and

$$N_\theta\bigg|_{\phi=\pi/2} = 2\gamma Rh\sin\theta$$

$N_{s\theta}$ and N_s are shown in Figures 8-29a and 8-29b, respectively.

The student is asked to show in the exercises that the resultants associated with these distributions are consistent with equilibrium of the hemisphere as a whole. ◆

The three examples presented above are relatively simple in that the loadings could be considered proportional to single $\sin\theta$ or $\cos\theta$ terms. This enabled the equations to be uncoupled and solved in a relatively straightforward fashion. In more complicated loading situations, it would likely be necessary to represent the loadings p_n, p_s and p_θ as a Fourier series in θ leading to systems of ordinary differential equations. Analysis of such systems is beyond the scope of this text.

8-6 Summary

Membrane theory for thin shells of revolution assumes that the primary mode of transmitting loads is by way of membrane force resultants, that is, resultants lying in the plane of the shell. For the membrane theory to be appropriate, the forces of constraint must be primarily membrane forces rather than transverse shear forces and bending moments, and the loading must be gradual so as not to introduce local bending. For such a shell the equilibrium equations in the circumferential, meridional and normal directions, respectively, are

$$\frac{1}{r}\frac{\partial N_\theta}{\partial \theta} + \frac{N_{s\theta}\cos\phi}{r} + \frac{1}{r_n}\frac{\partial N_{s\theta}}{\partial \phi} + \frac{N_{\theta s}\cos\phi}{r} + p_\theta = 0$$

$$\frac{1}{r_n}\frac{\partial N_s}{\partial \phi} + \frac{(N_s - N_\theta)\cos\phi}{r} + \frac{1}{r}\frac{\partial N_{\theta s}}{\partial \theta} + p_s = 0$$

$$\frac{N_\theta}{r_\theta} + \frac{N_s}{r_n} = p_n$$

When the shell is cylindrical, spherical or conical and when the loading and supports are such that the problem is statically determinate, it may be possible to solve these three equations for the three unknowns N_s, N_θ and $N_{s\theta}$ and to then determine the stresses using $\sigma_s = N_s/h$, $\sigma_\theta = N_\theta/h$ and $\tau_{s\theta} = N_{s\theta}$. It is then possible to express the strains in terms of the force resultants according to

$$\epsilon_\theta = \frac{N_\theta - \nu N_s}{Eh}$$

$$\epsilon_s = \frac{N_s - \nu N_\theta}{Eh}$$

$$\gamma_{\theta s} = \frac{N_{\theta s}}{Gh} = \frac{\tau_{s\theta}}{Gh}$$

and to then determine the displacements by integrating the equations

$$\epsilon_\theta = \frac{1}{r}\left(\frac{\partial v}{\partial \theta} + w\sin\phi + u\cos\phi\right) = \frac{N_\theta - \nu N_s}{Eh}$$

$$\epsilon_s = \frac{\partial u}{\partial s} + \frac{w}{r_n} = \frac{N_s - \nu N_\theta}{Eh}$$

$$\gamma_{s\theta} = \frac{\partial v}{\partial s} + \frac{1}{r}\left(\frac{\partial u}{\partial \theta} - v\cos\phi\right) = \frac{\tau_{s\theta}}{Gh}$$

With the stresses and displacements known, the problem can be considered solved.

REFERENCES

1. Timoshenko, S. P., and Woinowsky-Krieger, S. *Theory of Plates and Shells*, 2nd ed., McGraw-Hill, New York, 1959.
2. Langhaar, H. L. *Foundations of Practical Shell Analysis*. Department of Theoretical and Applied Mechanics, University of Illinois, 1962.
3. Kraus, H. *Thin Elastic Shells*. Wiley, New York, 1967.
4. Ugural, A. C. *Stresses in Plates and Shells*. McGraw-Hill, New York, 1981.
5. Roark, R. J., and Young, W. C., *Formulas for Stress and Strain*. 5th ed. McGraw-Hill, New York, 1975.
6. Flügge, W. *Stresses in Shells*. Springer-Verlag, Berlin, 1962.

EXERCISES

Section 8-2

1. Show that the derivatives of the unit vectors are given by

$$\frac{\partial \mathbf{e_n}}{\partial n} = \mathbf{0}, \qquad \frac{\partial \mathbf{e_\theta}}{\partial n} = \mathbf{0}, \qquad \frac{\partial \mathbf{e_s}}{\partial n} = \mathbf{0},$$

$$\frac{\partial \mathbf{e_n}}{\partial \theta} = \mathbf{e_\theta} \sin \phi, \qquad \frac{\partial \mathbf{e_\theta}}{\partial \theta} = -\mathbf{e_s} \cos \phi - \mathbf{e_n} \sin \phi, \qquad \frac{\partial \mathbf{e_s}}{\partial \theta} = \mathbf{e_\theta} \cos \phi,$$

$$\frac{\partial \mathbf{e_n}}{\partial s} = \frac{\mathbf{e_s}}{r_n}, \qquad \frac{\partial \mathbf{e_\theta}}{\partial s} = \mathbf{0}, \qquad \frac{\partial \mathbf{e_s}}{\partial s} = -\frac{\mathbf{e_n}}{r_n}.$$

(Hint: Since each of the unit vectors satisfies $\mathbf{e_i} \cdot \mathbf{e_i} = 1$, it follows by taking a derivative that $\mathbf{e_i} \cdot d\mathbf{e_i}/ds = 0$, that is, that $d\mathbf{e_i}/ds$ is perpendicular to $\mathbf{e_i}$.)

Section 8-3

2. Using the result of problem 1, show that the equilibrium equations are given by Equations (8-2).
3. Show that a moment equation about the n direction yields the result that $N_{s\theta} = N_{\theta s}$.
4. Using the definition

$$\epsilon_n(P) = \lim_{d\mathbf{r} \to 0} \frac{d\mathbf{r^*} \cdot d\mathbf{r^*} - d\mathbf{r} \cdot d\mathbf{r}}{2 d\mathbf{r} \cdot d\mathbf{r}}$$

where $d\mathbf{r} = \mathbf{e_n}\, dn + \mathbf{e_\theta} r\, d\theta + \mathbf{e_s}\, ds$, $d\mathbf{u} = d(w\mathbf{e_n} + u\mathbf{e_s} + v\mathbf{e_\theta})$ and $d\mathbf{r^*} = d\mathbf{r} + d\mathbf{u}$, carry out the details leading to

$$\epsilon_n(P) = n_2 n_2 \frac{1}{r}\left(\frac{\partial v}{\partial \theta} + w \sin \phi + u \cos \phi\right) + n_3 n_3 \left(\frac{\partial u}{\partial s} + \frac{w}{r_n}\right)$$
$$+ n_1 n_2 \frac{1}{r}\left(\frac{\partial w}{\partial \theta} - v \sin \phi\right) + n_1 n_3 \left(\frac{\partial w}{\partial s} - \frac{u}{r_n}\right) + n_2 n_3 \left(\frac{\partial v}{\partial s} + \frac{1}{r}\left(\frac{\partial u}{\partial \theta} - v \cos \phi\right)\right)$$

and thus verify that the membrane strains are given by Equations (8-4).

5. Verify that the combination of Equations (8-6) and (8-8) yields $\sigma_\theta = N_\theta/h$, $\sigma_s = N_\theta/h$ and $\tau_{s\theta} = N_{s\theta}/h$.

Section 8-4

6. Specialize Equations (8.9c), (8.10c) and (8.11c) for the case of a cylindrical shell. Determine $\tau_{s\theta}$ and v and show that they coincide with the results given by the equations from the corresponding elementary theory for a thin tube, that is, $\tau = Tr/J$ and $JG\theta' = T$.

7. Specialize Equations (8.9c), (8.10c) and (8.11c) for the case of a spherical shell and thus solve

EXERCISES

the problem shown below. How do these results compare to those corresponding to the use of the elementary theory to compute stresses according to $\tau = Tr/J$ and displacements according to $JG\phi' = T$?

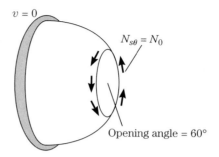

8. Repeat problem 7 with the opening angle = 45°.
9. Repeat problem 7 with the opening angle = 30°.
10. Specialize Equations (8.9c), (8.10c) and (8.11c) for the case of a conical shell and thus solve the problem shown below. Take $D = 2d$ and $L = 2D$. How do these results compare to those corresponding to the use of the elementary theory to compute stresses according to $\tau = Tr/J$ and displacements according to $JG\phi' = T$?

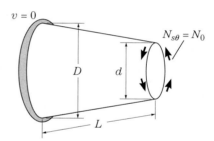

11. Repeat problem 10 with $D = 2d$ and $L = D$.
12. Repeat problem 10 with $D = 2d$ and $L = d$.
13. A cylindrical shell hangs under its own weight, as shown below. Determine the uniform internal pressure necessary to prevent any radial displacement at the top of the cylinder. Determine also the total change in length of the cylinder.

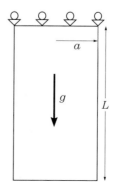

14. A cylindrical shell rotates about its axis, as shown below. Determine the stresses and displacements.

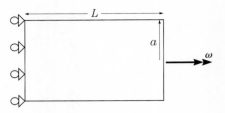

15. Repeat problem 13 if there are roller supports at the right end that prevent axial displacement.

16. A complete thin spherical shell is subjected to a uniform internal pressure p_0. Show that the governing equations can be solved to yield $N_s = p_0 R/2$, $w = p_0 R^2 (1 - \nu)/2Eh$ and $u = 0$.

17. A hemispherical shell hangs under its own weight, as shown below. Determine the stresses and displacements. Take the specific weight of the material composing the shell to be γ.

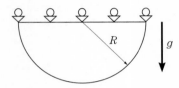

18. A hemispherical shell contains a liquid of specific weight γ, as shown below. Determine the stresses and displacements. Neglect the weight of the shell.

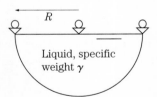

19. A conical frustrum rotates about its axis, as shown below. Determine the stresses and displacements. Take $D = 2d$, $L = D$ and the specific weight to be γ.

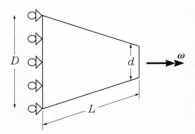

20. A conical shell hangs under its own weight, as shown below. Determine the stresses and displacements. Take $L = 2D$ and the specific weight to be γ.

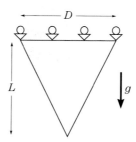

21. A water container is constructed from a cylindrical and a conical shell, as shown below. Determine (a) the stresses in the cylindrical and conical portions, (b) the displacements in the cylindrical and conical portions and (c) the difference between the horizontal displacements in the cylinder and cone at the junction. Take $D = H$ and the specific weight of the water to be γ.

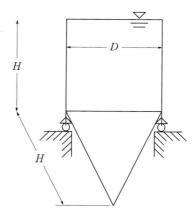

22. A water container is constructed from a cylindrical and a spherical shell, as shown below. Determine (a) the stresses in the cylindrical and spherical portions, (b) the displacements in the cylindrical and spherical portions and (c) the difference between the horizontal displacements in the cylinder and cone at the junction. Take $D = H$ and the specific weight of the water to be γ.

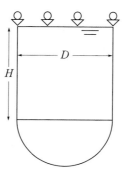

Section 8-5

23. Show in Example 8-4 that by integrating $N_{s\theta}(s=0)$ around the shell, the resultant force obtained is equal to the weight of the shell.

24. Show in Example 8-4 that by integrating $N_s(s=0)$ around the shell, the resultant moment obtained is equal to the moment of the weight of the shell.

25. Show in Example 8-5 that by integrating $N_{s\theta}(s=0)$ around the shell, the resultant force obtained is equal to the weight of the shell.

26. Show in Example 8-5 that by integrating $N_s(s=0)$ around the shell, the resultant moment obtained is equal to the moment of the weight of the shell.

27. Show in Example 8-6 that by integrating $N_{s\theta}(\phi=\pi/2)$ around the shell, the resultant force obtained is equal to the weight of the shell.

28. Show in Example 8-6 that by integrating $N_s(\phi=\pi/2)$ around the shell, the resultant moment obtained is equal to the moment of the weight of the shell.

Calculus of Variations

Recall that when dealing with discrete systems the potential energy V is a function of the form

$$V = V(x_1, x_2, \ldots, x_n)$$

The term *function* means that when the values of the x's are specified, the corresponding value of the scalar function V can be determined. The equations of equilibrium can then be generated by determining the values of the x's for which the function V is stationary, that is,

$$\frac{\partial V}{\partial x_i} = 0 \quad i = 1, 2, \ldots, n$$

Contrast this with a typical situation involving the total potential energy for the typical axial problem shown in Figure A-1. The total potential energy can be expressed as

$$V(u) = \int_0^L \left(\frac{AEu'^2}{2} - qu\right) dx - P_0 u(L) \tag{A-1}$$

with $u(0) = 0$ as the displacement boundary condition. If the *principle* of stationary potential energy is to be satisfied, it follows that it must be possible to determine what it means for the potential energy given by Equation (A-1) to be stationary.

The potential energy as given in (A-1) is expressed in terms of a *functional*, meaning roughly that the scalar value of V depends not on a discrete set of variables but

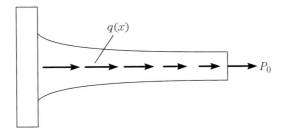

FIGURE A-1 Typical axial problem.

rather on an entire function $u(x)$, defined on the interval $0 \leq x \leq L$. Requiring that the functional (A-1) be stationary essentially asks the following question:

> With respect to which function $u(x)$ do small variations in that function produce corresponding vanishing variations in the value of $V(u)$?

In connection with the meaning of the principle of stationary potential energy, it is expected that the result of requiring that $V(u)$ be stationary will be a statement of equilibrium in terms of the function $u(x)$.

To make these ideas definite, consider Figure A-2. Shown are several permissible functions $u(x)$ each of which satisfies the condition $u(0) = 0$, dictated by the constraint indicated in Figure A-1. Also note that each $u_i(x)$ is a continuous function with a continuous derivative. For such a class of functions the integral portion of $V(u)$ is well defined. Strictly speaking, it turns out that the function $u(x)$ must be continuous but is only required to have a derivative that is piecewise-continuous.

For investigating what is meant by a stationary value of the functional $V(u)$, we refer to Figure A-3. The idea is to assume that the function $u(x)$ shown is the one that makes V stationary and to develop an algorithm that will then determine which $u(x)$ is actually the correct one. Consider then a nearby function $u(x) + \delta u(x)$, where $\delta u(x)$ is referred to as the variation of $u(x)$ and clearly vanishes at $x = 0$, that is, $\delta u(0) = 0$. Represent the variation as $\delta u(x) = \epsilon \eta(x)$, where ϵ is a small parameter and $\eta(x)$ is taken to be an *arbitrary* continuous function satisfying $\eta(0) = 0$. The potential energy associated with the function $u + \epsilon \eta$ is then

$$V^* = V(u + \epsilon \eta) = \int_0^L \left(\frac{AE(u' + \epsilon \eta')^2}{2} - q(u + \epsilon \eta) \right) dx - P_0(u(L) + \epsilon \eta(L))$$

The reader should show that this can be represented as

$$V^* = V(u) + \epsilon \left[\int_0^L (AEu'\eta' - q\eta) dx - P_0 \eta(L) \right] + \frac{\epsilon^2}{2} \int_0^L AE\eta'^2 \, dx$$

which, with u assumed known and η arbitrary, is a function of the small parameter ϵ.

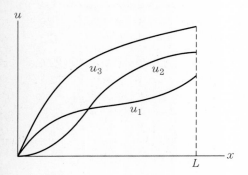

FIGURE A-2 Permissible u functions.

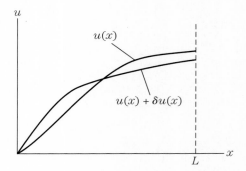

FIGURE A-3 Variation of u.

APPENDIX A CALCULUS OF VARIATIONS

It follows that for small ϵ, V^* can be represented as

$$V^* = V + \epsilon\left(\frac{dV^*}{d\epsilon}\bigg|_{\epsilon=0}\right) + \frac{\epsilon^2}{2}\left(\frac{d^2V^*}{d\epsilon^2}\bigg|_{\epsilon=0}\right) + \ldots$$

and that for V to be stationary when $\epsilon = 0$

$$\frac{dV}{d\epsilon}\bigg|_{\epsilon=0} = \int_0^L (AEu'\eta' - q\eta)\,dx - P_0\eta(L) = 0$$

Integration by parts yields

$$0 = AEu'\eta\bigg|_0^L - \int_0^L ((AEu')' + q)\eta\,dx - P_0\eta(L) = 0$$

which with $\eta(0) = 0$ can be expressed as

$$(AEu'(L) - P_0)\eta(L) - \int_0^L ((AEu')' + q)\eta\,dx = 0$$

With δu and hence η an arbitrary function, we conclude that the above equation is satisfied, and hence the stationary value of V occurs, when

$$(AEu')' + q = 0 \qquad 0 \leq x \leq L$$

and

$$AEu'(L) = P_0$$

which are, respectively, the equation of equilibrium and the force or mechanical boundary condition at $x = L$. Note that the quantity appearing on the left side of the boundary condition is the force transmitted, that is, $P = AEu'$.

Thus it is seen that the requirement for the potential energy functional $V(u)$ to be stationary, in the sense described above, leads to equilibrium in the interior given by $(AEu')' + q = 0$ and to what is essentially a local equilibrium condition requiring that the internal force transmitted assumes the boundary value P_0 at $x = L$.

In the context of the calculus of variations the displacement boundary condition $u(0) = 0$ is referred to as an *essential boundary condition*. The boundary condition $AEu'(L) = P_0$ is referred to as a *natural boundary condition*. In other settings an essential boundary condition can be referred to as a *geometric* boundary condition or a *forced* boundary condition, with a natural boundary condition being referred to as a *force* or *mechanical* boundary condition. These alternate designations have to do with the fact that a geometric boundary condition often refers to some aspect of the geometry of deformation, that is, displacement or slope, and that the force boundary condition refers to a force or moment at the boundary.

Another approach that leads formally to the stationary value of the functional Equation (A-1) is as follows. Beginning with

$$V(u) = \int_0^L \left(\frac{AEu'^2}{2} - qu\right)dx - P_0u(L)$$

form
$$\Delta V = V(u + \delta u) - V(u)$$
$$= \int_0^L (AEu'\, \delta u' - q\, \delta u)\, dx - P_0\, \delta u(L) + \frac{1}{2}\int_0^L AE(\delta u')^2\, dx$$
$$= \delta V + \frac{1}{2}\delta^2 V$$

where δu is an arbitrary virtual displacement satisfying $\delta u(0) = 0$. In the expression for ΔV, the first variation δV is the term linear in δu and its derivative while $\delta^2 V$, the second variation, is the quadratic term in the variations. The conditions for the stationary value of V are obtained by equating the first variation to zero, leading to

$$\int_0^L (AEu'\, \delta u - q\, \delta u)\, dx - P_0\, \delta u(L) = 0$$

which is precisely the same as $dV(0)/d\epsilon = 0$. Integration by parts would again lead to the differential equation of equilibrium and the mechanical boundary condition.

In a more general situation where the functional has the form

$$I(u) = \int_{x_1}^{x_2} F(x, u, u')\, dx \qquad\text{(A-2)}$$

with no essential boundary conditions on u, it turns out that carrying through the details of the algorithm described above leads to

$$0 = \int_{x_1}^{x_2} \left(\frac{\partial F}{\partial u}\, \delta u + \frac{\partial F}{\partial u'}\, \delta u'\right) dx$$

which after an integration by parts can be expressed as

$$0 = \left.\frac{\partial F}{\partial u'}\, \delta u\right|_{x_1}^{x_2} + \int_{x_1}^{x_2} \left[\frac{\partial F}{\partial u} - \frac{d}{dx}\left(\frac{\partial F}{\partial u'}\right)\right] \delta u\, dx$$

or

$$0 = \frac{\partial F}{\partial u'}(x_2)\, \delta u(x_2) - \frac{\partial F}{\partial u'}(x_1)\, \delta u(x_1) + \int_{x_1}^{x_2} \left[\frac{\partial F}{\partial u} - \frac{d}{dx}\left(\frac{\partial F}{\partial u'}\right)\right] \delta u\, dx$$

The arbitrariness of δu then allows us to conclude that

$$\frac{\partial F}{\partial u} - \frac{d}{dx}\left(\frac{\partial F}{\partial u'}\right) = 0 \qquad x_1 \le x \le x_2 \qquad\text{(A-3)}$$

is the differential or *Euler-Lagrange* equation and that

$$\frac{\partial F}{\partial u'}(x_1) = \frac{\partial F}{\partial u'}(x_2) = 0$$

are the natural boundary conditions that must be satisfied in order for a stationary value of the functional Equation (A-2).

Note that if u is prescribed at $x = x_1$, $\delta u(x_1) = 0$ and nothing need be said regarding $\partial F(x_1)/\partial u'$ with a similar observation about the boundary condition at $x = x_2$.

APPENDIX A CALCULUS OF VARIATIONS

Collectively, it can be stated that at a boundary

Essential		Natural
u is specified	or	$\dfrac{\partial F}{\partial u'} = 0$

If it turns out that there are additional boundary terms in the functional as in Equation (A-1), then it may be that at a boundary

Essential		Natural
u is specified	or	$\dfrac{\partial F}{\partial u'} = \text{const}$

For a functional of the form

$$I(u) = \int_{x_1}^{x_2} F(x, u, u', u'')\,dx$$

requiring I to be stationary leads to

$$0 = \left\{\delta u\left[\frac{\partial F}{\partial u'} - \frac{d}{dx}\left(\frac{\partial F}{\partial u''}\right)\right]\right\}\bigg|_{x_1}^{x_2} + \left\{\delta u'\left(\frac{\partial F}{\partial u''}\right)\right\}\bigg|_{x_1}^{x_2}$$
$$+ \int_{x_1}^{x_2}\left[\frac{\partial F}{\partial u} - \frac{d}{dx}\left(\frac{\partial F}{\partial u'}\right) + \frac{d^2}{dx^2}\left(\frac{\partial F}{\partial u''}\right)\right]\delta u\,dx$$

from which the Euler-Lagrange equation is

$$\frac{\partial F}{\partial u} - \frac{d}{dx}\left(\frac{\partial F}{\partial u'}\right) + \frac{d^2}{dx^2}\left(\frac{\partial F}{\partial u''}\right) = 0 \qquad x_1 \le x \le x_2$$

with boundary conditions as follows:

Essential		Natural
u is specified	or	$\dfrac{\partial F}{\partial u'} - \dfrac{d}{dx}\left(\dfrac{\partial F}{\partial u''}\right) = 0$, or a const
u' is specified	or	$\dfrac{\partial F}{\partial u''} = 0$, or a const

Generalizations to functionals containing higher derivatives are not difficult.

Another class of problems to which the calculus of variations is frequently applied are situations where the potential energy can be expressed as

$$I = \iint_D F(x, y, u, v, u_x, u_y, v_x, v_y)\,dA$$

where $u_x = \partial u/\partial x$, $u_y = \partial u/\partial y$, $v_x = \partial v/\partial x$, and $v_y = \partial v/\partial y$. The Euler-Lagrange equations are the coupled partial differential equations

$$\frac{\partial F}{\partial u} - \frac{\partial}{\partial x}\left(\frac{\partial F}{\partial u_x}\right) - \frac{\partial}{\partial y}\left(\frac{\partial F}{\partial u_y}\right) = 0 \qquad \text{in } D$$

and

$$\frac{\partial F}{\partial u} - \frac{\partial}{\partial x}\left(\frac{\partial F}{\partial v_x}\right) - \frac{\partial}{\partial y}\left(\frac{\partial F}{\partial v_y}\right) = 0 \quad \text{in } D$$

and the boundary conditions are as follows:

Essential		*Natural*	
u is specified	or	$n_x \dfrac{\partial F}{\partial u_x} + n_y \dfrac{\partial F}{\partial u_y}$	is specified
v is specified	or	$n_x \dfrac{\partial F}{\partial v_x} + n_y \dfrac{\partial F}{\partial v_y}$	is specified

where n_x and n_y are the direction cosines of the normal to the boundary of D.

B Finite-Element Models and MATLAB Codes

Chapter 3. As developed in Chapter 3, the torsion problem can be formulated in terms of the Prandtl stress function according to

$$\nabla^2 \Phi + 2G\theta = 0 \quad \text{in D}$$
$$\Phi = 0 \quad \text{on B}$$

with the torque given by

$$T = 2 \int\int_D \Phi \, dA$$

and the stresses by

$$\tau_{xz} = -\frac{\partial \Phi}{\partial y}$$

and

$$\tau_{xy} = \frac{\partial \Phi}{\partial z}$$

Except in relatively few instances—namely, those involving rectangles, circles and other special regions—analytical solutions of these equations are very difficult to obtain. One of the approaches for obtaining accurate numerical estimates is the use of the finite-element method. This appendix will outline the development of the finite-element model for the torsion problem.

As mentioned in Chapter 2, the finite-element method can be considered an application of the method of Ritz in those situations where there is a corresponding potential energy for the problem. For the torsion problem as developed, the potential energy can be represented as $V = U + \Omega$ where

$$U = \int_{Vol} \frac{(\tau_{xz}^2 + \tau_{xy}^2)}{2G} \, dVol = L \int_D \frac{\left(\frac{\partial \Phi}{\partial y}\right)^2 + \left(\frac{\partial \Phi}{\partial z}\right)^2}{2G} \, dA$$

and

$$\Omega = -T(L\theta) = -L\theta \left(2 \int\int_D \Phi \, dA \right)$$

Multiply by G and divide by L to obtain

$$V^* = \left(\frac{G}{L}\right)V = \int_D \left[\frac{\left(\frac{\partial \Phi}{\partial y}\right)^2 + \left(\frac{\partial \Phi}{\partial z}\right)^2}{2} - 2G\theta\Phi \right] dA$$

The reader should show that by requiring V^* to be stationary the governing differential equation $\nabla^2 \Phi + 2G\theta = 0$ results.

The first step in the development of a finite-element model is to choose a set of nodes and elements, as shown in Figure B-1a. This is referred to as the discretization process. The next step involves making an assumption for the form of the Φ function within a typical element, indicated in Figure B-1b. Consistent with the fact that each element is assumed to have three nodes, it is assumed that Φ is a linear function of y and z, that is,

$$\Phi_e = a + by + cz$$

as pictured in Figure B-2.

The constants a, b and c are determined by requiring

$$\Phi_i = a + by_i + cz_i$$
$$\Phi_j = a + by_j + cz_j$$
$$\Phi_k = a + by_k + cz_k$$

Solving for a, b and c and substituting back into $\Phi = a + by + cz$ gives

$$\Phi_e = N_i \Phi_i + N_j \Phi_j + N_k \Phi_k$$

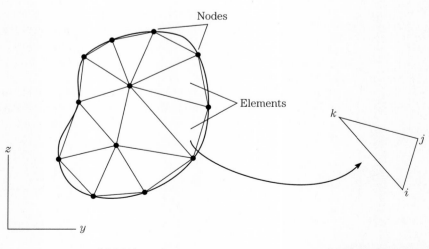

(a) Mesh   (b) Typical element

FIGURE B-1 Nodes and elements.

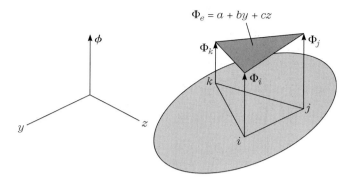

FIGURE B-2 Φ as a linear function on a typical element.

where

$$N_i = \frac{a_i + b_i y + c_i z}{2A_e} \quad i = 1, 2, 3$$

with

$$a_i = y_j z_k - y_k z_j$$
$$b_i = z_j - z_k$$
$$c_i = y_k - y_j$$

where $i, j,$ and k are to be permuted cyclically. $2A_e$ is the determinant of the coefficients, namely,

$$2A_e = \begin{vmatrix} 1 & y_i & z_i \\ 1 & y_j & z_j \\ 1 & y_k & z_k \end{vmatrix}$$

where A_e is the area of the element. The choice for the nodes defining the element, that is, the choice for i, j and k, must take place in a counterclockwise direction around the element. The functions N_i, N_j and N_k are referred to as the linear interpolation functions. They appear in Figure B-3.

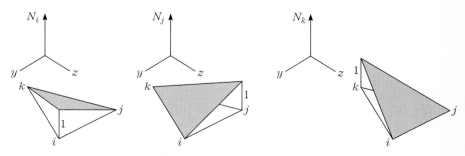

FIGURE B-3 Interpolation functions.

In matrix notation, the elemental representation can be expressed as

$$\Phi_e = \Phi_e^T \mathbf{N} = \hat{\mathbf{N}}^T \Phi_e$$

Additionally, the elemental representations of the derivatives are

$$\frac{\partial \Phi_e}{\partial y} = \Phi_e^T \frac{\partial \mathbf{N}}{\partial y} = \frac{\partial \mathbf{N}^T}{\partial y} \Phi_e$$

$$\frac{\partial \Phi_e}{\partial z} = \mathbf{u}_e^T \frac{\partial \mathbf{N}}{\partial z} = \frac{\partial \mathbf{N}^T}{\partial z} \mathbf{u}_e$$

Recalling the expressions for the components of $\mathbf{N}$, these can be represented as

$$\frac{\partial \mathbf{N}}{\partial y} = \frac{\mathbf{b_e}}{2A_e}$$

and

$$\frac{\partial \mathbf{N}}{\partial z} = \frac{\mathbf{c_e}}{2A_e}$$

resulting in

$$\frac{\partial \Phi_e}{\partial y} = \frac{\Phi_e^T \mathbf{b_e}}{2A_e} = \frac{\mathbf{b_e^T} \Phi_e}{2A_e}$$

and

$$\frac{\partial \Phi_e}{\partial z} = \frac{\Phi_e^T \mathbf{c_e}}{2A_e} = \frac{\mathbf{c_e^T} \Phi_e}{2A_e}$$

Note that both partial derivatives are constant within an element.

Return to the expression for the potential energy and write

$$V^* \approx \sum_e \int_{A_e} \left[\frac{\left(\frac{\partial \Phi_e}{\partial y}\right)^2 + \left(\frac{\partial \Phi_e}{\partial z}\right)^2}{2} - 2G\theta\Phi_e \right] dA$$

Substituting for Φ_e and its derivatives gives

$$V^* \approx \sum_e \int_{A_e} \left[\frac{\Phi_e^T \mathbf{b_e} \mathbf{b_e^T} \Phi_e + \Phi_e^T \mathbf{c_e} \mathbf{c_e^T} \Phi_e}{8A_e^2} - 2G\theta \Phi_e^T \mathbf{N} \right] dA$$

or, upon recognizing that the first term of the integrand is constant,

$$V^* = \sum_e \left[\frac{1}{2} \Phi_e^T \mathbf{k}_e \Phi_e - \Phi_e^T \mathbf{f}_e \right]$$

where each of the

$$\mathbf{k}_e = \frac{\mathbf{b_e} \mathbf{b_e^T} + \mathbf{c_e} \mathbf{c_e^T}}{4A_e}$$

APPENDIX B FINITE-ELEMENT MODELS AND MATLAB CODES

are three by three stiffness matrices and each of the

$$\mathbf{f_e} = 2G\theta \int_{A_e} \mathbf{N}\, dA = \frac{2G\theta A_e}{3}[1\ \ 1\ \ 1]^T$$

are three by one load vectors.

The sum in the expression for V^* indicates an assembly process whereby the individual stiffnesses and loads are assembled or loaded into their proper positions in a global stiffness matrix, leading to

$$V^* = \frac{\Phi_G^T \mathbf{k_G} \Phi_G}{2} - 2G\theta \Phi_G^T \mathbf{f_G}$$

The potential energy is now a function of the nodal Φ_i variables whose stationary value is given by

$$\frac{\partial V^*}{\partial \Phi_i} = 0$$

leading to

$$\mathbf{k_G}\Phi_G - \mathbf{f_G} = 0$$

The boundary conditions $\Phi = 0$ on B must now be enforced as constraints on all the nodal variables corresponding to nodes on the boundary. The equations are then solved for the remaining Φ_i, after which approximate values for the stresses and the torque transmitted can be computed.

EXAMPLE B-1 Consider the torsion problem for a square cross section, shown in Figure B-4a. Based on ideas discussed in section 3-3, only a quarter of the region is modeled, as shown in Figure B-4b. First form the stiffness matrices $\mathbf{k_e} = (\mathbf{b_e}\mathbf{b_e}^T + \mathbf{c_e}\mathbf{c_e}^T)/4A_e$ and the load matrices $\mathbf{f_e} = (2G\theta A_e/3)\,[1\ 1\ 1]^T$ for each of the elements and then assemble them into the global matrices.

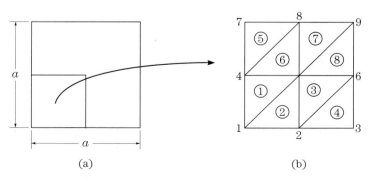

FIGURE B-4 Torsion of a square bar.

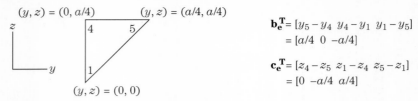

FIGURE B-5 Node numbers and $\mathbf{b_e}$ and $\mathbf{c_e}$ vectors.

Element 1: A permissible set of node numbers for this element is (1, 5, 4). Figure B-5 shows the corresponding $\mathbf{b_e}$ and $\mathbf{c_e}$ vectors. Then, with $4A_e = a^2/8$, the stiffness matrix is

$$\mathbf{k_e} = \frac{(\mathbf{b_e b_e^T} + \mathbf{c_e c_e^T})}{4A_e} = \left(\frac{1}{2}\right)\begin{bmatrix} 1 & 0 & -1 \\ 0 & 1 & -1 \\ -1 & -1 & 2 \end{bmatrix}$$

The corresponding load vector is $\mathbf{f_e} = (G\theta a^2/48)\,[1\ 1\ 1]^T$. The student should verify that the elemental stiffnesses for elements 3, 5 and 7 will be the same as the elemental stiffness for element 1 and that the load vectors will be the same for all elements.

Element 2: A permissible set of node numbers for this element is (1, 2, 5). Figure B-6 shows the corresponding $\mathbf{b_e}$ and $\mathbf{c_e}$ vectors. Then, with $4A_e = a^2/8$, the stiffness matrix is

$$\mathbf{k_e} = \frac{(\mathbf{b_e b_e^T} + \mathbf{c_e c_e^T})}{4A_e} = \left(\frac{1}{2}\right)\begin{bmatrix} 1 & -1 & 0 \\ -1 & 2 & -1 \\ 0 & -1 & 1 \end{bmatrix}$$

The student should verify that elements 4, 6 and 8 will have the same stiffness matrices as element 2.

Assembly: First, note that the size of the global stiffness matrix $\mathbf{K_G}$ will be nine by nine, that is, the number of nodes. Similarly, the size of the global load vector $\mathbf{F_G}$ is nine by one. The node numbers for element 1, namely, 1, 5 and 4, are essentially pointers for where the elements of $\mathbf{k_e}$ are to be loaded or assembled into the global stiffness matrix. This idea is indicated in Figure B-7.

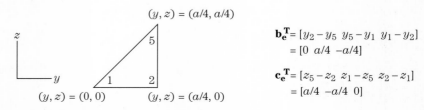

FIGURE B-6 Node numbers and $\mathbf{b_e}$ and $\mathbf{c_e}$ vectors.

APPENDIX B FINITE-ELEMENT MODELS AND MATLAB CODES

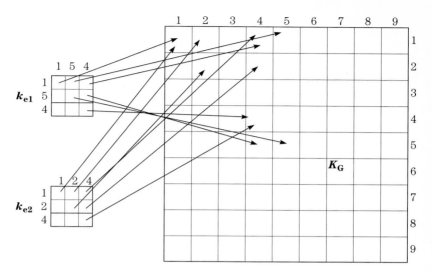

FIGURE B-7 Assembly from the elemental to the global.

Assembly for the load vectors is accomplished in an entirely similar fashion. Using these ideas, the student should verify that the global stiffness and load matrices are

$$\mathbf{K}_G = \frac{1}{2}\begin{bmatrix} 2 & -1 & 0 & -1 & 0 & 0 & 0 & 0 & 0 \\ -1 & 4 & -1 & 0 & -2 & 0 & 0 & 0 & 0 \\ 0 & -1 & 2 & 0 & 0 & -1 & 0 & 0 & 0 \\ -1 & 0 & 0 & 4 & -2 & 0 & -1 & 0 & 0 \\ 0 & -2 & 0 & -2 & 8 & -2 & 0 & -2 & 0 \\ 0 & 0 & -1 & 0 & -2 & 4 & 0 & 0 & -1 \\ 0 & 0 & 0 & -1 & 0 & 0 & 2 & -1 & 0 \\ 0 & 0 & 0 & 0 & -2 & 0 & -1 & 4 & -1 \\ 0 & 0 & 0 & 0 & 0 & -1 & 0 & -1 & 2 \end{bmatrix}; \quad \mathbf{F}_G = \frac{G\theta a^2}{48}\begin{bmatrix} 2 \\ 3 \\ 1 \\ 3 \\ 6 \\ 3 \\ 1 \\ 3 \\ 2 \end{bmatrix}$$

The boundary condition $\Phi = 0$ on the outer boundary translates to the constraints $\Phi_1 = 0$, $\Phi_2 = 0$, $\Phi_3 = 0$, $\Phi_4 = 0$ and $\Phi_7 = 0$. The student is encouraged to show that this essentially results in elimination of the rows and columns corresponding to the constrained degrees of freedom so that the final equations to be solved are

$$\begin{bmatrix} 8 & -2 & -2 & 0 \\ -2 & 4 & 0 & -1 \\ -2 & 0 & 4 & -1 \\ 0 & -1 & -1 & 2 \end{bmatrix}\begin{bmatrix} \Phi_5 \\ \Phi_6 \\ \Phi_8 \\ \Phi_9 \end{bmatrix} = \frac{G\theta a^2}{24}\begin{bmatrix} 6 \\ 3 \\ 3 \\ 2 \end{bmatrix}$$

Carrying out the solution yields $\Phi_5 = 0.0885\, G\theta a^2$, $\Phi_6 = 0.1146\, G\theta a^2$, $\Phi_8 = 0.1146\, G\theta a^2$ and $\Phi_9 = 0.1562\, G\theta a^2$. The torque-twist relation is then determined by evaluating

$$T = 2\iint \Phi\, dA = 8\sum \iint \Phi_e\, dA$$

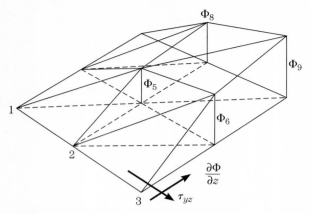

FIGURE B-8 Φ surface over the quarter region.

On a given element

$$\iint \Phi_e \, dA = A_e \frac{(\Phi_{ei} + \Phi_{ej} + \Phi_{ek})}{3}$$

Carrying out all the sums yields

$$T = 8\left(\frac{a^2}{32}\right)\frac{(6\Phi_5 + 3\Phi_6 + 3\Phi_8 + 2\Phi_9)}{3} = 0.1276 \, G\theta a^4$$

The Φ surface over the quarter region is shown in Figure B-8. The maximum shear stress occurs at node 3 (or node 7) and is equal to the slope of the Φ surface in the direction of the normal to the boundary, that is, from node 3 to node 6. Thus,

$$\tau_{yz} = \frac{(\Phi_6 - 0)}{\left(\dfrac{a}{4}\right)} = 0.4584 \, G\theta a$$

or, upon using the torque-twist relation to eliminate $G\theta$,

$$\tau_{max} = 3.592 \frac{T}{a^3}$$

The exact values are $T = 0.1406 \, G\theta a^3$ and $\tau_{max} = 4.808 \, T/a^3$ so that the error in the torque-twist relation is approximately 9% and the error in the maximum shear stress is approximately 25%. Results involving refinement of the mesh are presented in Chapter 3.

A MATLAB code for investigating the torsion problem using T3 elements follows.

```
% Prandtl.m
% This MATLAB code generates a finite-element model
% for the torsion of a simply connected region using
% T3 elements. T3 or "constant strain" elements are
```

```
% known to be slowly converging so that the user may
% need to use a mesh with many elements to obtain
% accurate results. The inputs required are ne = number
% of elements, nn = number of nodes, nc = number of
% constrained nodes, an array ndc containing the nodes
% to be constrained, an array x containing the x coordinates
% of the nodes, an array y containing the y coordinates of
% the nodes, and arrays ni, nj and nk containing the first,
% second and third node numbers defining the elements. Also
% required are the input torque T and a parameter nvol to
% account for any symmetries. Outputs are the shear
% stress in each of the elements and an estimate of the
% torsional rigidity. ALLOWANCES MUST BE MADE FOR ANY
% SYMMETRIES THAT ARE USED IN MODELING THE REGION.
%
clear
% The data that follows correspond to the problem
% of the torsion of the square with quarter symmetry
% discussed previously.
ne=8;    % Number of elements
nn=9;    % Number of nodes
nc=5;    % Number of constrained nodes
nvol=4;   % Factor because of quarter symmetry
T=1000;   % Torque applied
x=[0.0 0.5 1.0 0.0 0.5 1.0 0.0 0.5 1.0]/2;   % x nodal coords
y=[0.0 0.0 0.0 0.5 0.5 0.5 1.0 1.0 1.0]/2;   % y nodal coords
ni=[1 1 2 2 4 4 5 5];     % NI per element
nj=[5 2 6 3 8 5 9 6];     % NJ per element
nk=[4 5 5 6 7 8 8 9];     % NK per element
ndc=[1 2 3 4 7];          % Nodes to be constrained
% End of data
kg=zeros(nn,nn);fg=zeros(nn,1);
for i=1:ne
  n1=ni(i);n2=nj(i);n3=nk(i);
  b(1)=y(n2)-y(n3);b(2)=y(n3)-y(n1);b(3)=y(n1)-y(n2);
  c(1)=x(n3)-x(n2);c(2)=x(n1)-x(n3);c(3)=x(n2)-x(n1);
  a2=c(3)*b(2)-c(2)*b(3);
  ke=(b'*b+c'*c)/(2*a2);
  kg(n1,n1)=kg(n1,n1)+ke(1,1);kg(n1,n2)=kg(n1,n2)+ke(1,2);
  kg(n1,n3)=kg(n1,n3)+ke(1,3);
  kg(n2,n1)=kg(n1,n2);kg(n2,n2)=kg(n2,n2)+ke(2,2);
  kg(n2,n3)=kg(n2,n3)+ke(2,3);
  kg(n3,n1)=kg(n1,n3);kg(n3,n2)=kg(n2,n3);
  kg(n3,n3)=kg(n3,n3)+ke(3,3);
  fg(n1)=fg(n1)+a2/3;fg(n2)=fg(n2)+a2/3;fg(n3)=
         fg(n3)+a2/3;
```

```
      end
    for j=1:nc
      n1=ndc(j);
        for k=1:nn
          kg(k,n1)=0;kg(n1,k)=0;
        end
        kg(n1,n1)=1.0;fg(n1)=0;
    end
    u=kg\fg;
    pvol=0.0;
    for i=1:ne
      n1=ni(i);n2=nj(i);n3=nk(i);
      b(1)=y(n2)-y(n3);b(2)=y(n3)-y(n1);b(3)=y(n1)-y(n2);
      c(1)=x(n3)-x(n2);c(2)=x(n1)-x(n3);c(3)=x(n2)-x(n1);
      a2=c(3)*b(2)-c(2)*b(3);
      pvol=pvol+(a2/2)*(u(n1)+u(n2)+u(n3))/3;
      ue=[u(n1) u(n2) u(n3)];
      px=b*ue'/a2; py=c*ue'/a2;pn(i)=sqrt(px^2+py^2);
    end
      Vol=nvol*2*pvol;
    for i=1:ne
      ss(i)=T*pn(i)/(Vol);
    end
    fprintf('The torsional constant (J) is %6.4f\n',Vol)
    fprintf('\n')
    fprintf('The elemental shear stresses are\n')
    fprintf('    Element      Tau\n')
    for ip=1:ne
    fprintf('       %2.0f       %5.0f\n',ip,ss(ip))
    end
 »
OUTPUT FOR QUARTER SYMMETRY MODEL OF SQUARE

The torsional constant (J) is 0.1276

The elemental shear stresses are
    Element      Tau
         1       2776
         2       2776
         3       2893
         4       3592
         5       3592
         6       2893
         7       1540
         8       1540
 »
```

APPENDIX B   FINITE-ELEMENT MODELS AND MATLAB CODES

Chapter 6. The starting point for the development of the finite-element model for the plane axisymmetric elasticity problem will be taken to be the total potential energy, namely,

$$\frac{V}{2\pi} = \int_a^b \left[\frac{E}{2(1-\nu^2)}\left(u'^2 + \left(\frac{u}{r}\right)^2 + 2\nu u' \frac{u}{r}\right) - \frac{E\alpha T}{1-\nu}\left(u' + \frac{u}{r}\right) - \rho r \omega^2 u \right] tr\, dr$$
$$- p_i a t(a) u(a) + p_o b t(b) u(b)$$

As outlined in section 6-8, linear interpolation is used according to

$$u_e(r) = u_i\left(1 - \frac{r}{l_e}\right) + u_{i+1}\left(\frac{r}{l_e}\right)$$

resulting in

$$\frac{V}{2\pi} = \sum_e \frac{1}{2}\mathbf{u}_e^T \mathbf{k}_e \mathbf{u}_e - \sum_e \mathbf{u}_e^T \mathbf{f}_e + p_i a t(a) u_1 - p_o b t(b) u_{N+1}$$

where

$$\mathbf{k}_e = \frac{E}{1-\nu^2} \int_{r_i}^{r_{i+1}} \left[\mathbf{N}' rt \mathbf{N}'^T + \mathbf{N}\frac{t}{r}\mathbf{N}^T + \nu(\mathbf{N}' t \mathbf{N}^T + \mathbf{N} t \mathbf{N}'^T)\right] dr$$

and

$$\mathbf{f}_e = \int_{r_i}^{r_{i+1}} \left[\frac{E\alpha T}{1-\nu}(rt\mathbf{N}' + t\mathbf{N}) + \rho\omega^2 r^2 t \mathbf{N}\right] dr$$

In the MATLAB code named "**presves**," which is listed below, the integrations are carried out using a second-order Gauss quadrature. Outputs are displacements and stresses at the nodes. All required inputs are described in the listing.

```
% presves.m
% This MATLAB code solves the variable-thickness
% axisymmetric plane elasticity problem. The user
% specifies the geometry consisting of r1 the inside
% radius, r2 the outside radius, and an array t
% consisting of the nodal thicknesses, an array T
% consisting of the nodal temperatures, rh the mass
% density, om the angular velocity, ne the number
% of elements, Young's modulus E, Poisson's ratio nu,
% coefficient of thermal expansion alp, nc the number of
% constraints, an array cdf consisting of the
% constrained degrees of freedom, sgin the radial stress
% at the inner radius, and sgout the radial stress at the outer
% radius. Output consists of a plot of the
% displacement u(r), the radial stress sr(r) and the
% hoop stress st(r).
clear
```

```
% The data below correspond to an annular constant-thickness
% steel disk with radii 3 and 6 in rotating at 10000 rpm
% modeled with 10 elements.
E=3.e7;nu=0.3;ec=E/(1-nu*nu);
alp=6.5e-6;et=alp*(1+nu)*ec;
rh=0.000734;om=1000*pi/3;
rms=rh*om*om
ne=10;nn=ne+1;
r1=3;r2=6;ri=(r2-r1)/ne;
r=r1:ri:r2;
t=[1 1 1 1 1 1 1 1 1 1];
T=[0 0 0 0 0 0 0 0 0 0];
nc=0;      % no constraints
cdf=[1];
kg=zeros(nn);qg=zeros(nn,1);
sp=1/sqrt(3);gp=[-sp sp];wp=[1 1];
 for i=1:ne
   rp(i)=(r(i)+r(i+1))/2;
 end
sgin=0;
sgout=0;
for ie=1:ne
 ip=ie+1;
 t1=t(ie);t2=t(ip);dt=t2-t1;
 r1=r(ie);r2=r(ip);dr=r2-r1;
 T1=T(ie);T2=T(ip);dT=t2-t1;
 ke=zeros(2);qe=zeros(2,1);
 for j=1:2
  g=gp(j);w=wp(j);s=(1+g)/2;
    rc=r1+dr*s;
    tc=t1+dt*s;
    Tc=T1+dT*s;
    n=[1-s;s];
    np=[-1;1]/dr;
    ke=ke+2*pi*dr*w*ec*tc*(rc*np*np'+n*n'/rc+
         nu*(n*np'+np*n'))/2;
    qe=qe+2*pi*dr*w*tc*(rms*rc^2*n+et*Tc*(rc*np+n))/2;
end
for ii=1:2
 i1=ie+ii-1;
    qg(i1)=qg(i1)+qe(ii);
    for jj=1:2
     j1=ie+jj-1;
      kg(i1,j1)=kg(i1,j1)+ke(ii,jj);
    end
 end
```

```
end
qg(1)=qg(1)+2*pi*sgin*r(1)*t(1);
qg(nn)=qg(nn)-2*pi*sgout*r(nn)*t(nn);
if nc > 0
 for ic=1:nc
  nd=cdf(ic);
    qg(nd)=0;
     for j=1:nn
      kg(j,nd)=0;
        kg(nd,j)=0;
         end
   kg(nd,nd)=1;
  end
end
u=kg\qg;
plot(r,u)
xlabel('radius')
ylabel('displacement')
title('Radial displacement as a function of position')
grid
pause
for ie=1:ne
 ip=ie+1;
 r1=r(ie);r2=r(ip);dr=r2-r1;
 T1=T(ie);T2=T(ip);dT=t2-t1;
 u1=u(ie);u2=u(ip);
 sr(ie)=ec*((u2-u1)/dr+nu*(u1+u2)/(r1+r2))-et*(T1+T2)/2;
 st(ie)=ec*(nu*(u2-u1)/dr+(u1+u2)/(r1+r2))-et*(T1+T2)/2;
 end
 for i=2:ne
 srn(i)=(sr(i-1)+sr(i))/2;
 stn(i)=(st(i-1)+st(i))/2;
  end
 srn(1)=2*sr(1)-srn(2);
 stn(1)=2*st(1)-stn(2);
 srn(nn)=2*sr(ne)-srn(ne);
 stn(nn)=2*st(ne)-stn(ne);
 plot(r,srn,':g',r,stn,'r--',r,srn,'wo',r,stn,'c*')
 grid
 xlabel('radius')
 ylabel('stresses')
  title('Radial stress (o) and hoop stress (*)')
```

Shown in Figures B-9 and B-10 are the displacement and stresses corresponding to the data at the beginning of the MATLAB listing above.

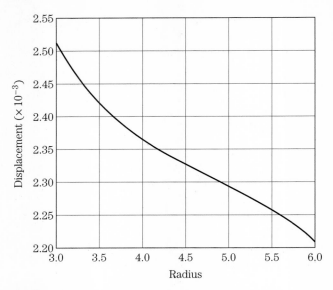

FIGURE B-9 Radial displacement as a function of position.

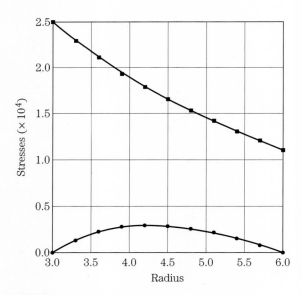

FIGURE B-10 Radial stress (•) and hoop stress (•).

Chapter 7. The starting point for the development of the finite-element model for the rectangular plate problem will be taken to be the total potential energy, namely,

$$V = \int_{\text{Area}} \left[\frac{D}{2} \{ (\nabla^2 w)^2 - 2(1-\nu)(w_{xx}w_{yy} - w^2_{xy}) \} - q(x,y)w \right] dA$$

plus terms to account for shears and moments applied at boundaries.

APPENDIX B FINITE-ELEMENT MODELS AND MATLAB CODES

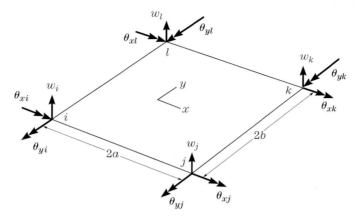

FIGURE B-11 Nodal degrees of freedom.

The element used for this model is due to Melosh [1], and to Zienkiewicz and Cheung [2], wherein the assumption made for the displacement w is the incomplete quartic equation

$$w = c_1 + c_2 x + c_3 y + c_4 x^2 + c_5 xy + c_6 y^2 + c_7 x^3 \\ + c_8 x^2 y + c_9 xy^2 + c_{10} y^3 + c_{11} x^3 y + c_{12} xy^3$$

As shown in Figure B-11, the 12 degrees of freedom for this element are the four nodal displacements w, the four nodal rotations $\theta_y = w_{,x}$ and the four nodal rotations $\theta_x = w_{,y}$, in terms of which the 12 constants $c_1, c_2, \ldots, c_{12}$ are determined. Also as shown, node numbering is in the counterclockwise direction.

The element that results from this assumption is called *nonconforming* because normal slope continuity is violated across inter-element boundaries. Constant strain state is contained within the assumption and convergence results. The MATLAB code given below describes the necessary inputs and resulting outputs. Examples and an explanation of the output are presented at the end of this appendix.

```
% Rectplate.m
% This MATLAB code uses the MZC element to analyze
% the displacements and moments in rectangular plates.
% Required inputs are arrays aa and bb containing the
% lengths of the sides of the rectangular elements, arrays
% n1a, n2a, n3a and n4a containing the first, second, third
% and fourth node number for the elements, Young's modulus E,
% Poisson's ratio nu, thickness h, an array containing the
% elemental pressure values, nc the number of constraints,
% an array cm containing the degrees of freedom to be
% constrained. Nodal displacements and moments are output.
% The values listed for the moments are obtained by
% evaluating the moments at the N = 2 Gauss points within
% the element and then extrapolating to the nodes.
%
```

```
clear
% The data below are for a four-element model of a square uniformly
loaded simply supported plate.
ne=4;nn=9;np=0;nc=15;
nu=0.3;E=1.e7;h=0.1;
aa=[1.25 1.25 1.25 1.25];bb=[1.25 1.25 1.25 1.25];
n1a=[1 2 4 5];
n2a=[2 3 5 6];
n3a=[5 6 8 9];
n4a=[4 5 7 8];
qz=1*[2 2 2 2];cm=[1 2 3 4 6 7 9 10 11 18 19 20 23 26 27];
% End of four-element data
% Beginning of 16-element data
ne=16;nn=25;np=0;nc=27;
nu=0.3;E=1.e7;h=0.1;
aa=[5/8 5/8 5/8 5/8 5/8 5/8 5/8 5/8 5/8 5/8 5/8 5/8 5/8
     5/8 5/8 5/8];
bb=[5/8 5/8 5/8 5/8 5/8 5/8 5/8 5/8 5/8 5/8 5/8 5/8 5/8
     5/8 5/8 5/8];
n1a=[1 2 3 4 6 7 8 9 11 12 13 14 16 17 18 19];
n2a=[2 3 4 5 7 8 9 10 12 13 14 15 17 18 19 20];
n3a=[7 8 9 10 12 13 14 15 17 18 19 20 22 23 24 25];
n4a=[6 7 8 9 11 12 13 14 16 17 18 19 21 22 23 24];
qz=1*[2 2 2 2 2 2 2 2 2 2 2 2 2 2 2 2];
cm=[1 2 3 4 6 7 9 10 12 13 15 16 17 30 31 32 45 46 47 60 61
     62 65 68 71 74 75];
% End of 16-element data
fr=E*h^3/(12*(1-nu^2));la=(1-nu)/2;
em=(E/(1-nu^2))*[1 nu 0;nu 1 0;0 0 la];
ndf=3*nn;
r3=sqrt(3);
se=[r3 -r3 -r3 r3];
te=[r3 r3 -r3 -r3];
kg=zeros(ndf);qg=zeros(ndf,1);kes=zeros(12,ndf);
                         % assembly
for i=1:ne
 a=aa(i);b=bb(i);as=a*a;bs=b*b;
 n1=n1a(i);n2=n2a(i);n3=n3a(i);n4=n4a(i);
 m3=3*n1;m2=m3-1;m1=m2-1;
 m6=3*n2;m5=m6-1;m4=m5-1;
 m9=3*n3;m8=m9-1;m7=m8-1;
 m12=3*n4;m11=m12-1;m10=m11-1;
 k1=(b/(6*a*a*a))*[6 0 -6*a -6 0 -6*a -3 0 -3*a 3 0 -3*a;
    0 0 0 0 0 0 0 0 0 0 0 0;
    -6*a 0 8*as 6*a 0 4*as 3*a 0 2*as -3*a 0 4*as;
    -6 0 6*a 6 0 6*a 3 0 3*a -3 0 3*a;
    0 0 0 0 0 0 0 0 0 0 0 0;
```

```
            -6*a 0 4*as 6*a 0 8*as 3*a 0 4*as -3*a 0 2*as;
            -3 0 3*a 3 0 3*a 6 0 6*a -6 0 6*a;
            0 0 0 0 0 0 0 0 0 0 0 0;
            -3*a 0 2*as 3*a 0 4*as 6*a 0 8*as -6*a 0 4*as;
            3 0 -3*a -3 0 -3*a -6 0 -6*a 6 0 -6*a;
            0 0 0 0 0 0 0 0 0 0 0 0;
            -3*a 0 4*as 3*a 0 2*as 6*a 0 4*as -6*a 0 8*as];
k2=(a/(6*b*b*b))*[6 6*b 0 3 3*b 0 -3 3*b 0 -6 6*b 0;
           6*b 8*bs 0 3*b 4*bs 0 -3*b 2*bs 0 -6*b 4*bs 0;
           0 0 0 0 0 0 0 0 0 0 0 0;
           3 3*b 0 6 6*b 0 -6 6*b 0 -3 3*b 0;
           3*b 4*bs 0 6*b 8*bs 0 -6*b 4*bs 0 -3*b 2*bs 0;
           0 0 0 0 0 0 0 0 0 0 0 0;
           -3 -3*b 0 -6 -6*b 0 6 -6*b 0 3 -3*b 0;
           3*b 2*bs 0 6*b 4*bs 0 -6*b 8*bs 0 -3*b 4*bs 0;
           0 0 0 0 0 0 0 0 0 0 0 0;
           -6 -6*b 0 -3 -3*b 0 3 -3*b 0 6 -6*b 0;
           6*b 4*bs 0 3*b 2*bs 0 -3*b 4*bs 0 -6*b 8*bs 0;
           0 0 0 0 0 0 0 0 0 0 0 0];
k3=nu/(2*a*b)*[1 b -a -1 -b 0 1 0 0 -1 0 a;
      b 0 -2*a*b -b 0 0 0 0 0 0 0 0;
      -a -2*a*b 0 0 0 0 0 0 0 a 0 0;
      -1 -b 0 1 b a -1 0 -a 1 0 0;
      -b 0 0 b 0 2*a*b 0 0 0 0 0 0;
      0 0 0 a 2*a*b 0 -a 0 0 0 0 0;
      1 0 0 -1 0 -a 1 -b a -1 b 0;
      0 0 0 0 0 0 -b 0 -2*a*b b 0 0;
      0 0 0 -a 0 0 a -2*a*b 0 0 0 0;
      -1 0 a 1 0 0 -1 b 0 1 -b -a;
      0 0 0 0 0 0 b 0 0 -b 0 2*a*b;
      a 0 0 0 0 0 0 0 0 -a 2*a*b 0];
k4=la/(15*a*b)*[21 3*b -3*a -21 -3*b -3*a 21 -3*b 3*a -21
      3*b 3*a;
           3*b 8*bs 0 -3*b -8*bs 0 3*b 2*bs 0 -3*b -2*bs 0;
           -3*a 0 8*as 3*a 0 -2*as -3*a 0 2*as 3*a 0 -8*as;
           -21 -3*b 3*a 21 3*b 3*a -21 3*b -3*a 21 -3*b -3*a;
           -3*b -8*bs 0 3*b 8*bs 0 -3*b -2*bs 0 3*b 2*bs 0;
           -3*a 0 -2*as 3*a 0 8*as -3*a 0 -8*as 3*a 0 2*as;
           21 3*b -3*a -21 -3*b -3*a 21 -3*b 3*a -21 3*b 3*a;
           -3*b 2*bs 0 3*b -2*bs 0 -3*b 8*bs 0 3*b -8*bs 0;
           3*a 0 2*as -3*a 0 -8*as 3*a 0 8*as -3*a 0 -2*as;
           -21 -3*b 3*a 21 3*b 3*a -21 3*b -3*a 21 -3*b -3*a;
           3*b -2*bs 0 -3*b 2*bs 0 3*b -8*bs 0 -3*b 8*bs 0;
           3*a 0 -8*as -3*a 0 2*as 3*a 0 -2*as -3*a 0 8*as];
ke=fr*(k1+plk2+k3+k4);
ij=12*(i-1)+1;ik=12*i;
kes(1:12,ij:ik)=ke;
```

```
kes;
qe=qz(i)*a*b*[1;b/3;-a/3;1;b/3;a/3;1;-b/3;a/3;1;
      -b/3;-a/3];
kg(m1:m3,m1:m3)=kg(m1:m3,m1:m3)+ke(1:3,1:3);
kg(m1:m3,m4:m6)=kg(m1:m3,m4:m6)+ke(1:3,4:6);
kg(m1:m3,m7:m9)=kg(m1:m3,m7:m9)+ke(1:3,7:9);
kg(m1:m3,m10:m12)=kg(m1:m3,m10:m12)+ke(1:3,10:12);
kg(m4:m6,m1:m3)=kg(m1:m3,m4:m6)';
kg(m7:m9,m1:m3)=kg(m1:m3,m7:m9)';
kg(m10:m12,m1:m3)=kg(m1:m3,m10:m12)';
kg(m4:m6,m4:m6)=kg(m4:m6,m4:m6)+ke(4:6,4:6);
kg(m4:m6,m7:m9)=kg(m4:m6,m7:m9)+ke(4:6,7:9);
kg(m4:m6,m10:m12)=kg(m4:m6,m10:m12)+ke(4:6,10:12);
kg(m7:m9,m4:m6)=kg(m4:m6,m7:m9)';
kg(m10:m12,m4:m6)=kg(m4:m6,m10:m12)';
kg(m7:m9,m7:m9)=kg(m7:m9,m7:m9)+ke(7:9,7:9);
kg(m7:m9,m10:m12)=kg(m7:m9,m10:m12)+ke(7:9,10:12);
kg(m10:m12,m7:m9)=kg(m7:m9,m10:m12)';
kg(m10:m12,m10:m12)=kg(m10:m12,m10:m12)+ke(10:12,10:12);
qg(m1:m3)=qg(m1:m3)+qe(1:3);
qg(m4:m6)=qg(m4:m6)+qe(4:6);
qg(m7:m9)=qg(m7:m9)+qe(7:9);
qg(m10:m12)=qg(m10:m12)+qe(10:12);
qg;
end
for ij=1:np
 qg(npdf(ij))=qg(npdf(ij))+p(ij);
 end
kgm=kg-kg';
for i=1:nc
 ic=cm(i);
 qg(ic)=0;
 for j=1:ndf
    kg(ic,j)=0;
    kg(j,ic)=0;
 end
 kg(ic,ic)=1.0;
end
kg;
qg;
fprintf('         displacements\n')
d=kg\qg;
fprintf(' node    w      rotx     roty\n')
for iu=1:nn
i1=3*iu-2;i2=i1+1;i3=i2+1;
fprintf('   %2.0f  %8.4f   %8.4f   %8.4f\n',iu,
      d(i1),d(i2),d(i3))
```

```
end
fprintf('\n')
for i=1:ne
n1=n1a(i);n2=n2a(i);n3=n3a(i);n4=n4a(i);
d1=3*n1;d2=3*n2;d3=3*n3;d4=3*n4;
e1=d1-2;e2=d2-2;e3=d3-2;e4=d4-2;
ij=12*(i-1)+1;ik=12*i;
km=kes(1:12,ij:ik);
u(1:3)=d(e1:d1);
u(4:6)=d(e2:d2);
u(7:9)=d(e3:d3);
u(10:12)=d(e4:d4);fm=km*u';
ef(i,:)=fm';
end
              %         fprintf('elemental forces\n')

fprintf('       Moments extrapolated to the nodes\n')
for i=1:ne
 a=aa(i);b=bb(i);as=a*a;bs=b*b;
n1=n1a(i);n2=n2a(i);n3=n3a(i);n4=n4a(i);
d1=3*n1;d2=3*n2;d3=3*n3;d4=3*n4;
e1=d1-2;e2=d2-2;e3=d3-2;e4=d4-2;
u(1:3)=d(e1:d1);
u(4:6)=d(e2:d2);
u(7:9)=d(e3:d3);
u(10:12)=d(e4:d4);
for j=1:4
s=se(j);t=te(j);p=1/s;q=1/t;
B(1,1)=3*p*(1-q)*bs;
B(2,1)=3*q*(1-p)*as;
B(3,1)=(4-3*(p^2+q^2))*a*b;
B(1,2)=0;
B(2,2)=-(1-p)*(1-3*q)*as*b;
B(3,2)=(1-q)*(1+3*q)*a*bs;
B(1,3)=(1-3*p)*(1-q)*a*bs;
B(2,3)=0;
B(3,3)=-(1-p)*(1+3*p)*as*b;
B(1,4)=-3*p*(1-q)*bs;
B(2,4)=3*(1+p)*q*as;
B(3,4)=-(4-3*(p*p+q*q))*a*b;
B(1,5)=0;
B(2,5)=-(1+p)*(1-3*q)*as*b;
B(3,5)=-(1-q)*(1+3*q)*a*bs;
B(1,6)=-(1+3*p)*(1-q)*a*bs;
B(2,6)=0;
B(3,6)=-(1+p)*(1-3*p)*as*b;
B(1,7)=-3*p*(1+q)*bs;
```

```
B(2,7)=-3*(1+p)*q*as;
B(3,7)=(4-3*(p*p+q*q))*a*b;
B(1,8)=0;
B(2,8)=(1+p)*(1+3*q)*as*b;
B(3,8)=-(1+q)*(1-3*q)*a*bs;
B(1,9)=-(1+3*p)*(1+q)*a*bs;
B(2,9)=0;
B(3,9)=(1-3*p)*(1+p)*as*b;
B(1,10)=3*p*(1+q)*bs;
B(2,10)=-3*(1-p)*q*as;
B(3,10)=-(4-3*(p*p+q*q))*a*b;
B(1,11)=0;
B(2,11)=(1-p)*(1+3*q)*as*b;
B(3,11)=(1+q)*(1-3*q)*a*bs;
B(1,12)=(1-3*p)*(1+q)*a*bs;
B(2,12)=0;
B(3,12)=(1-p)*(1+3*p)*as*b;
B=B/(4*as*bs);
EB=em*B*h^3/12;
mom=EB*u';
mmo(:,j)=mom;
end       % Gauss point moment loop
for k=1:4     % Extrapolate Gauss point moments to nodes
s=se(k);t=te(k);
n=[(1+s)*(1+t)/4 (1-s)*(1+t)/4 (1-s)*(1-t)/4
    (1+s)*(1-t)/4];
momex(:,k)=(n*mmo')';
end     %end extrapolation loop
fprintf('            For element %2.0f\n',i)
fprintf('        node I    node II    node III   node IV\n')
fprintf('Mxx  %8.6f   %8.6f   %8.6f   %8.6f\n',momex(1,:))
fprintf('Myy  %8.6f   %8.6f   %8.6f   %8.6f\n',momex(2,:))
fprintf('Mxy  %8.6f   %8.6f   %8.6f   %8.6f\n',momex(3,:))
fprintf('\n')
end     %end ne loop
```

EXAMPLE B-2 Consider the problem of a square uniformly loaded plate simply supported on all edges, as shown in Figure B-12. Take the plate to be 10 in on a side and 0.1 in in thickness with a 2 psi loading upwards. Let $E = 10^7$ psi and $\nu = 0.3$.

On the basis of symmetry, only a quarter of the plate will be modeled. On the sides corresponding to the original boundaries of the plate, the constraints are on w and what amounts to the tangential derivative $\partial w/\partial t$. On the interior boundaries, the conditions of symmetry dictate that the normal derivative be zero, as shown. The node and element numbers for the two meshes that will be investigated are shown in Figure B-13. The input data for the four-element mesh is so indicated in the listing of the MATLAB code presented above. Note that the dimensions of an element are 2 aa by 2 bb where aa and bb are the arrays specified in MATLAB.

APPENDIX B   FINITE-ELEMENT MODELS AND MATLAB CODES

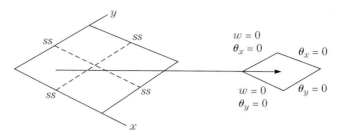

FIGURE B-12 Use of symmetry in modeling square plate.

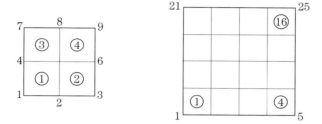

FIGURE B-13 Node and element numbers.

For the four-element mesh, the displacements are given in Table B-1. Also presented are moments that are computed at the Gauss points and then extrapolated to the nodes. For each element, the roman numerals correspond to those indicated in Figure B-14. The values are presented in Table B-2.

For the 16-element model, the displacements are given in Table B-3.

The values of the moments extrapolated to the nodes are given in Table B-4.

The results for displacement and moment for the 4- and 16-element cases are summarized in Table B-5, where the corresponding exact results are also presented for comparison.

TABLE B-1 Nodal Displacements and Rotation

	Displacements		
Node	w	rotx	roty
1	0.0000	0.0000	0.0000
2	0.0000	0.0232	0.0000
3	0.0000	0.0314	0.0000
4	0.0000	0.0000	−0.0232
5	0.0497	0.0147	−0.0147
6	0.0684	0.0204	0.0000
7	0.0000	0.0000	−0.0314
8	0.0684	0.0000	−0.0204
9	0.0945	0.0000	0.0000

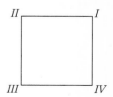

FIGURE B-14
Numbering for Gauss Points.

TABLE B-2 Moments Extrapolated to the Nodes

	For Element 1			
	Node I	Node II	Node III	Node IV
Mxx	−6.751441	−1.009703	0.000000	−0.302911
Myy	−6.751441	−0.302911	0.000000	−1.009703
Mxy	2.926198	5.097290	7.268381	5.097290

	For Element 2			
	Node I	Node II	Node III	Node IV
Mxx	−7.822106	−6.663549	−0.302911	−0.244825
Myy	−8.907099	−6.725073	−1.009703	−0.816084
Mxy	−0.286593	3.481101	4.118895	0.351202

	For Element 3			
	Node I	Node II	Node III	Node IV
Mxx	−8.907099	−0.816084	−1.009703	−6.725073
Myy	−7.822106	−0.244825	−0.302911	−6.663549
Mxy	−0.286593	0.351202	4.118895	3.481101

	For Element 4			
	Node I	Node II	Node III	Node IV
Mxx	−10.433852	−8.631926	−6.637182	−7.739554
Myy	−10.433852	−7.739554	−6.637182	−8.631926
Mxy	−0.702793	0.765573	2.233939	0.765573

TABLE B-3 Nodal Displacements and Rotations

	Displacements		
Node	w	rotx	roty
1	0.0000	0.0000	0.0000
2	0.0000	0.0123	0.0000
3	0.0000	0.0219	0.0000
4	0.0000	0.0279	0.0000
5	0.0000	0.0299	0.0000
6	0.0000	0.0000	−0.0123
7	0.0147	0.0108	−0.0108
8	0.0263	0.0194	−0.0077
9	0.0336	0.0249	−0.0039
10	0.0360	0.0268	0.0000
11	0.0000	0.0000	−0.0219
12	0.0263	0.0077	−0.0194
13	0.0474	0.0140	−0.0140
14	0.0607	0.0181	−0.0072
15	0.0652	0.0194	0.0000
16	0.0000	0.0000	−0.0279
17	0.0336	0.0039	−0.0249
18	0.0607	0.0072	−0.0181
19	0.0779	0.0094	−0.0094
20	0.0838	0.0101	0.0000
21	0.0000	0.0000	−0.0299
22	0.0360	0.0000	−0.0268
23	0.0652	0.0000	−0.0194
24	0.0838	0.0000	−0.0101
25	0.0902	0.0000	0.0000

TABLE B-4 Moments Extrapolated to the Nodes

For Element 1

	Node I	Node II	Node III	Node IV
Mxx	−2.565081	−0.261458	0.000000	−0.078437
Myy	−2.565081	−0.078437	0.000000	−0.261458
Mxy	5.233871	6.015980	6.798089	6.015980

For Element 2

	Node I	Node II	Node III	Node IV
Mxx	−3.598093	−2.558727	−0.078437	−0.068174
Myy	−4.133508	−2.563175	−0.261458	−0.227247
Mxy	3.730154	5.325468	5.797541	4.202228

For Element 3

	Node I	Node II	Node III	Node IV
Mxx	−4.087367	−3.607916	−0.068174	−0.071279
Myy	−4.976006	−4.136455	−0.227247	−0.237595
Mxy	1.900464	3.808750	4.079955	2.171669

For Element 4

	Node I	Node II	Node III	Node IV
Mxx	−4.233046	−4.090661	−0.071279	−0.071011
Myy	−5.249226	−4.976994	−0.237595	−0.236704
Mxy	−0.040090	1.981388	2.069351	0.047874

For Element 5

	Node I	Node II	Node III	Node IV
Mxx	−4.133508	−0.227247	−0.261458	−2.563175
Myy	−3.598093	−0.068174	−0.078437	−2.558727
Mxy	3.730154	4.202228	5.797541	5.325468

For Element 6

	Node I	Node II	Node III	Node IV
Mxx	−6.093338	−4.090151	−2.556821	−3.585085
Myy	−6.093338	−3.585085	−2.556821	−4.090151
Mxy	2.681879	3.886551	5.091223	3.886551

For Element 7

	Node I	Node II	Node III	Node IV
Mxx	−7.019461	−6.080239	−3.594909	−4.075649
Myy	−7.551405	−6.089409	−4.093098	−4.936944
Mxy	1.340341	2.896461	3.619481	2.063362

For Element 8

	Node I	Node II	Node III	Node IV
Mxx	−7.292151	−7.017057	−4.078943	−4.220707
Myy	−8.029511	−7.550684	−4.937932	−5.208097
Mxy	−0.115453	1.568896	1.808871	0.124521

For Element 9

	Node I	Node II	Node III	Node IV
Mxx	−4.976006	−0.237595	−0.227247	−4.136455
Myy	−4.087367	−0.071279	−0.068174	−3.607916
Mxy	1.900464	2.171669	4.079955	3.808750

For Element 10

	Node I	Node II	Node III	Node IV
Mxx	−7.551405	−4.936944	−4.093098	−6.089409
Myy	−7.019461	−4.075649	−3.594909	−6.080239
Mxy	1.340341	2.063362	3.619481	2.896461

continues

TABLE B-4 *Continued*

	For Element 11			
	Node I	Node II	Node III	Node IV
Mxx	−8.794370	−7.527700	−6.076309	−7.012349
Myy	−8.794370	−7.012349	−6.076309	−7.527700
Mxy	0.618237	1.607698	2.597159	1.607698

	For Element 12			
	Node I	Node II	Node III	Node IV
Mxx	−9.158648	−8.784843	−7.009946	−7.284146
Myy	−9.385822	−8.791512	−7.526979	−8.002830
Mxy	−0.164308	0.933823	1.270653	0.172522

	For Element 13			
	Node I	Node II	Node III	Node IV
Mxx	−5.249226	−0.236704	−0.237595	−4.976994
Myy	−4.233046	−0.071011	−0.071279	−4.090661
Mxy	−0.040090	0.047874	2.069351	1.981388

	For Element 14			
	Node I	Node II	Node III	Node IV
Mxx	−8.029511	−5.208097	−4.937932	−7.550684
Myy	−7.292151	−4.220707	−4.078943	−7.017057
Mxy	−0.115453	0.124521	1.808871	1.568896

	For Element 15			
	Node I	Node II	Node III	Node IV
Mxx	−9.385822	−8.002830	−7.526979	−8.791512
Myy	−9.158648	−7.284146	−7.009946	−8.784843
Mxy	−0.164308	0.172522	1.270653	0.933823

	For Element 16			
	Node I	Node II	Node III	Node IV
Mxx	−9.783654	−9.373857	−8.781984	−9.155059
Myy	−9.783654	−9.155059	−8.781984	−9.373857
Mxy	−0.187892	0.191188	0.570267	0.191188

TABLE B-5 Summary of Results

Number of Elements	w_{max} (in)	M_{max} (in-lbf/in)
4	0.0945	10.43
16	0.0902	9.78
exact	0.0886	9.58

REFERENCES

1. Melosh, R. J. "Basis of Derivation of Matrices for the Direct Stiffness Method." *AIAA Journal*, vol. 1, no. 7, July 1963, pp. 1631–1637.
2. Zienkiewicz, O. C., and Cheung, Y. K. "The Finite Element Method for Analysis of Elastic Isotropic and Orthotropic Slabs." *Proc. Inst. Civ. Eng.*, vol. 28, 1964, pp. 471–488.

Circular Plate Model. The starting point for the development of the finite-element model for the axisymmetric circular plate problem will be taken to be the total potential

APPENDIX B FINITE-ELEMENT MODELS AND MATLAB CODES

energy, namely,

$$V = 2\pi \int_a^b \left[\frac{D}{2}\left(w''^2 + \left(\frac{w'}{r}\right)^2 + 2\nu w'' \frac{w'}{r}\right) - q(r)w \right] r\, dr$$

plus terms to account for line loads and/or line moments and for shears and moments applied at boundaries.

Hermite polynomials are used for interpolation according to

$$w_e(r) = w_1 N_1 + \theta_1 N_2 + w_3 N_3 + \theta_2 N_4$$

where

$$N_1 = 1 - 3\left(\frac{s}{\ell_e}\right)^2 + 2\left(\frac{s}{\ell_e}\right)^3$$

$$N_2 = \ell_e\left(\left(\frac{s}{\ell_e}\right) - 2\left(\frac{s}{\ell_e}\right)^2 + \left(\frac{s}{\ell_e}\right)^3\right)$$

$$N_3 = 3\left(\frac{s}{\ell_e}\right)^2 - 2\left(\frac{s}{\ell_e}\right)^3$$

$$N_4 = \ell_e\left(\left(\frac{s}{\ell_e}\right)^3 - \left(\frac{s}{\ell_e}\right)^2\right)$$

and where the degrees of freedom w_1, θ_1, w_2 and θ_2 are as shown in Figure B-15. The potential energy function can then be expressed as

$$\frac{V}{2\pi} = \sum_e \frac{1}{2} \mathbf{w}_e^T \mathbf{k}_e \mathbf{w}_e - \sum_e \mathbf{w}_e^T \mathbf{f}_e$$

where

$$\mathbf{k}_e = D \int_{r_i}^{r_{i+1}} \left[\mathbf{N}'' r \mathbf{N}''^T + \mathbf{N}\frac{1}{r}\mathbf{N}^T + \nu(\mathbf{N}''\mathbf{N}'^T + \mathbf{N}'\mathbf{N}''^T) \right] dr$$

and

$$\mathbf{f}_e = \int_{r_i}^{r_{i+1}} \mathbf{N} q\, dr$$

In the MATLAB code named "circplate," which is listed below, the integrations are carried out using a third-order Gauss quadrature. Outputs are displacements w and bending moments M_r and M_θ at the nodes. All required inputs are described in the listing.

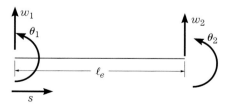

FIGURE B-15 Nodal displacements and rotations.

```
% circplate.m
% This code can be used to investigate displacements and
% stresses for the symmetric circular-plate problem of
% Chapter 7. Beam functions are used for interpolation.
% Three-point Gauss integration is used to evaluate stiffness
% and load matrices. Nodal values for moments are
% obtained by computing N = 2 Gauss point values per element
% and then extrapolating those to the nodes.
% Inputs are r1 the inner radius, r2 the outer radius,
% E Young's modulus, nu Poisson's ratio, h the thickness,
% ne the number of elements, nc the number of constraints,
% ncd an array containing the constrained degrees of freedom,
% P an array containing the nodal line loads (F/L),
% M an array containing the nodal line moments (FL/L), and
% qz an array containing the nodal distributed load values (F/LL).
%
% Test Case 1: uniformly loaded, simply supported plate
% Specify elastic and geometric constants
E=1.e7;nu=0.3;h=0.1;
flexrig=Eh³/(12*(1-nu*nu));
% Specify nodes, elements
ne=8;nn=ne+1;ndf=2*nn;
r2=10;r1=0.0;le=(r2-r1)/ne;
r=r1:le:r2
P=zeros(nn,1);M=zeros(nn,1);
kg=zeros(ndf);qg=zeros(ndf,1);qz=zeros(nn,1)
% Specify Loads
%qz=[1 1 1 1 1 1 1 1 1 ]/10;
% Specify constraints
nc=1;
cdf=[2*ne+1];
% End of test case 1 input
% Gauss data
ng=3;s=sqrt(0.6);gp=[-s 0 s];w=[5 8 5]/9;
%ng=2;s=sqrt(1/3);gp=[-s s];w=[1 1];
% Computation of stiffness and load matrices
for i=1:ne
 ls=le*le;
 ke=zeros(4);qe=zeros(4,1);
 n1=2*(i-1);
 qel=qz(i);
 radi=r(i);
  for i=1:ng
   g1=gp(i);w1=w(i);
   rg=radi+le*(1+g1)/2;
```

```
        nf=[(2+g1)*(1-g1)^2/4 le*(1+g1)*(1-g1)^2/8 (2-g1)*(1+g1)^2/4
            le*(-1+g1)*(1+g1)^2/8];
        np=(2/le)*[3*(g1*g1-1)/4 le*(g1-1)*(3*g1+1)/8
              -3*(g1*g1-1)/4
                le*(g1+1)*(3*g1-1)/8];
        nd=[6*g1 le*(3*g1-1) -6*g1 le*(3*g1+1)]/ls;
        ke=ke+flexrig*(le*w1/2)*(nd'*nd*rg+np'*np/
             rg+nu*(np'*nd+nd'*np));
        qe=qe+(le/2)*w1*qel*rg*nf';
     end
     for ii=1:4
      i1=n1+ii;
      qg(i1)=qg(i1)+qe(ii);
       for jj=1:4
        j1=n1+jj;
        kg(i1,j1)=kg(i1,j1)+ke(ii,jj);
       end
      end
end
kg;
% Add line and boundary loadings
for i=1:nn
 i1=2*i-1;i2=i1+1;
 qg(i1)=qg(i1)+r(i)*P(i);
 qg(i2)=qg(i2)+M(i)*r(i);
end
% Enforce constraints
for i=1:nc
  c=cdf(i);
 for j=1:ndf
  kg(c,j)=0;
  kg(j,c)=0;
  qg(c)=0;
  kg(c,c)=1;
 end       %end j
end            %end i
% Solution
w=kg\qg;
% Print displacements
fprintf('     Displacements\n')
fprintf('\n')
fprintf(' node    disp     rotation\n')
for i=1:nn
 wn(i)=w(2*i-1);tn(i)=w(2*i);
 fprintf(' %2.0f  %8.6f  %8.6f\n',i,wn(i),tn(i))
```

```
end
fprintf('\n')
% Calculations of moments @ N = 2 Gauss points
s2=sqrt(1/3);gp=[-s2 s2];wg=[1 1];s3=1/s2;
for im=1:ne        %loop on number of elements
r1=r(im);ls=le*le;
 ue=zeros(4,1);
 ii=2*(im-1);
  for jj=1:4
    ip=ii+jj;
   ue(jj)=w(ip);
  end
 mr=zeros(2,1);mth=zeros(2,1);
  for ig=1:2
   g1=gp(ig);w1=wg(ig);
   rg=r(im)+le*(1+g1)/2;
   np=(2/le)*[3*(g1*g1-1)/4 le*(g1-1)*(3*g1+1)/8
       -3*(g1*g1-1)/4
               le*(g1+1)*(3*g1-1)/8];
   nd=[6*g1 le*(3*g1-1) -6*g1 le*(3*g1+1)]/ls;
   mr(ig)=mr(ig)+flexrig*ue'*(nd'+np'*nu/rg);
   mth(ig)=mth(ig)+flexrig*ue'*(np'/rg+nu*nd');
  end
% Extrapolation to nodes
 mre(1)=mr(1)*(1+s3)/2+mr(2)*(1-s3)/2;
 mthe(1)=mth(1)*(1+s3)/2+mth(2)*(1-s3)/2;
 mre(2)=mr(1)*(1-s3)/2+mr(2)*(1+s3)/2;
 mthe(2)=mth(1)*(1-s3)/2+mth(2)*(1+s3)/2;
 mmr(:,im)=mre';
 mmt(:,im)=mthe';
end
% Averages at interior nodes
MR(1)=mmr(1,1);MR(nn)=mmr(2,ne);
MT(1)=mmt(1,1);MT(nn)=mmt(2,ne);
for i=2:nn-1
MR(i)=(mmr(1,i)+mmr(2,i-1))/2;
MT(i)=(mmt(1,i)+mmt(2,i-1))/2;
end
% Print moments
fprintf('   Nodal moments\n')
fprintf('\n')
fprintf(' node   mom r   mom theta\n')
for i=1:nn
fprintf(' %2.0f   %8.6f   %8.6f\n',i,MR(i),MT(i))
end
```

```
%for i=1:nn
% rn(i)=r1+(i-1)*le;
%end
for i=1:nn
wd(i)=w(2*i-1);
end
plot(r,wd)
xlabel('radius')
 ylabel('displacement')
 title('Transverse displacement as a function of position')
 grid
pause
plot(r,MR,':g',r,MT,'r--',r,MR,'wo',r,MT,'c*')
 grid
 xlabel('radius')
 ylabel('Moments')
 title('Radial moment (o) and circumferential moment (*)')
```

Test Case 1. The first test case is for a simply supported solid plate with $a = 0$, $b = 10, h = 0.1, E = 1.\mathrm{e}7$ and $q_0 = 0.1$. A model using eight elements is used. The nodal displacements and rotations are shown in Table B-6. The maximum displacement at the center compares well with the theoretical value of 0.06956 in. The radial and circumferential moments are shown in Table B-7. The maximum values at the center compare well with the exact values of $M_r = M_\theta = -2.0625$ in-lbf/in. The value of M_θ at the outer edge compares well with the exact value of 0.875. Plots of these data are shown in Figures B-16 and B-17.

Test Case 2. The second test case is for an annular plate with $a = 3$ in, $b = 6$ in, $h = 0.1, E = 1.\mathrm{e}7$ and an edge load of 1 lbf/in at the outer edge. A model using eight elements is used. The nodal displacements and rotations are shown in Table B-8. The maximum displacement at the outer edge compares well with the exact value of 0.1190 in. The radial and circumferential moments are shown in Table B-9. The maxi-

TABLE B-6 Displacements

Node	Displacement (in)	Rotation
1	0.069563	0.000001
2	0.068213	−0.002152
3	0.064215	−0.004225
4	0.057718	−0.006137
5	0.048973	−0.007809
6	0.038328	−0.009162
7	0.026235	−0.010114
8	0.013242	−0.010587
9	0.000000	−0.010500

TABLE B-7 Nodal Moments (in-lbf/in)

Node	M_r	M_θ
1	−2.071976	−2.067975
2	−2.034954	−2.046760
3	−1.938855	−1.991288
4	−1.777785	−1.898564
5	−1.552221	−1.768699
6	−1.262192	−1.601714
7	−0.907705	−1.397617
8	−0.488763	−1.156408
9	−0.005214	−0.877951

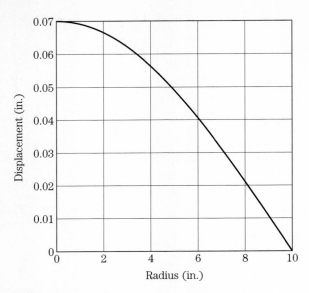

FIGURE B-16 Transverse displacement as a function of position.

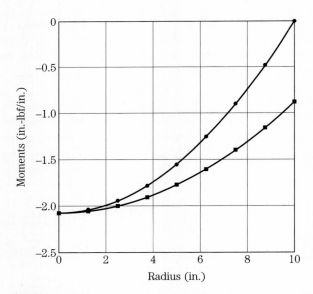

FIGURE B-17 Radial moment (•) and circumferential moment (■).

TABLE B-8 Displacements

Node	Displacement (in)	Rotation
1	0.000000	0.000000
2	0.000331	0.001689
3	0.001214	0.002959
4	0.002510	0.003901
5	0.004107	0.004579
6	0.005916	0.005035
7	0.007860	0.005304
8	0.009874	0.005411
9	0.011901	0.005376

TABLE B-9 Nodal Moments (in-lbf/in)

Node	M_r	M_θ
1	4.710535	1.432808
2	3.688371	1.538157
3	2.876555	1.529666
4	2.207575	1.456346
5	1.641770	1.344317
6	1.153148	1.209272
7	0.723981	1.060926
8	0.341760	0.905429
9	−0.002759	0.746826

mum value at the center gives a bending stress of magnitude $\sigma_r = 6M_r/h^2 = 2826$ psi, which compares with the exact value of 2839 psi. Plots of these data are shown in Figures B-18 and B-19.

Chapter 8. The starting point for the development of the finite-element model for the axisymmetric shell bending problem will be taken to be the total potential energy, namely,

$$V = 2\pi \int_0^L \frac{1}{2}\left[\frac{Eh}{(1-\nu^2)}\{\epsilon_s^2 + \epsilon_\theta^2 + 2\nu\epsilon_s\epsilon_\theta\} + \frac{Eh^3}{12(1-\nu^2)}\{\chi_s^2 + \chi_\theta^2 + 2\nu\chi_s\chi_\theta\}r\,ds$$

$$- 2\pi \int_0^L (up_s + wp_n)r\,ds$$

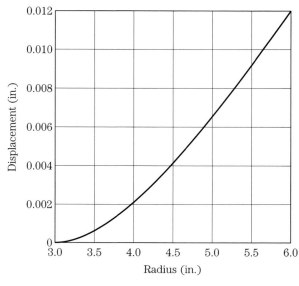

FIGURE B-18 Transverse displacement as a function of position.

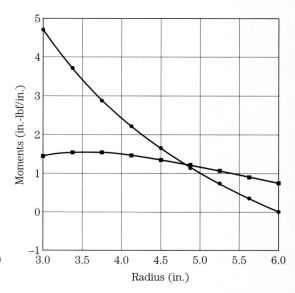

FIGURE B-19 Radial moment (•) and circumferential moment (■).

where

$$\epsilon_s = \frac{du}{ds}$$

$$\epsilon_\theta = \frac{w\cos\phi + u\sin\phi}{r}$$

$$\chi_s = -\frac{d^2w}{ds^2}$$

$$\chi_\theta = -\frac{\sin\phi}{r}\frac{dw}{ds}$$

The meridional displacement u is assumed to depend linearly on the meridional position according to

$$u_e(s) = u_1\left(1 - \frac{s}{\ell_e}\right) + u_2\frac{s}{\ell_e}$$

and the transverse displacement is assumed to be

$$w_e(s) = w_1 N_1 + \theta_1 N_2 + w_2 N_3 + \theta_2 N_4$$

where N_1, N_2, N_3 and N_4 are the Hermite polynomials commonly used in bending situations.

The MATLAB code listed below constructs a finite-element model using conical frustra to represent the axisymmetric shell.

```
% axshbend.m
% This MATLAB code analyzes the displacements, membrane
% forces and moments in an axisymmetric shell modeled
% as a collection of conical frustra. Inputs required
% are ne the number of elements, an array r containing the
% r coordinates of the nodes, an array z containing the
% z coordinates of the nodes, E Young's modulus, nu
% Poisson's ratio, h the thickness, pn an array of the
% normal pressures at the nodes, ps an array of the
% meridional pressures at the nodes, nc the number of
% constraints and cdf an array constraining the GLOBAL
% constrained degrees of freedom. Line loads and moments
% must be multiplied by the total circumferential length
% over which they act. Output consists of global
% displacements, local membrane forces and bending moments.
% The student can use the code to investigate the influence of
% geometry, loading and support conditions on membrane and
% bending stresses.
% The data is for Example 8.3. r1 = 72 in, H = 144 in
% and h = 0.625 in. The fluid is water. Twelve elements
% are used. For the finite-element model, the shell is assumed
% to be pinned at the top.
%
```

```
clear
ne=12;nn=ne+1-;ndf=3*nn;
pn=zeros(nn,1);ps=zeros(nn,1);r=zeros(nn,1);z=zeros(nn,1);
ns=zeros(1,ne);nt=zeros(1,ne);sp=zeros(nn,1);
ms=zeros(1,ne);mt=zeros(1,ne);
nsn=zeros(1,nn);ntn=zeros(1,nn);msn=zeros(1,nn);
     mtn=zeros(1,nn);
kg=zeros(ndf);qg=zeros(ndf,1);MF=zeros(4,ne);
ym=1.e7;nu=0.3;h=0.625;
ra1=0.0;ra2=72.0;dr=(ra2-ra1)/ne;r=ra1:dr:ra2;
z1=0.0;z2=144;dz=(z2-z1)/ne;z=z1:dz:z2;
pn=5.2*[12 11 10 9 8 7 6 5 4 3 2 1 0]/12;
nc=2;cdf=[ndf-2 ndf-1];
mr=ym*h/(1-nu*nu);fr=mr*h*h/12;
E=[mr nu*mr 0 0;nu*mr mr 0 0;0 0 fr nu*fr;0 0 nu*fr fr];
plot(r,z),pause
% Gauss data
ngb=3;s=sqrt(0.6);gpb=[-s 0 s];wb=[5 8 5]/9;
% Generate and assemble stiffness and load matrices
% Loop on elements
for i=1:ne
 km=zeros(6);qu=zeros(6,1);
 ip=i+1;
 n1=3*i-3;n2=3*i+1;
 p1=pn(i);p2=pn(ip);
 q1=ps(i);q2=ps(ip);
 r1=r(i);r2=r(ip);dr=r2-r1;
 z1=z(i);z2=z(ip);dz=z2-z1;
 le=sqrt(dr*dr+dz*dz);sg=dr/le;cg=dz/le;
 sp(ip)=sp(i)+le;
 % Define rotation matrix
 R=[cg sg 0 0 0 0;
   -sg cg 0 0 0 0;
    0 0 1 0 0 0;
    0 0 0 cg sg 0;
    0 0 0 -sg cg 0;
    0 0 0 0 0 1];
 % Loop on Gauss
   for j=1:ngb
        g1=gpb(j);w1=wb(j);
        s=(1+g1)/2;
     rg=r1+dr*s;
        bb(1,1)=-1/le;bb(1,2)=0;bb(1,3)=0;
        bb(1,4)=1/le;bb(1,5)=0;bb(1,6)=0;
        bb(2,1)=(1-s)*sg/rg;bb(2,2)=cg*(1-s^2*(3-2*s))/rg;
        bb(2,3)=s*le*(1-s)^2*cg/rg;
```

```
              bb(2,4)=s*sg/rg;bb(2,5)=s^2*cg*(3-2*s)/rg;
              bb(2,6)=s^2*(s-1)*le*cg/rg;
              bb(3,1)=0;bb(3,2)=6*(1-2*s)/le^2;bb(3,3)=
                  2*(2-3*s)/le;
              bb(3,4)=0;bb(3,5)=-6*(1-2*s)/le^2;bb(3,6)=
                  2*(1-3*s)/le;
              bb(4,1)=0;bb(4,2)=6*s*(1-s)*sg/(le*rg);
              bb(4,3)=(-1+s*(4-3*s))*sg/rg;
              bb(4,4)=0;bb(4,5)=-6*s*(1-s)*sg/(le*rg);
              bb(4,6)=s*(2-3*s)*sg/rg;
              ms=E*bb;
              kt=bb'*ms;
              km=km+w1*rg*kt;
              qu(1)=qu(1)+w1*pi*le*rg*(q1*(1-s)+q2*s)*(1-s);
              qu(2)=qu(2)+w1*pi*le*rg*(p1*(1-s)+p2*s)*
                  (1-s^2*(3-2*s));
              qu(3)=qu(3)+w1*pi*le*rg*(p1*(1-s)+p2*s)*s*(1-s)^2*le;
              qu(4)=qu(4)+w1*pi*le*rg*(q1*(1-s)+q2*s)*s;
              qu(5)=qu(5)+w1*pi*le*rg*(p1*(1-s)+p2*s)*s^2*(3-2*s);
              qu(6)=qu(6)+w1*pi*le*rg*(p1*(1-s)+p2*s)*
                  (-s^2*(1-s))*le;
      end     % Gauss
      km=pi*km*le;
      kv=R'*km;
      ke=kv*R;
      qe=R'*qu;
      for kl=1:6
       kr=n1+kl;
       qg(kr)=qg(kr)+qe(kl);
       for jl=1:6
        jr=n1+jl;
          kg(kr,jr)=kg(kr,jr)+ke(kl,jl);
       end
      end
end
% Enforce constraints
for i=1:nc
 ic=cdf(i);
 for j=1:ndf
      kg(ic,j)=0;kg(j,ic)=0;
 end
 kg(ic,ic)=1;
 qg(ic)=0;
end
u=kg\qg;
% Output displacements
```

```
for ip=1:nn
 j=3*ip-1;
 wu(ip)=u(j-1);
 ww(ip)=u(j);
 wt(ip)=u(j+1);
 end
 plot(sp,wu,':g',sp,ww,'r--',sp,wu,'wo',sp,ww,'c*')
 xlabel('Meridional position')
 ylabel('Displacements')
 title('Global z (o) and r (*) components of displacement')
 grid
 pause
% Force resultant computations: Force resultants
% are computed at the middle of each element,
% averaged at the interior nodes and extrapolated to
% the nodes at the boundaries.
for ie=1:ne
 ue=zeros(6,1);
 ip=ie+1;
 r1=r(ie);r2=r(ip);dr=r2-r1;ra=(r1+r2)/2;
 z1=z(ie);z2=z(ip);dz=z2-z1;
 le=sqrt(dr²+dz²);sg=dr/le;cg=dz/le;
 R=[cg sg 0 0 0 0;
   -sg cg 0 0 0 0;
    0 0 1 0 0 0;
    0 0 0 cg sg 0;
    0 0 0 -sg cg 0;
    0 0 0 0 0 1];
 n1=3*ie-3;
  for ii=1:6
   ij=n1+ii;
   ve(ii)=u(ij);
   end      %ii
    ue=R*ve';
    s=0.5;
  rg=r1+dr*s;
  bb(1,1)=-1/le;bb(1,2)=0;bb(1,3)=0;
    bb(1,4)=1/le;bb(1,5)=0;bb(1,6)=0;
    bb(2,1)=(1-s)*sg/rg;bb(2,2)=cg*(1-s²*(3-2*s))/rg;
    bb(2,3)=s*le*(1-s)²*cg/rg;
    bb(2,4)=s*sg/rg;bb(2,5)=s²*cg*(3-2*s)/rg;
    bb(2,6)=s²*(s-1)*le*cg/rg;
    bb(3,1)=0;bb(3,2)=6*(1-2*s)/le²;bb(3,3)=2*(2-3*s)/le;
    bb(3,4)=0;bb(3,5)=-6*(1-2*s)/le²^;bb(3,6)=
          2*(1-3*s)/le;
    bb(4,1)=0;bb(4,2)=6*s*(1-s)*sg/(le*rg);
```

```
        bb(4,3)=(-1+s*(4-3*s))*sg/rg;
        bb(4,4)=0;bb(4,5)=-6*s*(1-s)*sg/(le*rg);
        bb(4,6)=s*(2-3*s)*sg/rg;
        bb;
  tmp=bb*ue;
  fm=E*tmp;
  MF(:,ie)=fm;
end
ns=MF(1,:);nt=MF(2,:);ms=MF(3,:);ms=&MF(4,:);
nsn(1)=(3*ns(1)-ns(2))/2;
ntn(1)=(3*nt(1)-nt(2))/2;
msn(1)=(3*ms(1)-ms(2))/2;
mtn(1)=(3*mt(1)-mt(2))/2;
nsn(nn)=(3*ns(ne)-ns(ne-1))/2;
ntn(nn)=(3*nt(ne)-nt(ne-1))/2;
msn(nn)=(3*ms(ne)-ms(ne-1))/2;
mtn(nn)=(3*mt(ne)-mt(ne-1))/2;
for i=2:ne
 nsn(i)=(ns(i-1)+ns(i))/2;
 ntn(i)=(nt(i-1)+nt(i))/2;
 msn(i)=(ms(i-1)+ms(i))/2;
 mtn(i)=(mt(i-1)+mt(i))/2;
end
plot(sp,nsn,':g',sp,ntn,'r--',sp,nsn,'wo',sp,ntn,'c*')
     xlabel('Meridional position')
      ylabel('Membrane forces')
     title('Axial (o) and transverse (*) membrane forces')
     grid
     pause
plot(sp,msn,':g',sp,mtn,'r--',sp,msn,'wo',sp,mtn,'c*')
     xlabel('Meridional position')
     ylabel('Membrane moments')
     title('Meridional (o) and transverse (*) moments')
     grid
```

For Example 8-3 the value of the axial displacement at the apex of the cone is given by

$$u(0) = -\frac{\gamma r_1 L^3}{EtL}\frac{5-6\nu}{36} = -0.000958 \text{ in}$$

From the plot shown in Figure B-20, the global value is approximately -0.00108 in. Multiplying this by the cosine of the angle gives $u(0) = -0.000966$ in. The actual value that can be output from the code is -0.000960, showing the excellent agreement between the theory and the finite-element model.

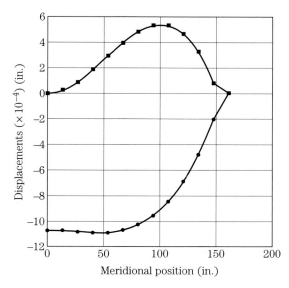

FIGURE B-20 Global z (■) and r (●) components of displacement.

FIGURE B-21 Axial (■) and transverse (●) membrane forces.

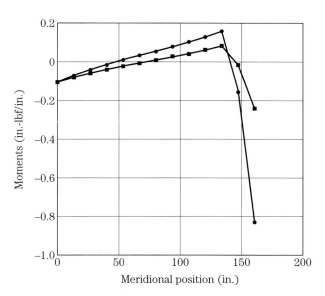

FIGURE B-22 Meridional (■) and transverse (●) moments.

From Example 8-3 the maximum value of the meridional membrane force is at $s = 3L/4$ where $N_s = 3\gamma r_1 L/16 = 78.5$ lbf/in. Similarly, the maximum hoop stress occurs at $s = L/2$ with the value $N_\theta = \gamma r_1 L/4 = 104.7$ lbf/in. Both values are seen to compare well with those given in the plots shown in Figure B-21.

Shown in Figure B-22 are plots of the meridional and circumferential bending moments. The assumption that bending stresses are negligible is seen to be a reasonable one since the maximum bending stress is of the order of $6(1)/h^2 \approx 15$ psi compared to membrane stress of the order of 170 psi. Many more elements would be necessary to resolve the character of the bending at the upper support.

C Sectorial Area

It was seen in Section 3-7 that the quantity $\omega = \int r\,ds$, called the sectorial area, arises naturally in formulating the torsional problem for thin-walled sections, including the effect of warping. It is the purpose of this appendix to investigate some of the properties of the sectorial area.

In connection with Figure C-1, sectorial area is defined according to $d\omega = r\,ds = 2\,dA$, where r is the perpendicular distance from the pole P to the directed line segment ds. Thinking of ds as a force and r as the moment arm, $d\omega$ is positive when the resulting direction of the moment is counterclockwise.

Integration of $d\omega = r\,ds$ yields

$$\omega = \omega_0 + \int_I^s r\,ds$$

where ω_0 is a constant of integration and I indicates the point at which the integration is begun. Also implicit in evaluating the sectorial area is the location of the pole P. The sectorial area will depend on the choice of the points P and I as well as the constant ω_0.

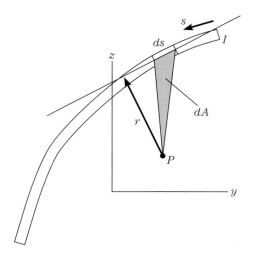

FIGURE C-1 Definition of sectorial area.

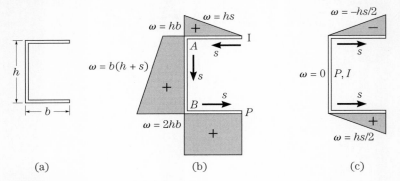

FIGURE C-2 Channel sectorial area.

EXAMPLE C-1 Consider the channel section shown in Figure C-2. For the choice of P and I as per Figure C-2b, the sectorial area is calculated as follows, assuming that $\omega_0 = 0$.

On IA: $r = h$ and $\omega = hs$ (counterclockwise direction), where $0 \leq s \leq b$
On AB: $r = b$ and $\omega = hb + bs$, $0 \leq s \leq h$
On BP: $r = 0$ and $\omega = 2bh$, $0 \leq s \leq b$

The function ω is shown as shaded in Figure C-2b. The reader should verify that for $\omega_0 = 0$, the sectorial area is as shown in Figure C-2c for that choice of I and P.

Clearly, ω depends on P and I, both of which are arbitrary. In the context of using the sectorial area in formulating and solving the torsion problem, the points P and I as well as the constant ω_0 must be chosen so as to satisfy certain equilibrium conditions on the normal stress. In this regard, refer to section 3-7 for the normal stress due to the warping as $\sigma_x = -E\omega\phi''$. For the torsion problem the only force resultant transmitted is the torque T so that the normal stress distribution must be such that the equilibrium equations

$$P = \int_A \sigma_x \, dA = 0, \quad M_z = -\int_A y\sigma_x \, dA = 0 \quad \text{and} \quad M_y = \int_A z\sigma_x \, dA = 0$$

are satisfied. This leads to

$$\int_A \omega \, dA = 0, \quad \int_A y\omega \, dA = 0 \quad \text{and} \quad \int_A z\omega \, dA = 0$$

As mentioned above, the points P and I as well as the constant ω_0 must be chosen so that these equations are satisfied.

EXAMPLE C-2 Consider the I-section shown in Figure C-3. With P and I chosen as shown in Figure C-3a, it follows with $\omega(I) = \omega_0$ that

APPENDIX C SECTORIAL AREA

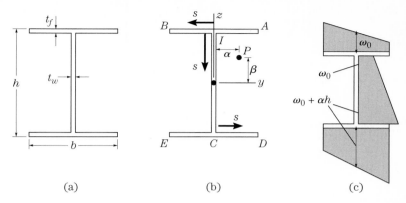

FIGURE C-3 Sectorial area for the I-section.

On AB: $r = \dfrac{h}{2} - \beta$, $\omega = \omega_0 + \left(\dfrac{h}{2} - \beta\right)s$ $\qquad -\dfrac{b}{2} \leq s \leq \dfrac{b}{2}$

On IC: $r = \alpha$, $\omega = \omega_0 + \alpha s$ $\qquad 0 \leq s \leq h$

On ED: $r = \dfrac{h}{2} + \beta$, $\omega = \omega_0 + \alpha h + \left(\dfrac{h}{2} + \beta\right)s$ $\qquad -\dfrac{b}{2} \leq s \leq \dfrac{b}{2}$

as shown in Figure C-3b. Evaluating the three integrals and simplifying leads to

$$\int_A \omega \, dA = (t_w h + 2t_f b)\left(\omega_0 + \dfrac{\alpha h}{2}\right) = 0$$

$$\int_A y\omega \, dA = \dfrac{t_f \beta h^3}{12} = 0$$

$$\int_A y\omega \, dA = \dfrac{-t_f \alpha h^2 b}{2} - \dfrac{t_w \alpha h^3}{12} = 0$$

that is, three equations for ω_0, α and β with the solution $\alpha = \beta = \omega_0 = 0$. The final sectorial area distribution is shown in Figure C-4. For other choices of I, ω_0 would in

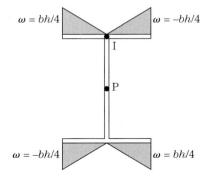

FIGURE C-4 Sectorial area for the I-section.

general be nonzero. However, the same result would obtain for α and β irrespective of the choice for I. It turns out that in general the pole P coincides with the shear center, which is discussed in Chapter 4. Note that the manner in which the normal stress σ_x depends on position over the cross section coincides with that of the sectorial area diagram shown in Figure C-3c, as it must by virtue of the relation $\sigma_x = -E\phi''\omega$.

If there is an axis of symmetry for the cross section, the pole and hence the shear center will lie somewhere on the axis of symmetry. If there are two axes of symmetry, the shear center lies at the intersection of the two axes of symmetry, which also turns out to be the centroid.

Note that using the values for ω indicated in Figure C-4, the integral J_ω can be evaluated to yield

$$J_\omega = \int_A \omega^2 \, dA = 4 \int_0^{b/2} \left(\frac{hs}{2}\right)^2 t_f \, ds = \frac{b^3 h^2 t_f}{24}$$

which is the same as the torsional constant resulting from the first development given in section 3-6.

EXAMPLE C-3 Consider the C-section shown in Figure C-5a. Select the pole on the axis of symmetry and the initial point I, as shown in Figure C-5b. The sectorial area then becomes

$$\text{On } IA: r = e, \; \omega = \omega_0 + es, \qquad 0 \leq s \leq \frac{h}{2}$$

$$\text{On } IC: r = e, \; \omega = \omega_0 + es, \qquad -\frac{h}{2} \leq s \leq 0$$

which can be consolidated to

$$\text{On } CA: r = e\omega = \omega_0 + es \qquad -\frac{h}{2} \leq s \leq \frac{h}{2}$$

$$\text{On } AB: r = \frac{h}{2}, \; \omega = \omega_0 + \frac{eh}{2} - \frac{hs}{2} \qquad 0 \leq s \leq b$$

$$\text{On } CD: r = \frac{h}{2}, \; \omega = \omega_0 - \frac{eh}{2} + \frac{hs}{2} \qquad 0 \leq s \leq b$$

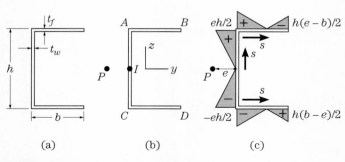

FIGURE C-5 Pole location and sectorial area for a C-section.

APPENDIX C SECTORIAL AREA

Evaluating the three required integrals yields

$$\int_A \omega \, dA = (t_w h + 2t_f b)\omega_0 = 0 \quad \Rightarrow \omega_0 = 0$$

$$\int_A y\omega \, dA \equiv 0$$

$$\int_A z\omega \, dA = e\left[\frac{t_w h^3}{12} + \frac{h^2 t_f b}{2}\right] - \frac{h^2 b^2 t_f}{4} = 0$$

from which

$$e = \frac{3b^2 t_f}{h t_w + 6b t_f}$$

as the distance from the center of the web to the pole P. The reader should show that the warping constant is given by

$$J_\omega = \int_A \omega^2 \, dA = \frac{t_f b^3 h^2}{12} \frac{2h t_w + 3b t_f}{h t_w + 6b t_f}$$

INDEX

Admissible functions in Ritz method, 85–86
Allowable stress, 47–50
Annular plates, 351–59, 437, 439
Approximate formulations, 50–55
Approximate methods
 finite element method, 87–98
 method of Ritz. *See* Method of Ritz
Approximate theory, 55
Assembly, defined, 93
Asymmetrically loaded membrane shells
 conical, 390–92
 cylindrical, 386–90
 spherical, 393–96
Axial deformation problem
 finite element methods in, 87–98
 one-dimensional, 74
 potential energy for, 403–4
 prismatic bar, 51–56
 uniform elastic bar, 69–72
Axial loading of a bar
 complementary energies in, 100
 one-dimensional, 74, 82
 segmented, 104–5
 strain energies in, 72, 78–79
Axial stress in thick-walled cylinders, 297
Axis representation
 neutral, 194, 197
 principal, 193–94, 195–97
Axisymmetric
 defined, 366
 elasticity problem, 419–22
 regions, 277, 278, 313
 shell, 367
 shell bending problem, 439–46
Axisymmetrically loaded membrane shells, 377–79
 conical, 384–86
 cylindrical, 379–81
 spherical, 381–84

Bar. *See also* Axial loading of a bar; Cross-section bars, torsion of
 prismatic, 51–56
Basic Ideas. *See also* Boundary value problems; Equilibrium and stress
 deformation and strain, 15–17, 18–24
 material behavior: stress-strain relations, 25–28
Beams
 plane-curved. *See* Plane-curved beams
 plates. *See* Plates
 transverse loading of. *See* Transverse loading of unsymmetrical beams
Behavior of material. *See* Material behavior
Bending
 of beams. *See* Transverse loading of unsymmetrical beams
 deformations, strain energy in, 73
 of a load, complementary energies in, 101–3
 shell, 367–69
 single thin-walled tube, 218–22
 stiffness of circular plates, 344
 symmetric, 183
 thin-walled multitube, 222–27
 and torsion, 227–32
 transverse, 101–3
 unsymmetrical. *See* Unsymmetrical bending
Bending moment
 of annular plates, 353, 355
 of rectangular plates, 325, 336
Bending stress
 multitube beam, 229
 plane-curved beam, 249–53
 plate, 329, 351
 warping restraint, 159
Biharmonic equation, 330

Blocks, deformations of, 24
Boundary conditions
 axial deformation problem, 55
 calculus of variations, 405–8
 displacements, 260–66
 force or nonessential, 260
 geometric or essential, 260
 for rectangular plates, 330–34
 in Ritz method, 85–86
 stress, 15
Boundary value problems
 equations for deformation plane stress case, 28
 stress and displacement of, 29–30
 stress formulations, 30–31
Brittle material
 failure by fracture, 36
 failure criteria of, 39, 45–47
 ultimate stress of, 48
 uniaxial tension test of, 37–38, 45–47

C-section. *See* Channel section (C-section)
Calculus of variations, 403–8
Cantilever beam
 end-loaded, 79–80, 194–95, 200–202
 loaded, 202–4
 shear center of, 215–18
 T-section, 183–88, 252–53
 transverse bending of, 102–3
Castigliano's first theorem
 to derive governing equations, 99
 in structural analysis, 77–81
Castigliano's second theorem
 generally, 104
 in thin-walled tube, torsion of, 149
Centroidal coordinate system, 209, 218, 222–23
Chalk, torsion test experiment, 37–39

INDEX

Channel section (C-section)
  loaded in shear, 211–15
  sectorial area, 448, 450–51
  under torque, 141–42
  warping function, 169–71
Circplate MATLAB code, 434–37
Circular plates, 339–40
  annular plate problems, 351–59
  assumptions regarding geometry and loading, 340
  deformations of, 342–44
  finite element-model for, 432–39
  force resultants and equilibrium, 340–42
  solid plate problems, 348–51
  solutions to problems, 347–48
  stress-strain and combination, 344–47
Circumferential moment, 438, 439, 446
Clamped edge, boundary conditions for, 331–32, 347
Clamped solid plate problems, 350–51
Closed cylinder, 290–91
Combination
  axial deformation problem, 54–56
  circular plate, 344–47
  membrane shell, 375–77
  plane-curved beam, 248–49
  rectangular plate, 329–30
  thick-walled cylinders, 279–80, 281–82
  thin-walled open section, 160–64
  thin-walled tube, torsion of, 148–49
  unsymmetrical bending, 190–91
Compatibility equation, 31
Complementary energy, 98–99
  density, 99
  for dummy loads, 105–7
  statistically indeterminate systems, 107–12
  for structural members, 100–5
  of thin-walled tube, torsion of, 149
Components of strain, 19
Compression test. *See* Uniaxial tension test
Conical shells, 384–86, 390–92
Conservative force, 65–66
Conservative mechanical system, 67–68
Constant forces and moments, 66
Constants, elastic and thermal, 26
Constraint, defined, 93
Cosines, direction. *See* Direction cosines
Coulomb-Mohr failure criterion
  of brittle materials, 39, 46–47
  factor of safety, 50
  working stress of brittle materials, 48–49
Cross-section bars, torsion of
  circular, 124–27
  noncircular, 128–36
  rectangular, 134–36, 142–44
  square, 128
  thin rectangular, 137–39
Cubic approximation, 299
Cylinders
  closed, 290–91
  long, 288–92
  open, 288–89
  thick-walled. *See* Thick-walled cylinders
Cylindrical shell, 379–81, 386–90

Deformation
  axial. *See* Axial deformation problem
  bending, 73
  blocks along principal strain directions, 24
  circular plate, 342–44
  defined, 2
  elastic, 36
  equation, 28
  extensional, 17
  flat panel, 324
  kinematics for, 16
  membrane shell, 373–75
  multitube construction, 153–54
  plane-curved beam, 246–47
  radial, in closed tube, 290–91
  rectangle, 22
  rectangular plate, 326–28
  shear, 17
  and strain, 15–17, 18–24
  stress equilibrium, 14
  thick-walled cylinders, 278–79
  thin-walled open sections, 160, 166–67
  thin-walled tube, torsion of, 148
  in torsion, 125
  transverse, of a beam, 76
  two-dimensional, 18–19
Design constraints, 33
Design criteria, 33
Design process
  defined, 31
  iterative, 34
  and modeling. *See* Modeling and the design process
Direction cosines
  in strain transformation equations, 21
  for stress transformations, 6–7
Directions, principal, geometry of, 10
Discrete problem, linear, 74
Discrete system, strain energy of, 69
Discretization
  defined, 87
  finite-element model, 410
  finite element, 309–10
Disks
  hollow, 294–97, 300–1, 307–9
  rotating, 299–309
  solid, 298–99, 301–4
  variable-thickness rotating, 305–9
Displacement
  boundary, 29–30
  degrees of freedom, 81
  formulations, 56
  gradients, defined, 16
  plane-curved beam, 260–66
  and potential energy, 77
  nodal, 429, 430, 433, 437
  radial, in a closed tube, 290–91
  shell, 374
  solid plate, 437–439
  two- and six-element model, 96
  in unsymmetrical bending, 198–204
Distribution, force, 2
Divergence theorem, two-dimensional form of, 131
Ductile material
  failure by excessive deformation, 36
  failure criteria of, 39, 40–45
  tensile test of, 37
  ultimate stress of, 48
Dummy load, 105–7, 263–64
Dummy moment, 105, 107

Edge, boundary conditions for
  in circular plates, 346–47
  in rectangular plates, 330–34
Edge moments on a circular plate, 348–49, 357
Eigenvalue, 8–9
Eigenvector, 8–9
Elastic
  body loaded by forces and couples, 77
  constant, 26
  deformation, 36
Element
  defined, 87
  in finite-element model, 410
Elemental
  formulation, 91
  load matrix, 90–91
  stiffness matrix, 89–90, 95

Elliptical torsion problem, 132–33
End-loaded cantilever beam, 79–80, 194–95, 200–2
Energy. See Complementary energy; Potential energy; Stationary potential energy; Strain energy
Engesser's first theorem, 103–5, 113
  in determination of displacements, 112, 263, 266
  in dummy loads, 106, 107
  in statically indeterminate systems, 108–9
Engesser's second theorem (principle of least work), 110
Engineering shear strain, 21
Environment of modeling process, 32
Equation
  biharmonic, 330
  compatibility, 31
  deformation, 28
  equilibrium. See Equilibrium equations
  Euler-Lagrange, 406, 407
  governing, 245–49, 371–77
  integrability, 31
  material behavior, 28
  membrane shells, 371–77
  strain-displacement, 376, 377
  strain transformation, 21
  stress transformation, 8
Equilibrium
  axial deformation problem, 52, 53
  membrane shell, 371–73
  multicelled tube, torsion of, 151–53
  plane-curved beam, 245–46
  requiring of, 2
  tetrahedron, 5
  thick-walled cylinder, 277–78
  thin-walled open section, 158–59, 166
  thin-walled tube, torsion of, 146–47
  unsymmetrical bending, 186–88
Equilibrium equations
  conical shell, 384, 390
  cylindrical shell, 386
  and displacement degrees of freedom, 81
  and force resultants, 325–26, 340–42
  membrane theory, 376, 377
  obtained by sectioning, 3
  single thin-walled tube, torsion of, 146–47
  small deformation plane stress case, 28–29
  spherical shell, 381, 393

Equilibrium and stress, 2–3
  plane stress, 10–13
  principal stresses, 8–10
  stress at a point, 3–4
  stress equilibrium, 14–15
  stress notation, 4–5
  stress transformation, 5–8
Essential boundary conditions, 260, 405, 407–8
Euler-Lagrange equation, 406, 407
Excessive deformations, types of, 36
Extensional deformation, 17
Extensional strain, of line segments, 18–20
External pressures, 282–88
External loading in axial deformation problem, 51–52

Face, negative and positive, 4
Factor of safety (FS), 47–50
Failure criteria, 35–39
  for brittle materials, 45–47
  for ductile materials, 40–45
Failure of structure
  due to fatigue, 36
  excessive deformations, 36
  fracture, 36
  and stress, 38–39
Fatigue, fracture due to, 36
Fatigue life, 36
FBD (free body diagram). See Free body diagram (FBD)
FEM. See Finite-element method (FEM)
Finite element method (FEM), 87–98. See also Finite-element model
  assumption of variable, 87–88
  axial problems, 88–95, 95–98
  model, generation of, 87
  solutions in thick-walled cylinders, 309–13
  in torsion of rectangular cross-section bar, 142–44
Finite-element model. See also Finite element method (FEM)
  axisymmetric shell bending problem, 439–46
  circular plate model, 432–39
  development of, 410–16
  noncircular cross-section bars, 142–44
  plane axisymmetric elasticity problem, 419–22
  Prandtl stress function, 130, 409
  rectangular plate problem, 422–32
First law of thermodynamics, 70

First moments of the sectorial area, 211
Fixed edge, boundary conditions for, 331–32
Flange, shear stress in, 257
Flat plate, 322, 323
Flexibility formulations, 65, 113–14
Flexural rigidity, in plates, 329, 344
Flexure formula, 244–45
Force
  boundary conditions, 260, 405
  constant, 66
  distribution, 2
  gravitational, 66
  work of, 65–69
Force-displacement relation
  conical shell, 384
  defined, 54
  membrane theory, 376, 377
  spherical shell, 382
Force resultants
  beam, 323–24
  circular plate, 340–42
  equilibrium equations, 325–26, 340–42
  and external loading, 325
  flat panel, 323–24
  shell, 367, 387, 390
  stress relations, 323–24
Forced boundary condition, 405
Forces, loading by, 77–81
Form between forces and deformations, 26
Formulations
  approximate, 50–55
  displacement, 56
  elemental, 91
  flexibility, 65, 113–14
  neutral axis, 197–98
  plane stress, 314
  principal axis, 195–97
  stiffness, 65, 81, 112–14
  stress, 56
Fracture, failure by, 36
Free body diagram (FBD)
  annular plate, 355
  beams, 263
  curved beam segment, 246
  for determining shear center, 213, 217, 226
  for determining transverse shear stresses, 254, 257
  differential element, 341
  drawing of, 3
  element in bending, 159

Free body diagram (*continued*)
   for equilibrium of thick-walled cylinders, 277
   force resultants and external loading, 325
   membrane shell, 378
   for shear flows in the web, 258
   for torque equation, 161
Free edge, boundary conditions for, 332, 347
FS (factor of safety), 47–50
Function, defined, 403

Gauss quadrature, 419, 429
Geometric boundary condition, 260, 405
Geometry
   axial deformation problem, 51
   circular plate, 340
   conical shell, 384
   cylindrical shell, 379
   flat plate, 323
   multicelled tube, torsion of, 151, 152
   plane-curved beam, 245
   plane strain, 280–81
   plane stress, 11, 279
   principal directions, 10
   shaft and disk, 303
   shells of revolution, 369–71
   shrink-fit, 285
   spherical shell, 381
   symmetrical problem, 310
   thick-walled, 277
   thin-walled tube, torsion of, 146
   variable-thickness rotating disk, 311
Global displacement vector, 91
Global stiffness matrix, 92–93
Governing equations
   development of, 245–49
   for membrane theory, 371–77
Gravitational force of a mass near earth, 66

Hermite polynomials, 433
Hollow disks
   rotating, 300–1
   and temperature, 294–97
   variable-thickness, 307–9
Homogeneous body, 27
Hooke's law, 26
Hoop stress
   rotating disks, 304, 308
   immersed cylinder, 299
   plane axisymmetric elasticity problem, 419–22
   shrink-fit problems, 287
   thick-walled cylinder, 284–87, 296, 297
   variable-thickness rotating disk, 312

I-section
   sectorial area, 448–50
   transverse shear stresses in, 257–60
Idealized structural components, 72–74
Initial configuration, 31
Inputs in modeling process environment, 32
Integrability equation, 31
Interference, 285
Internal force distribution, 2
Internal-friction theory. *See* Coulomb-Mohr failure criterion
Internal moment, resultant, 2
Internal pressures, 282–88
Interpolation
   defined, 87
   of finite elements, 310
   linear functions, 87–88, 411
Isotropic material, 26–27

Kinematics, 15
Kinetics, 3
Kirchoff assumption
   in circular plates, 343, 345
   in unsymmetrical bending, 189
ksi units, 4

Laplace's equation
   in circular plates, 346
   in hollow disks, 295
   in noncircular cross-section bars, 129–30
Levy solution to rectangular-plate problems, 336–38
Liebnitz rule for differentiation of an integral, 255
Linear algebraic eigenvalue problem, 8–9
Linear discrete problem, 74
Linear interpolation function, 87–88, 411
Linear relationship of stress and strain, 26
Linear strain-displacement relations, 19
Linearly interpolated (T3) triangular element, 143, 416–18
Load, dummy, 105–7
Loading
   by forces, 77–80
   of a bar, 73. *See also* Axial loading of a bar
   linear representation of, 90–91
   of plane-curved beams, 245
   shear, 101–3
   single plane of, 193, 199–200
   torsional, 72–73, 100–1
Loads in modeling process environment, 32
Luder bands, in tensile test of ductile material, 37

Material
   constants, 26
   isotropic, 26–27
   mechanics of, 51
Material behavior
   axial deformation problem, 53–54
   defined, 25, 26
   equation, 28
   membrane shells, 375–77
   multitube construction, 153–54
   stress-strain relations, 25–28
   thick-walled cylinders, 279
   thin-walled open sections, 160, 167–68
   thin-walled tube, torsion of, 148
   unsymmetrical bending, 190
Mathematical model
   defined, 31–32
   qualitative changes in, 34–35
   quantative changes in, 34–35
MATLAB code
   axisymmetric shell bending problem, 440–44
   circplate, 433–37
   rectangular plate problem, 423–28
   presves, 419–21
   using T3 elements, 416–18
Matrix
   elemental load, 90–91
   elemental stiffness, 89–90, 95
   global stiffness, 92–93
   stiffness, 81
Maximum distortion strain-energy theory. *See* von Mises failure criterion
Maximum normal-stress failure criterion for brittle materials, 45–47
Maximum shear-stress failure criterion, 40–44, 48, 49
Mechanical boundary condition, 405
Mechanics of materials, characteristics of, 51
Membrane analogy
   for circular tube, 145
   soap film used in, 136–37

in single thin-walled tube, 144–45
in thin-walled open sections, 139
Membrane force, 366–69. *See also*
    Shells of revolution
  axisymmetric shell bending problem, 445
  resultants, 372
Membrane shells of revolution. *See*
    Shells of revolution
Membrane theory, governing equations of
  deformations, 373–75
  equilibrium, 371–73
Meridional
  line, 369
  membrane force, 445
  moment, 445–46
Method of Ritz. *See also* Finite element methods
  admissible functions in, 85–86
  approximation, 309
  boundary conditions, 85–86
  exact solutions, 84, 86
  in finite-element models, 409
  in linear solid mechanics problem, 86–87
  one-dimensional axial problem, 82–83
  solutions, 83–85
Middle surface of plate, 322, 323
Moan, and interval midpoint, 94
Model
  circular plate, 432–39
  finite-element. *See* Finite-element model
  mathematical. *See* Mathematical model
  physical, 31
Modeling and the design process. *See also* Mathematical model
  defined, 31
  environment in, 32
  flowchart, 34
  laws and principles to include in, 33
  modeling process, 31–33
  steps in, 31
  structure in, 31, 32
  testing phase, 34–35
Modulus of elasticity, in axial deformation problem, 54
Mohr's circle
  in Coulomb-Mohr criterion, 47
  in plane stress, 12–13
  for strain, 23
  for strain transformation equations, 21
  for stress, 40–43

Moment
  bending. *See* Bending moment
  circumferential, 438, 439, 446
  constant, 66
  dummy, 105, 107
  edge, on a circular plate, 348–49, 357
  extrapolated to the nodes, 430, 431–32
  first, of the sectorial area, 211
  meridional, in axisymmetric shell bending problem, 445
  nodal, 430, 431–32, 437, 439
  to prevent warping, 158–59
  radial, 438, 439
  transverse, in axisymmetric shell bending problem, 445
  twisting, 325, 336
Multicelled tube, torsion of. *See* Tube, multicelled, torsion of
Mutually orthogonal planes, 9, 10

N term Ritz approximation, 82
Natural boundary condition, 405, 407–8
Navier solution to rectangular-plate problems, 334–36
Negative face, 4
Negative of work done, 66
newton/m^2, 4
Newton's laws
  of equilibrium, 33
  of kinetics (second law), 2–3, 64
  stress vector (third law), 4, 210
Nodal degrees of freedom, 423
Nodal displacement
  circular plate model, 433, 437
  four-element mesh, 429
  16-element mesh, 430
Nodal moments in solid plates, 437, 439
Nodes
  for concentrated loads, 91, 93
  defined, 87
  in finite-element model, 410, 423
  moments extrapolated to, 430, 431–32
Nonconforming element, 423
Nonessential boundary conditions, 260
Normal stress
  in plane stress, 11
  in stress transformations, 6

Octahedral plane, 44
Octahedral shear-stress theory, 44
Open cylinder, 288–89
Open sections. *See* Thin-walled open sections

Orthogonal coordinate system, 6
Orthogonal plane, 9, 10

Pascals, as units of stress, 4
Path
  work done independent of, 66, 67
  work of a force in, 65
Physical model, in modeling process, 31
Piecewise constant function, 94
Plane-curved beams, 244–45
  bending stresses, 249–53
  displacements, 260–66
  governing equations, 245–49
  transverse shear stresses, 254–60
Plane strain in thick-walled cylinders, 280–81
Plane stress
  defined, 10
  differential equations of equilibrium for, 14
  formulations, 314
  geometry for, 279
  geometry and loading, 11
  Mohr's circle for, 12–13
  stress transformation, 12
Plastic deformation, 36
Plates
  Circular. *See* Circular Plates
  Rectangular. *See* Rectangular plates
Poisson's equation, 130
Poisson's ratio, 27, 279, 375
Pole, defined, 166
Positive face, 4
Potential energy, 65–66. *See also* Stationary potential energy
  axial problem, 403–4
  constant forces and moments, 66
  gravitational force, 66
  linear spring, 67–69
Power-law assumption, 306, 309
Prandtl stress function, 130, 409
Pressure vessel, thin-walled, 276
Pressures, internal and external, 282–88
Presves MATLAB code, 419–21
Principal sectorial properties, 167
Principal stresses, 8–9
Principle of least work, 107–12, 266
Principle of stationary potential energy. *See* Stationary potential energy
Principle of superposition, 227, 233, 356–58
Prismatic bar, axial deformation of, 51–56

Problem statement in solid mechanics, 1–2
psi (lbf/in^2), 4

Quadratically interpolated (T6) element, 143–44

Radial displacements
  deformations with, 278–79
  closed tube, 290–91
  plane axisymmetric elasticity problem, 419–22
Radial moment
  annular plate, 439
  solid plate, 438
Radial stress
  hollow variable-thickness disk, 307–9
  immersed cylinder, 299
  plane axisymmetric elasticity problem, 419–22
  between shaft and disk, 303
  shrink-fit problems, 286–87
  thick-walled cylinders, 284–87, 296, 297
  variable-thickness rotating disk, 312
Radial temperature, 294–95
Rankine, William, 45
Rectangular coordinates, notation for stresses, 5
Rectangular cross section, 255–56
Rectangular plates
  constants for determining displacement and stresses, 339
  deformations of, 22–23, 326–28
  finite-element model for, 422–32
  flat, 322
  force resultants and equilibrium, 323–26
  Levy solution to problems, 336–38
  middle surface of, 322, 323
  Navier solution to problems, 334–36
  stress-strain and combination, 328–34
  thickness of, 322, 323
Redundants, 107–11
Resultant internal force distribution, 2
Resultant internal moment, 2
Rigid body motion, 15, 16
Ritz, method of. *See* Method of Ritz
Rotating disks
  constant-thickness, 299–304
  hollow, 300–1
  variable-thickness, 305–9
Rotation, rigid body, 15–16

Safety, factor of, 47–50
Saint Venant torque, 158, 161–65, 166

Sectorial area
  channel, 448, 450–51
  definition of, 447
  I-section, 448–50
Shafts, attached to rotating disks, 302–4
Shear
  deformation, 17
  loading, 101–3
  modulus, 27
Shear center
  cantilever beam, 215–18
  thin-walled multitubes, 226–27
  thin-walled open sections, 209–18
  thin-walled tubes, 218–22
Shear flow
  multicelled tube, torsion of, 152–53
  multitube section, 156–57, 229–32
  thin-walled multitubes, 222–27
  single thin-walled tube, 218–22
  thin-walled open sections, 161, 168, 205, 212
  single thin-walled tube, torsion of, 146–47
  two-tube section, 154–55
  in the web, 258
Shear force resultant in shell of revolution, 367–69
Shear strain
  engineering, 21
  geometric interpretation of, 20
  tensorial component of, 21
Shear stress
  extreme values, 11
  maximum, in plane stress, 13
  multitube beam, 229–31
  multitube section, 157–58
  notation, 4
  plane-curved beam, 254–60
  in plane stress, 11
  in thin rectangular cross-section bar, 138
  thin-walled open sections, 161, 204–8
  at a point, 4
  in stress transformations, 6, 7
Shells of revolution, 366–69, 397
  asymmetrically loaded, 386–96
  axisymmetrically loaded, 377–86
  conical, 384–86, 390–92
  cylindrical, 379–81, 386–90
  defined, 366
  geometry of, 369–71
  membrane theory equations, 371–77
  spherical, 381–84, 393–96
Ship cross sections, in torsion of multicelled tubes, 152

Shrink-fit
  problem, 285–88
  of rotating disks and shafts, 302
SI system, units of stress in, 4
Simply supported edge, boundary conditions of, 330–31, 346
Small displacement theory of thin, flat plates, 333
Soap film used in membrane analogy, 136–37
Solid disks, 298–99, 301–4
Solid plates, 348–51
Solution, 83–84, 93
Spherical shells, 381–84, 393–96
Springs, potential energy of, 67–69
Statically indeterminate problem, 264–66
Statically indeterminate systems, 107–12
Stationary potential energy, 67, 74, 403–5
  in axial deformation problem, 74–75
  in complementary energy ideas, 99
  displacements and, 77
  in finite-element models, 309–13
  in transverse deformations of a beam, 75–77
Stiffness formulations
  defined, 81
  generally, 65, 112–14
  and strain energy. *See* Strain energy
Stiffness matrix, 81
Stiffness method
  defined, 81
  in springs, potential energy of, 69
Strain
  components of, 19
  extensional, 18–20
  Mohr's circle for, 23–24
  shear, 20–21
Strain-displacement equations
  conical shells, 384, 390
  cylindrical shells, 387
  membrane theory, 376, 377
  spherical shells, 381
Strain-displacement relations
  circular plate, 342–44
  defined, 16–17
  linear, 19
  in rectangular plates, 326–28
  spherical shell, 393
  three-dimensional case, 21–22
  two-dimensional deformation, 18–19
Strain-displacement in unsymmetrical bending, 188–90

INDEX

Strain energy
  in axial deformation problem, 69–72
  defined, 65, 69, 70
  density, 71–72
  stationary potential energy, 74–77
  and structural analysis, 77–81
  for structural members, 72–74
Strain transformation equations, 21
Strength of materials, characteristics of, 51
Stress
  axial, 297
  bending. *See* Bending stress
  boundary, 15, 29–31
  and failure of the material, 38–39
  formulations, 30–31, 56
  hollow rotating disk, 300–1, 309
  hoop. *See* Hoop stress
  and internal force resultants, 340–41
  measurement of, in SI system, 4
  normal, 6
  notation, 4–5
  plane. *See* Plane stress
  plate, 323
  at a point, 3–4
  radial. *See* Radial stress
  shear. *See* Shear stress
  solid rotating disk, 302–4
  torsional, 232
  transformations, 5–8
  in variable-thickness rotating disk, 309
  in unsymmetrical bending, 191–98
  working, 47–50
Stress equilibrium, 14–15
Stress tensor
  defined, 5, 8
  in principal stresses, 9
  symmetry of, 5, 14–15
Stress-strain diagram, of one-dimensional conservative elastic solid, 98
Stress-strain relations
  axial deformation problem, 54
  circular plates, 344–47
  linear, 26
  material behavior in, 25–28
  plane-curved beams, 247
  spherical shell, 393
  for thin-plate theory, 328–34
Stress transformation equation, 8
Stress and twist coefficients, 136
Structure, 31, 32
Structural analysis, 77–81
Structural members
  complementary energies for, 100–5
  strain energy for, 72–74

Supports, boundary conditions for, 260–61
Superposition
  combined bending and torsion of multitube structure, 227, 233
  of a uniformly loaded solid circular plate, 356–58
Symmetry of the stress tensor
  defined, 5
  in stress equilibrium, 14–15

T-section cantilever beam, 183–88, 252–53
T3 element, 143, 416–18
T6 element, 143–44
Taylor series, in rectangular plate, 327
Temperature in thick-walled cylinders, 298–99. *See also* Thermal problems
Tensile test
  of chalk, 37–38
  of a ductile material, 37
Tensorial component of shear strain, 21
Tests
  in modeling and the design process, 34–35
  tensile test of ductile material, 37
  uniaxial tension, of brittle material, 37–38
Tetrahedron, equilibrium of, 5
Thermal constant, 26
Thermal problems, 292–94
  in hollow disks, 294–97
  in solid disks, 298–99
Thermodynamics, first law of, in axial deformation of elastic bar, 70
Thick-walled cylinders, 276–77
  closed, 290–91
  finite-element solutions, 309–13
  geometry, 277
  internal and external pressures, 282–88
  long cylinders, 288–92
  open, 288–89
  rotating disks, 299–304, 305–9
  theory, 277–82
  thermal problems, 292–99
Thickness of a plate, 322, 323
Thin rectangular cross-section bar, torsion of, 137–39
Thin-walled multitube bending, 222–32
Thin-walled open sections
  alternate formulation for warping restraint, 166–71
  beam with warping restraint, 169

  sectorial area, 447
  shear center for, 209–18
  shear stress in, 161, 204–8
  torsion of, 139–42
  warping restraint in, 158–65, 166–71
Thin-walled tube
  bending of, 218–22
  pressure vessel, 276
  torsion of. *See* Tube, single thin-walled, torsion of
Torsion
  and bending, 227–32
  chalk, testing of, 37–39
  circular cross-section bar, 124–27
  deformations in, 125
  elliptical torsion problem, 132–33
  finite-element model. *See* Finite-element model
  membrane analogy, 136–37
  multicelled tubes, 151–58
  noncircular cross-section bar, 128–36
  rectangular cross-section bar problem, 134–36
  single thin-walled tube, 144–51
  of square cross-section bar, 128, 413
  stress and twist coefficients, 136
  of thin rectangular cross section, 137–39
  of thin-walled sections. *See* Sectorial area
  of thin-walled open sections, 139–42
  of thin-walled rectangular tube, 149–51
  torque, 124
  warping restraint in thin-walled open sections, 158–65
  warping restraint in thin-walled open sections, alternate formulation, 166–71
Torsional loading of a bar
  complementary energies in, 100–1
  strain energies in, 72–73
Torsionless axisymmetric membrane shell equations, 378
Translation, 15, 16
Transmission of force, 2
Transverse
  bending of a cantilever, 102–3
  bending problem, 101–2
  deformations of a beam, 76
  deflections of a thin membrane, 136–37
  displacement in annular plates, 439
  displacement in circular plates, 349, 359

Transverse (*continued*)
  displacement in solid plates, 438
  loading of a bar, 73
  shear stresses in multitube beam, 229–31
  shear stresses in plane-curved beams, 254–60
Transverse loading of unsymmetrical beams
  assumptions in bending of beams theory, 182–83
  bending of single thin-walled tube, 218–22
  combined bending and torsion, 227–32
  shear center for thin-walled open sections, 209–18
  shear stresses in thin-walled open sections, 204–8
  T-section cantilever beam, 183–88
  thin-walled multitube bending, 222–27
  unsymmetrical bending. *See* Unsymmetrical bending
Tube. *See also* Cylinders
  bending of single thin-walled, 218–22
  closed, 290–91
  open, 288–89
Tube, multicelled, torsion of
  deformations and stress strain, 153–54
  equilibrium, 151–53
  geometry, 151, 152
  multitube section example, 155–58
  two-tube section example, 154–55
Tube, single thin-walled, torsion of, 144–46
  combination, 148–49
  deformations, 148
  equilibrium, 146–47
  geometry of region, 146
  with longitudinal slit, 150–51
  material behavior, 148
  rectangular, 149–50
Twisting moment, 325, 336

Uniaxial tension test
  of brittle material, 37–38, 45–47
  maximum shear-stress failure criterion, 40–44
  in octahedral shear-stress theory, 44–45
Uniform elongation, 17
Uniform extension, 17
Units of stress
  in SI system, 4
  in US customary system, 4
Unsymmetrical beams, transverse loading of. *See* Transverse loading of unsymmetrical beams
Unsymmetrical bending, 185–86
  combination, 190–91
  defined, 185
  displacements in, 198–204
  equilibrium, 186–88
  material behavior, 190
  multiplane loading of a beam, 197–98
  neutral axis coordinate representation, 194, 197
  principal axis representation, 193–94, 195–97
  single plane of loading, 193, 199–200
  strain-displacement, 188–90
  stresses in, 191–92
US customary system, units of stress in, 4

Variable-thickness rotating disks, 305–9
Variational approach, 64
Variations, calculus of, 403–8
Vectorial approach, 64
von Mises failure criterion
  diagram, 49
  of ductile materials, 39, 44–45, 47
  ellipse for plane stress, 45
  factor of safety, 49
  ultimate stress, 48

Warping constant, 161
Warping restraint, in thin-walled open sections
  combination, 160–64
  deformations, 160, 166–67
  equilibrium, 158–59, 166
  material behavior, 160, 167–68
  sectorial area, 447–51
  thin-walled beam, 169
  warping function data table, 170
Web, shear flows in, 258
Wing cross sections, in torsion of multicelled tubes, 152
Work, principle of least, 107–12, 266
Work of a force, 65–69
Working stress, 47–50

XYZ coordinate system, in stress transformations, 6, 8

Yielding of material, 36
Young's modulus, 27
  in axial deformation problem, 54
  in membrane shells, 375
  in thick-walled cylinders, 279
  in unsymmetrical bending, 190

Z-section, shear flows in, 207–8